Industrial Machinery Repair: Best Maintenance Practices Pocket Guide

Industrial Machinery Repair: Best Maintenance Practices Pocket Guide

Ricky Smith and
R. Keith Mobley

BUTTERWORTH
HEINEMANN

An imprint of Elsevier Science
www.bh.com

Amsterdam Boston London New York Oxford Paris San Diego
San Francisco Singapore Sydney Tokyo

Butterworth–Heinemann is an imprint of Elsevier Science.

∞ Recognizing the importance of preserving what has been written, Elsevier-Science prints its books on acid-free paper whenever possible.

Library of Congress Cataloging-in-Publication Data
Smith, Ricky.
Industrial machinery repair : best maintenance practices pocket guide / Ricky Smith and Keith Mobley.
p. cm.
ISBN 0-7506-7621-3 (pbk : alk. paper)
1. Machinery–Maintenance and repair. 2. Industrial equipment–Maintenance and repair.
I. Mobley, Keith. II. Title.

TJ153.S6355 2003
621.8′16–dc21

2003040435

British Library Cataloguing-in-Publication Data
A catalogue record for this book is available from the British Library.

The publisher offers special discounts on bulk orders of this book.
For information, please contact:

Elsevier Science
Manager of Special Sales
200 Wheeler Road, 6th Floor
Burlington, MA 01803
Tel: 781-313-4700
Fax: 781-313-4882

For information on all Butterworth–Heinemann publications available, contact our World Wide Web home page at: http://www.bh.com

10 9 8 7 6 5 4 3 2 1

Transferred to Digital Printing 2009

Contents

Acknowledgments

Ricky Smith wants to offer his thanks to the following individuals who contributed to the writing of this book. Bruce Hawkins, Life Cycle Engineering; Darryl Meyers, former U.S. Army Warrant Officer; Steve Lindborg, Chemical Lime Company; Robby Smith (his brother), International Paper Corporation; and J.E. Hinkel, Lincoln Electric Company. Ricky also wants to thank Life Cycle Engineering, where he is currently employed, for the opportunity to write this book; Alumax–Mt. Holly—currently Alcoa–Mt. Holly—where he worked as a maintenance technician, for all the training and the chance to expand his knowledge; and Dr. John Williams, who always believed in him.

1 Introduction: Why Use Best Maintenance Repair Practices?

"Only Permanent Repairs Made Here"

This book addresses, in a simplistic manner, the proper principles and techniques in "Best Maintenance Practices—Mechanical."

If these principles and techniques are followed, they will result in a serious reduction in "self-induced failures." This book is a tool that should be carried and referenced by all mechanical maintenance personnel.

A number of surveys conducted in industries throughout the United States have found that 0 of equipment failures are self-induced.

Maintenance personnel who are not following what are termed "Best Maintenance Repair Practices" substantially affect these failures. Between 30% and 50% of the self-induced failures are the result of maintenance personnel not knowing the basics of maintenance. Maintenance personnel who, although skilled, choose not to follow best maintenance repair practices potentially cause another 20% to 30% of those failures. The existence of this problem has been further validated through the skills assessment process performed in companies throughout the state of Georgia. This program evaluated the knowledge of basic maintenance fundamentals through a combination of written, identification, and performance assessments of thousands of maintenance personnel from a wide variety of industries. The results indicated that over 90% lacked complete basic fundamentals of mechanical maintenance. This book focuses on the "Best Maintenance Repair Practices" necessary for maintenance personnel to keep equipment operating at peak reliability and companies functioning more profitably through reduced maintenance costs and increased productivity and capacity.

The potential cost savings can often be beyond the understanding or comprehension of management. Many managers are in a denial state regarding maintenance. The result is that they do not believe that repair practices directly impact an organization's bottom line or profitability.

More enlightened companies have demonstrated that, by reducing self-induced failures, they can increase production capacity by as much as 0 .

Other managers accept lower reliability standards from maintenance efforts because they either do not understand the problem or they choose to ignore this issue. A good manager must be willing to admit to a maintenance problem and actively pursue a solution.

You may be asking, what are the Best Maintenance Repair Practices Here are a few that maintenance personnel must know. (See Table 1.1.)

Looking through this abbreviated Best Maintenance Repair Practices table, try to determine whether your company follows these guidelines.

The results will very likely surprise you. You may find that the best practices have not been followed in your organization for a long time. In order to fix the problem you must understand that the culture of the organization is at the bottom of the situation. Everyone may claim to be a maintenance expert but the conditions within a plant generally cannot often validate that this is true. In order to change the organization's basic beliefs, the reasons why an organization does not follow these best practices in the repair of their equipment must be identified.

"Only Permanent Repairs Made Here"

A few of the most common reasons that a plant does not follow best maintenance repair practices are:

1 Maintenance is totally reactive and does not follow the definition of maintenance, which is to protect, preserve, and prevent from decline (reactive plant culture).

2 Maintenance personnel do not have the requisite skills.

3 The maintenance workforce lacks either the discipline or direction to follow best maintenance repair practices.

4 Management is not supportive, and/or does not understand the consequences of not following the best practices (real understanding must involve a knowledge of how much money is lost to the bottom line).

Table 1.1 *Best maintenance repair practices*

Maintenance task	Standard	Required best practices	Consequences for not following best practices	Probability of future failures—number of self-induced failures vs. following best practices
Lubricate Bearing	Lubrication interval: time based ± 10% variance.	1 Clean fittings. 2 Clean end of grease gun. 3 Lubricate with proper amount and right type of lubricant. 4 Lubricate within variance of frequency.	Early bearing failure: reduced life by 20–80%.	100% ⟩ 20 vs. 1
Coupling Alignment	Align motor couplings utilizing dial indicator or laser alignment procedures. (Laser is preferred for speed and accuracy.) Straightedge method is unacceptable.	1 Check runout on shafts and couplings. 2 Check for soft foot. 3 Align angular. 4 Align horizontal. 5 Align equipment specifications, not coupling specifications.	Premature coupling failure. Premature bearing and seal failure in motor and driven unit. Excessive energy loss.	100% ⟩ 7 vs. 1
V-Belts	Measure the tension of V-belts through tension and deflection utilizing a belt tension gauge.	1 Identify the proper tension and deflection for the belt. 2 Set tension to specifications.	Premature belt failures through rapid belt wear or total belt failure. Premature bearing failure of driven and driver unit. Belt creeping or slipping causing speed variation without excessive noise. Motor shaft breakage.	100% ⟩ 20 vs. 1

Continued

Table 1.1 *continued*

Maintenance task	Standard	Required best practices	Consequences for not following best practices	Probability of future failures—number of self-induced failures vs. following best practices
Hydraulic Components	Hydraulic fluid must be conditioned to component specifications.	1 Hydraulic fluid must be input into the hydraulic reservoir utilizing a filter pumping system only. 2 Filters must be rated to meet the needs of the component reliability and not equipment manufacturer's specification. 3 Filters must be changed on a timed basis based on filter condition. 4 Oil samples must be taken on a set frequency, and all particles should be trended in order to understand the condition and wear of the hydraulic unit.	Sticking hydraulic. Premature or unknown hydraulic pump life. Sustaining hydraulic competency by maintenance personnel. Length of equipment breakdown causes lost production.	100% ⟩ 30 vs. 1

Figure 1.1 *Maintenance*

In order to solve the problem of not following Best Maintenance Repair Practices, a sequential course of action should be taken:

First, identify whether a problem exists (i.e., track repetitive equipment failures, review capacity losses in production and identify causes for these losses, and measure the financial losses due to repair issues). (See Figure 1.2, "The Maintenance Cost Iceberg.")

Second, identify the source of the problem (this could be combination of issues):

- Maintenance skill level: Perform skills assessment (written and performance based) to evaluate whether skill levels are adequate to meet "Best Maintenance Repair Practices" for your specific maintenance organization.
- Maintenance culture: Provide training to all maintenance and management relative to a change in maintenance strategy and how it will impact them individually (e.g., increase in profit for the plant, less overtime

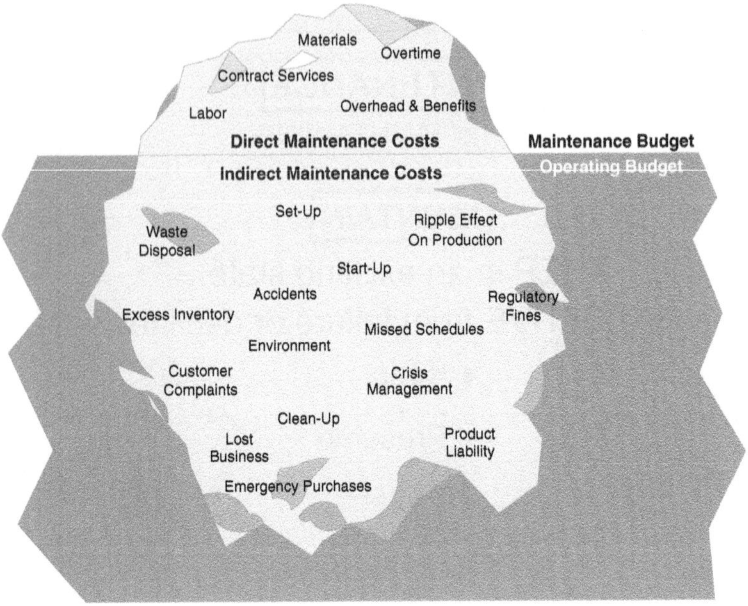

Figure 1.2 *The maintenance cost iceberg*

resulting from fewer equipment breakdowns, etc.). Track and measure the changes and display the results to everyone.

- Maintenance strategy: Develop a plan to introduce a proactive maintenance model with "Preventive and Planned Maintenance" at the top of planned priorities. This will provide more time for performing maintenance utilizing the "Best Maintenance Repair Practices."

Third, implement the changes needed to move toward following "Best Maintenance Repair Practices" and measure the financial gains.

Everyone should be aware that financial rewards can be great but we must understand why they can also be hard to achieve.

Several of the reasons why implementing a program of change, such as the one discussed, can be doomed to failure include:

- Management not committed;
- Lack of discipline and direction;

- Lack of management commitment and accountability;
- Momentum becomes slowed or changes direction;
- Lack of an adequately skilled workforce;
- No gap analysis or specific action plan to guide the effort to close the gaps;
- Conflict between emergencies and performing maintenance following "Best Maintenance Repair Practices" (this does not mean all "emergent" repairs must be performed to "as built" specifications the first time, but it does mean that the repair, especially a temporary fix, will be corrected during the next outage of the equipment).

To conclude, as many as 90% of companies in the United States do not follow "Best Maintenance Repair Practices." The percent that do follow these practices are realizing the rewards of a well run, capacity-driven organization that can successfully compete in today's and tomorrow's marketplace. Remember that use of the "Best Maintenance Repair Practices" might just become a mandatory requirement for the future success of an organization in today's economy.

Utilize this book as a resource to:

1 Write corrective maintenance procedures to attach to specific work order tasks in a computerized maintenance management system (CMMS).

2 Train personnel regarding new and existing maintenance repair procedures.

3 Be used as a tool by a current maintenance staff. Have all maintenance personnel use this book on the job site in order to follow "Best Maintenance Practices." (This book can be set up in a storeroom in order to be replaced as the book is worn and damaged on the job site.)

Preventive and Predictive Maintenance (PPM)

PPM is more than the regular cleaning, inspection, tightening, lubrication, and other actions intended to keep durable equipment in good operating condition and to avoid failures. It is an investment in the future—a future without major, disruptive breakdowns of critical equipment. It is, at times, an investment without an immediate return on that investment. Here the philosophy must be, "Pay me now or pay me later," because that is exactly what happens when it comes to PPM. If preventive maintenance is not accomplished in the proper way and in a timely manner, then the pay-me-later clause will occur at the most inopportune time. That is the premise that must be established and promoted by management. While the maintenance department has the day-to-day responsibility of plant PPM, the plant manager is ultimately responsible for setting the expectations concerning plant preventive maintenance.

In general, the preventive and predictive maintenance effort is not focused. To spotlight some of the weaknesses, review the following points:

1 Many of the components that make up a good predictive maintenance program have not been developed;

2 Thermography;

3 Oil sampling for some gearboxes and hydraulic units.

Preventive Maintenance (PM)

Some of the PM procedures have been developed; however, they lack details to make them efficient and safe, and to reinforce sound maintenance practices.

1 No mobile equipment has written PMs.

2 Logbooks are used, but it can't be determined who was looking at and using the data.

3 All preventive maintenance regarding lubrication should be reviewed for detail and accuracy.

4 The forecasting and generation of PM tasks because of the state of the CMMS is not part of the normal maintenance routine.

Recommendations

Review all lubrication-related task procedures for detail and content.

- Provide training in effective plant lubrication procedures and techniques. The lubricator, maintenance personnel, and maintenance supervisors should attend.

- The maintenance manager needs to conduct monthly spot checks on randomly selected PMs and repairs for quality assurance.
- The present quantity and content of preventive maintenance tasks currently established are too generic in content. Task information should be detailed enough to help build consistency and training for maintenance—for example, "Line 1, Tire Stacker: On the full monthly PM, job #01 states: Lubricate, Check Photo Eyes, Clean/Sweep/Pickup, etc." For the experienced maintenance person that has done the task many times, they most likely know what to look for. But the maintenance person with less experience does not. Again, provide enough information to build consistency and training.
- Make the production associates aware of the fact that they are the eyes and ears of maintenance—the first line of defense—and that they are also an important part of the predictive maintenance process. They are the ones who will see or hear a problem first. Make sure to praise workers who contribute to the process.
- Ensure all machines have operator checklists, that these checks are done properly, and that the results are turned in to the responsible supervisor.
- Have the operators (in the future) perform routine clean-and-inspection tasks on equipment.
- Encourage operator involvement concerning equipment PPM.
- Train plant management and workforce in the importance of preventive and predictive maintenance.

Summary

Preventive Maintenance is the most important routine function that maintenance personnel can accomplish. The reactive, breakdown maintenance mode will never be gotten away from if PMs are not performed consistently and properly on a regularly scheduled basis.

2 Fundamental Requirements of Effective Preventive/Predictive Maintenance

When most people think of preventive maintenance, they visualize scheduled, fixed interval maintenance that is done every day, every month, every quarter, every season, or at some other predetermined intervals. Timing may be based on days, or on intervals such as miles, gallons, activations, or hours of use. The use of performance intervals is itself a step toward basing preventive tasks on actual need, instead of just on a generality.

The two main elements of fixed interval preventive maintenance are procedure and discipline. Procedure means that the correct tasks are done, the right lubricants applied, and consumables replaced at the best interval. Discipline requires that all the tasks are planned and controlled so that everything is done when it should be done. Both these areas deserve attention. The topic of procedures is covered in detail in following sections.

Discipline is a major problem in many organizations. This is obvious when one considers the fact that many organizations do not have an established program. Further, organizations that do claim to have a program often fail to establish a good planning and control procedure to assure accomplishment. Elements of such a procedure include:

1. Listing of all equipment and the intervals at which it must receive PMs;
2. A master schedule for the year that breaks down tasks by month, week, and possibly even by the day;
3. Assignment of responsible persons to do the work;
4. Inspection by the responsible supervisor to make sure that quality work is done on time;
5. Updating of records to show when the work was done and when the next preventive task is due;
6. Follow-up as necessary to correct any discrepancies.

Fundamental Requirements of Effective Maintenance

Effective maintenance is not magic, nor is it dependent on exotic technologies or expensive instruments or systems. Instead, it is dependent on doing simple, basic tasks that will result in reliable plant systems. These basics include:

Inspections

Careful inspection, which can be done without "tearing down" the machine, saves both technician time and exposure of the equipment to possible damage. Rotating components find their own best relationship to surrounding components. For example, piston rings in an engine or compressor cylinder quickly wear to the cylinder wall configuration. If they are removed for inspection, chances are that they will not easily fit back into the same pattern. As a result, additional wear will occur, and the rings will have to be replaced much sooner than if they were left intact and performance-tested for pressure produced and metal particles in the lubricating oil.

Human Senses

We humans have a great capability for sensing unusual sights, sounds, smells, tastes, vibrations, and touches. Every maintenance manager should make a concerted effort to increase the sensitivity of his own and that of his personnel's human senses. Experience is generally the best teacher. Often, however, we experience things without knowing what we are experiencing. A few hours of training in what to look for could have high payoff.

Human senses are able to detect large differences but are generally not sensitive to small changes. Time tends to have a dulling effect. Have you ever tried to determine if one color was the same as another without having a sample of each to compare side by side? If you have, you will understand the need for standards. A standard is any example that can be compared to the existing situation as a measurement. Quantitative specifications, photographs, recordings, and actual samples should be provided. The critical parameters should be clearly marked on them with displays as to what is good and what is bad.

As the reliability-based preventive maintenance program develops, samples should be collected that will help to pinpoint with maximum accuracy how much wear can take place before problems will occur. A display where craftsmen gather can be effective. A framed 4′ × 4′ pegboard works well since shafts, bearings, gears, and other components can be easily wired to it or hung on hooks for display. An effective, but little used, display area where notices can be posted is above the urinal or on the inside of the toilet stall door. Those are frequently viewed locations and allow people to make dual use of their time.

Sensors

Since humans are not continually alert or sensitive to small changes and cannot get inside small spaces, especially when operating, it is necessary to use sensors that will measure conditions and transmit information to external indicators.

Sensor technology is progressing rapidly; there have been considerable improvements in capability, accuracy, size, and cost. Pressure transducers, temperature thermocouples, electrical ammeters, revolution counters, and a liquid height level float are examples found in most automobiles.

Accelerometers, eddy-current proximity sensors, and velocity seismic transducers are enabling the techniques of motion, position, and expansion analysis to be increasingly applied to large numbers of rotating equipment. Motors, turbines, compressors, jet engines, and generators can use vibration analysis. The normal pattern of operation, called its "signature," is established by measuring the performance of equipment under known good conditions. Comparisons are made at routine intervals, such as every thirty days, to determine if any of the parameters are changing erratically, and further, what the effect of such changes may be.

The *spectrometric oil analysis* process is useful for any mechanical moving device that uses oil for lubrication. It tests for the presence of metals, water, glycol, fuel dilution, viscosity, and solid particles. Automotive engines, compressors, and turbines all benefit from oil analysis. Most major oil companies will provide this service if you purchase lubricants from them.

The major advantage of spectrometric oil analysis is early detection of component wear. Not only does it evaluate when oil is no longer lubricating properly and should be replaced, it also identifies and measures small quantities of metals that are wearing from the moving surfaces. The metallic elements found, and their quantity, can indicate what components are wearing and to what degree so that maintenance and overhaul can be carefully planned. For example, presence of chrome would indicate cylinder-head wear, phosphor bronze would probably be from the main bearings, and stainless steel would point toward lifters. Experience with particular equipment naturally leads to improved diagnosis.

Thresholds

Now that instrumentation is becoming available to measure equipment performance, it is still necessary to determine when that performance is "go" and when it is "no go." A human must establish the threshold point, which can then be controlled by manual, semiautomatic, or automatic means. First, let's decide how the threshold is set and then discuss how to control it.

To set the threshold, one must gather information on what measurements can exist while equipment is running safely and what the measurements are just prior to or at the time of failure. Equipment manufacturers, and especially their experienced field representatives, will be a good starting source of information. Most manufacturers will run equipment until failure in their laboratories as part of their tests to evaluate quality, reliability, maintainability, and maintenance procedures. Such data are necessary to determine under actual operating conditions how much stress can be put on a device before it will break. Many devices, such as nuclear reactors and flying airplanes, should not be taken to the breaking point under operating conditions, but they can be made to fail under secure test conditions so that the knowledge can be used to keep them safe during actual use.

Once the breaking point is determined, a margin of safety should be added to account for variations in individual components, environments, and operating conditions. Depending on the severity of failure, that safety margin could be anywhere from one to three standard deviations before the average failure point. One standard deviation on each side of the mean will include 68% of all variations, two standard deviations include 95%, and three standard deviations are 98.7%. Where our mission is to prevent failures, however, only the left half of the distribution is applicable. This single-sided distribution also shows that we are dealing with probabilities and risk.

The earlier the threshold is set and effective preventive maintenance done, the greater is the assurance that it will be done prior to failure. If the mean time between failures (MTBF) is 9,000 miles with a standard deviation of 1,750 miles, then proper preventive maintenance at 5,500 miles could eliminate almost 98% of the failures. Note the word "proper," meaning that no new problems are injected. That also means, however, that costs will be higher than need be since components will be replaced before the end of their useful life, and more labor is required.

Once the threshold set point has been determined, it should be monitored to detect when it is exceeded. The investment in monitoring depends on the period over which deterioration may occur, means of detection, and benefit value. If failure conditions build up quickly, a human may not easily detect the condition, and the relatively high cost of automatic instrumentation will be repaid.

Lubrication

Friction of two materials moving relative to each other causes heat and wear. Friction-related problems cost industries over $1 billion per annum. Technology intended to improve wear resistance of metal, plastics, and other surfaces in motion has greatly improved over recent years, but planning, scheduling, and control of the lubricating program is often reminiscent of a plant handyman wandering around with his long-spouted oil can.

Anything that is introduced onto or between moving surfaces in order to reduce friction is called a lubricant. Oils and greases are the most commonly used substances, although many other materials may be suitable. Other liquids and even gases are being used as lubricants. Air bearings, for example, are used in gyroscopes and other sensitive devices in which friction must be minimal. The functions of a lubricant are to:

1. Separate moving materials from each other in order to prevent wear, scoring, and seizure;
2. Reduce heat;
3. Keep out contaminants;
4. Protect against corrosion;
5. Wash away worn materials.

Good lubrication requires two conditions: sound technical design for lubrication and a management program to assure that every item of equipment is properly lubricated.

Lubrication Program Development

Information for developing lubrication specifications can come from four main sources:

1 Equipment manufacturers;

2 Lubricant vendors;

3 Other equipment users;

4 Individuals' own experience.

Like most other preventive maintenance elements, initial guidance on lubrication should come from manufacturers. They should have extensive experience with their own equipment both in their test laboratories and in customer locations. They should know what parts wear and are frequently replaced. Therein lies a caution: a manufacturer could, in fact, make short-term profits by selling large numbers of spare parts to replace worn ones. Over the long term, however, that strategy will backfire, and other vendors, whose equipment is less prone to wear and failure, will replace them.

Lubricant suppliers can be a valuable source of information. Most major oil companies will invest considerable time and effort in evaluating their customers' equipment to select the best lubricants and intervals for change. Naturally, these vendors hope that the consumer will purchase their lubricants, but the total result can be beneficial to everyone. Lubricant vendors perform a valuable service of communicating and applying knowledge gained from many users to their customers' specific problems and opportunities.

Experience gained under similar operating conditions by other users or in your own facilities can be one of the best teachers. Personnel, including operators and mechanics, have a major impact on lubrication programs.

A major step in developing the lubrication program is to assign specific responsibility and authority for the lubrication program to a competent maintainability or maintenance engineer. The primary functions and steps involved in developing the program are to:

1 Identify every piece of equipment that requires lubrication;

2 Assure that all major equipment is uniquely identified, preferably with a prominently displayed number;

3 Assure that equipment records are complete for manufacturer and physical location;

4 Determine locations on each piece of equipment that needs to be lubricated;

5 Identify lubricant to be used;

6 Determine the best method of application;

7 Establish the frequency or interval of lubrication;

8 Determine if the equipment can be safely lubricated while operating, or if it must be shut down;

9 Decide who should be responsible for any human involvement;

10 Standardize lubrication methods;

11 Package the above elements into a lubrication program;

12 Establish storage and handling procedures;

13 Evaluate new lubricants to take advantage of state of the art;

14 Analyze any failures involving lubrication and initiate necessary corrective actions.

An individual supervisor in the maintenance department should be assigned the responsibility for implementation and continued operation of the lubrication program. This person's primary functions are to:

1 Establish lubrication service actions and schedules;

2 Define the lubrication routes by building, area, and organization;

3 Assign responsibilities to specific persons;

4 Train lubricators;

5 Assure supplies of proper lubricants through the storeroom;

6 Establish feedback that assures completion of assigned lubrication and follows up on any discrepancies;

7 Develop a manual or computerized lubrication scheduling and control system as part of the larger maintenance management program;

8 Motivate lubrication personnel to check equipment for other problems and to create work requests where feasible;

9 Assure continued operation of the lubrication system.

It is important that a responsible person who recognizes the value of thorough lubrication be placed in charge. As with any activity, interest diminishes over time, equipment is modified without corresponding changes to the lubrication procedures, and state-of-the-art advances in lubricating technology may not be undertaken. A factory may have thousands of lubricating points that require attention. Lubrication is no less important to computer systems, even though they are often perceived as electronic. The computer field engineer must provide proper lubrication to printers, tape drives, and disks that spin at 3,600 rpm. A lot of maintenance time is invested in lubrication. The effect on production uptime can be measured nationally in billions of dollars.

Calibration

Calibration is a special form of preventive maintenance whose objective is to keep measurement and control instruments within specified limits. A "standard" must be used to calibrate the equipment. Standards are derived from parameters established by the National Bureau of Standards (NBS). Secondary standards that have been manufactured to close tolerances and set against the primary standard are available through many test and calibration laboratories and often in industrial and university tool rooms and research labs. Ohmmeters are examples of equipment that should be calibrated at least once a year and before further use if subjected to sudden shock or stress.

The government sets forth calibration system requirements in MIL-C-45662 and provides a good outline in the military standardization handbook MIL-HDBK-52, *Evaluation of Contractor's Calibration System.* The principles are equally applicable to any industrial or commercial situation. The purpose of a calibration system is to provide for the prevention of tool inaccuracy through prompt detection of deficiencies and timely application of corrective action. Every organization should prepare a written description of its calibration system. This description should cover the measuring of test equipment and standards, and should:

1 Establish realistic calibration intervals;

2 List all measurement standards;

3 List all environmental conditions for calibration;

4 Ensure the use of calibration procedures for all equipment and standards;

5 Coordinate the calibration system with all users;

6 Assure that equipment is frequently checked by periodic system or cross-checks in order to detect damage, inoperative instruments, erratic readings, and other performance degrading factors that cannot be anticipated or provided for by calibration intervals;

7 Provide for timely and positive correction action;

8 Establish decals, reject tags, and records for calibration labeling;

9 Maintain formal records to assure proper controls.

The checking interval may be in terms of time—hourly, weekly, monthly—or based on amount of use—every 5,000 parts, or every lot. For electrical test equipment, the *power-on* time may be the critical factor and can be measured through an electrical elapsed-time indicator.

Adherence to the checking schedule makes or breaks the system. The interval should be based on stability, purpose, and degree of usage. If initial records indicate that the equipment remains within the required accuracy for successive calibrations, then the intervals may be lengthened. On the other hand, if equipment requires frequent adjustment or repair, the intervals should be shortened. Any equipment that does not have specific calibration intervals should be (1) examined at least every six months, and (2) calibrated at intervals of no longer than one year. Adjustments or assignment of calibration intervals should be done in such a way that a minimum of 95% of equipment, or standards of the same type, is within tolerance when submitted for regularly scheduled recalibration. In other words, if more than 5% of a particular type of equipment is out of tolerance at the end of its interval, then the interval should be reduced until less than 5% is defective when checked.

A record system should be kept on every instrument, including:

1 History of use;

2 Accuracy;

3 Present location;

4 Calibration interval and when due;

5 Calibration procedures and necessary controls;

6 Actual values of latest calibration;

7 History of maintenance and repairs.

Test equipment and measurement standards should be labeled to indicate the date of last calibration, by whom it was calibrated, and when the next calibration is due. When the size of the equipment limits the application of labels, an identifying code should be applied to reflect the serviceability and due date for the next calibration. This provides a visual indication of the calibration serviceability status. Both the headquarters calibration organization and the instrument user should maintain a two-way check on calibration. A simple means of doing this is to have a small form for each instrument with a calendar of weeks or months (depending on the interval required) across the top, which can be punched and noticed to indicate the calibration due date.

Planning and Estimating

Planning is the heart of good inspection and preventive maintenance. As described earlier, the first thing to establish is what items must be maintained and what the best procedure is for performing that task. Establishing good procedures requires a good deal of time and talent. This can be a good activity for a new graduate engineer, perhaps as part of a training process that rotates him or her through various disciplines in a plant or field organization. This experience can be excellent training for a future design engineer.

Writing ability is an important qualification, along with pragmatic experience in maintenance practices. The language used should be clear and concise, using short sentences. Who, what, when, where, why, and how should be clearly described. The following points should be noted from this typical procedure:

1 Every procedure has an identifying number and title;

2 The purpose is outlined;

3 Tools, reference documents, and any parts are listed;

4 Safety and operating cautions are prominently displayed;

5 A location is clearly provided for the maintenance mechanic to indicate performance as either satisfactory or deficient. If deficient,

details are written in the space provided at the bottom for planning further work.

The procedure may be printed on a reusable, plastic-covered card that can be pulled from the file, marked, and returned when the work order is complete; on a standard preprinted form; or on a form that is uniquely printed by computer each time a related work order is prepared. Whatever the medium of the form, it should be given to the preventive maintenance craftsperson together with the work order so that he has all the necessary information at his fingertips. The computer version has the advantage of single-point control that may be uniformly distributed to many locations. This makes it easy for an engineer at headquarters to prepare a new procedure or to make any changes directly on the computer and have them instantly available to any user in the latest version.

Two slightly different philosophies exist for accomplishing the unscheduled actions that are necessary to repair defects found during inspection and preventive maintenance. One is to fix them on the spot. The other is to identify them clearly for later corrective action. If a "priority one" defect that could hurt a person or cause severe damage is observed, the equipment should be immediately stopped and "46 red tagged" so that it will not be used until repairs are made. Maintenance management should establish a guideline such as, "Fix anything that can be corrected within ten minutes, but if it will take longer, write a separate work request." The policy time limit should be set based on:

1 Travel time to that work location;

2 Effect on production;

3 Need to keep the craftsperson on a precise time schedule.

The inspector who finds them can effect many small repairs the most quickly. This avoids the need for someone else to travel to that location, identify the problem, and correct it. And it provides immediate customer satisfaction. More time-consuming repairs would disrupt the inspector's plans, which could cause other, even more serious problems to go undetected. The inspector is like a general practitioner, who performs a physical exam and may give advice on proper diet and exercise but who refers any problems he may find to a specialist.

The inspection or preventive maintenance procedure form should have space where any additional action required can be indicated. When the procedure is completed and turned into maintenance control, the planner or scheduler should note any additional work required and see that it gets done according to priority.

Estimating Time

Since inspection or preventive maintenance is a standardized procedure with little variation, the tasks and time required can be accurately estimated. Methods of developing time estimates include consideration of such resources as:

1 Equipment manufacturers' recommendations;

2 National standards such as *Chilton's* on automotive or *Means'* for facilities;

3 Industrial engineering time-and-motion studies;

4 Historical experience.

Experience is the best teacher, but it must be carefully critiqued to make sure that the "one best way" is being used and that the pace of work is reasonable.

The challenge in estimating is to plan a large percentage of the work (preferably at least 90%) so that the time constraints are challenging but achievable without a compromise in high quality. The trade-off between reasonable time and quality requires continuous surveillance by experienced supervisors. Naturally, if a maintenance mechanic knows that his work is being time studied, he will follow every procedure specifically and will methodically check off each step of the procedure. When the industrial engineer goes away, the mechanic will do what he feels are necessary items, in an order that may or may not be satisfactory. As discussed earlier in regard to motivation, an experienced preventive maintenance inspector mechanic can vary performance as much as 50% either way from the standard without most maintenance supervisors recognizing a problem or opportunity for improvement. Periodic checking against national or time-and-motion standards, as well as trend analysis of repetitive tasks, will help keep preventive task times at a high level of effectiveness.

Estimating Labor Cost

Cost estimates follow from time estimates simply by multiplying the hours required by the required labor rates. Beware of coordination problems where multiple crafts are involved. For example, one Fortune 100 company has trade jurisdictions that require the following personnel in order to remove an electric motor: a tinsmith to remove the cover, an electrician to disconnect the electrical supply, a millwright to unbolt the mounts, and one or more laborers to remove the motor from its mound. That situation is fraught with inefficiency and high labor costs, since all four trades must be scheduled together, with at least three people watching while the fourth is at work. The cost will be at least four times what it could be, and is often greater if one of the trades does not show up on time. The best a scheduler can hope for is if he has the latitude to schedule the cover removal at, say, 8:00 A.M., and the other functions at reasonable time intervals thereafter: electrician at 9:00, millwright at 10:00, and laborers at 11:00.

It is recommended that estimates be prepared on "pure" time. In other words, the exact hours and minutes that would be required under perfect scheduling conditions should be used. Likewise, it should be assumed that equipment would be immediately available from production. Delay time should be reported, and scheduling problems should be identified so that they can be addressed separately from the hands-on procedure times. Note that people think in hours and minutes, so one hour and ten minutes is easier to understand than 1.17 hours.

Estimating Materials

Most parts and materials that are used for preventive maintenance are well known and can be identified in advance. The quantity of each item planned should be multiplied by the cost of the item in inventory. The sum of those extended costs will be the material cost estimate. Consumables such as transmission oil should be enumerated as direct costs, but grease and other supplies used from bulk should be included in overhead costs.

Scheduling

Scheduling is, of course, one of the advantages to doing preventive maintenance over waiting until equipment fails and then doing emergency repairs. Like many other activities, the watchword should be "PADA," which stands for "Plan a Day Ahead." In fact, the planning for inspections and preventive activities can be done days, weeks, and even months in advance in order to assure that the most convenient time for production is chosen, that maintenance parts and materials are available, and that the maintenance workload is relatively uniform.

Scheduling is primarily concerned with balancing demand and supply. Demand comes from the equipment's need for preventive maintenance. Supply is the availability of the equipment, craftspeople, and materials that are necessary to do the work. Establishing the demand is partially covered in the chapters on on-condition, condition monitoring, and fixed interval preventive maintenance tasks. Those techniques identify individual equipment as candidates for PM.

Coordination with Production

Equipment is not always available for preventive maintenance just when the maintenance schedulers would like it to be. An overriding influence on coordination should be a cooperative attitude between production and maintenance. This is best achieved by a meeting between the maintenance manager and production management, including the foreman level, so that what will be done to prevent failures, how this will be accomplished, and what production should expect to gain in uptime may all be explained.

The cooperation of the individual machine operators is of prime importance. They are on the spot and most able to detect unusual events that may indicate equipment malfunctions. Once an attitude of general cooperation is established, coordination should be refined to monthly, weekly, daily, and possibly even hourly schedules. Major shutdowns and holidays should be carefully planned so any work that requires "cold" shutdown can be done during those periods. Maintenance will often find that they must do this kind of work on weekends and holidays, when other persons are on vacation. Normal maintenance should be coordinated according to the following considerations:

1 Maintenance should publish a list of all equipment that is needed for inspections, preventive maintenance, and modifications, and the amount of cycle time that such equipment will be required from production.

2 A maintenance planner should negotiate the schedule with production planning so that a balanced workload is available each week.

3 By Wednesday of each week, the schedule for the following week should be negotiated and posted where it is available to all concerned; it should be broken down by days.

4 By the end of the day before the preventive activity is scheduled, the maintenance person who will do the PM should have seen the first-line

production supervisor in charge of the equipment to establish a specific time for the preventive task.

5 The craftsperson should make every effort to do the job according to schedule.

6 As soon as the work is complete, the maintenance person should notify the production supervisor so that the equipment may be put back into use.

Overdue work should be tracked and brought up-to-date. Preventive maintenance scheduling should make sure that the interval is maintained between preventive actions. For example, if a preventive task for May is done on the thirtieth of the month, the next monthly task should be done during the last week of June. It is foolish to do a preventive maintenance task on May 30th and another June 1st, just to be able to say one was done each month. In the case of preventive maintenance, the important thing is not the score but how the game was played.

Assuring Completion

A formal record is desirable for every inspection and preventive maintenance job. If the work is at all detailed, a checklist should be used. The completed checklist should be returned to the maintenance office on completion of the work. Any open preventive maintenance work orders should be kept on report until the supervisor has checked the results for quality assurance and signed off approval. Modern computer technology with handheld computers and pen-based electronic assistants permits paperless checklists and verification. In many situations, a paper work order form is still the most practical media for the field technician. The collected data should then be entered into a computer system for tracking.

Record Keeping

The foundation records for preventive maintenance are the equipment files. In a small operation with less than 200 pieces of complex equipment, the records can easily be maintained on paper. The equipment records provide information for purposes other than preventive maintenance. The essential items include:

- Equipment identification number;
- Equipment name;

- Equipment product/group/class;
- Location;
- Use meter reading;
- PM interval(s)
- Use per day;
- Last PM due;
- Next PM due;
- Cycle time for PM;
- Crafts required, number of persons, and time for each;
- Parts required.

Back to Basics

Obviously, effective maintenance management requires much more than these fundamental tasks. However, these basic tasks must be the foundation of every successful maintenance program. The addition of other tools, such as CMMS, predictive maintenance, etc., cannot replace them.

3 Maintenance Skills Assessment

Introduction

A maintenance skills assessment is a valuable tool in determining the strengths and weaknesses of an individual or a given group of employees in order to design a high-impact training program that targets those documented needs. Maintenance personnel have often found it difficult to upgrade their technical skills because much that is available is redundant or does not take their current skill level into consideration. An assessment is designed to eliminate those problems by facilitating the construction of customized training paths for either individuals or the group based upon demonstrated existing knowledge and skills. When used in conjunction with a job task analysis, a gap analysis can be performed to determine both what skills are needed in order to perform the job effectively and what skills those in the workforce presently have.

Definition of a Skills Assessment

A maintenance skills assessment consists of a series of written tests, performance exercises, and identification activities selected from a listing of mechanical basic skill areas. In this chapter maintenance mechanics will be able to assess their maintenance knowledge only because skills can only be assessed through a hands-on assessment. However, the knowledge assessment is the prerequisite for all skills. The written assessment in this chapter is written at an eighth-grade level (maintenance mechanics, in most industries, must be able to read proficiently at least at the 12-year twelfth-grade level). A maintenance person without the knowledge required for a specific skill can be assured mistakes will be made in mechanical judgment and ability and thus will cause equipment failures. This knowledge assessment will not cover all skill areas covered in this book but will cover chapters that are considered the mechanical basics.

Knowledge Assessment

This knowledge assessment is directed to the following skills. The answers will be provided in Appendix A at the end of the book. A minimum score of 90% in each skill area should be considered acceptable in most cases. However, some skill areas could require a higher score if the risk of failure due to a knowledge gap is high. In all areas of maintenance, a maintenance person must know the risk.

The knowledge assessment will be performed in the following skills areas:

- Safety
- Lubrication
- Bearings
- Chain Drives
- Belt Drives
- Hydraulics
- Couplings

Knowledge Assessment

The assessment is multiple choice. Select the best answer. Do not look at the answers until you have answered all the questions.

Knowledge Area: Safety

1 What term is used to describe places where moving parts meet or come near each other?

A. Guard
B. Closer

C. Pinch points
D. Assembly point

2 What is another name for back-and-forth motion?

A. Reciprocating motion
B. Away
C. Advancing lateral
D. None of the above

3 What is the term "point of operation?"

A. The section of the process where the machine centers meet
B. The main focus of process
C. The place where the raw material or work-piece is processed by a machine
D. A point where operators cannot see

4 If a bench grinder is equipped with safety guards, is it necessary for you to wear personal protective equipment?

A. Yes
B. No

5 What type of machine guard limits the operator's access to the danger zone?

A. Safety chain
B. E-stop
C. A barrier guard
D. None of the above

6 What type of machine guard prevents access to the danger zone altogether?

A. An enclosure guard
B. Safety chain
C. Barrier guard
D. None of the above

7 What kind of guards cannot be moved when a machine is in operation?

A. Barrier guards
B. Fixed guards
C. E-stop guards
D. None of the above

8 What type of guard prevents a machine from operating when the guard is opened or removed?

A. Barrier guard
B. Tapeless guard
C. An interlocking guard
D. None of the above

9 What type machine guard is capable of physically pulling an operator's hands out of the danger zone?

A. An automatic guard
B. Barrier guard
C. Restrictive guard
D. None of the above

10 When will a presence-sensing guard stop a machine?

A. When a person is located outside of the danger zone
B. When a light signals a safety alert
C. There is no such item
D. When a person or object enters the danger zone

11 What kind of controls does a machine have if the operator must remove both hands from the danger zone in order to start the machine?

A. Hand-free controls
B. Two-hand trip controls
C. Standard controls
D. Automatic controls

12 What kinds of tools make it unnecessary for an operator to reach into the danger zone?

A. Hand-extraction tools
B. Local guarding tools
C. Feeding and extracting tools
D. None of the above

13 Under what conditions would you remove someone else's lock from a lockout device?

A. When plant manager or maintenance manager approves
B. When the person that installed the lockout cannot be found after 30 minutes
C. When you think it is OK to do so safely
D. According to your plant's lockout procedure

14 What must your employer provide in addition to the appropriate PPE?

A. Training in its use
B. Safety bulletins
C. Training material and trainers
D. None of the above

15 What is your responsibility before using PPE?

A. None
B. You must inspect it
C. Locate all documents controlling it
D. You must report its condition to your supervisor

16 Why should you avoid loose-fitting clothing in the plant?

A. It can create a barrier from sharp parts
B. It can get caught in moving parts
C. It is unprofessional
D. None of the above

17 What should you do if you accidentally come in contact with a dangerous chemical?

A. It depends on the chemical
B. Report to the safety director
C. Shower for at least 15 minutes to rinse thoroughly
D. None of the above

18 How much clearance should hard-hat webbing provide between your head and the top of the shell?

A. Close as possible
B. $\frac{1}{2}$"
C. Does not matter
D. 1"

19 How can you keep dust and liquids from reaching your eyes from any direction?

A. By wearing safety goggles
B. By wearing safety glasses
C. By wearing a face shield
D. Any of the above

20 What units are used to measure noise?

A. Trebels
B. Decibels
C. Milibars
D. None of the above

Knowledge Area: Lubrication

1 A lubricant's viscosity is rated by what type of unit?

A. SSU
B. SAE
C. ISA
D. LVU

2 A lubricant with high viscosity has a:

A. High speed.
B. High temperature.
C. High resistance to flow.
D. High resistance to breakdown.

3 A low-viscosity lubricant:

A. Provides good cushioning for machine shock loads.
B. Can flow into tight spaces for better lubrication.
C. Does not carry heat away as well as a high-viscosity lubricant.
D. Costs less than a high-viscosity lubricant.

4 What are two disadvantages of high-viscosity lubricants?

A. They are expensive and cannot be used on high-speed motors.
B. They break down quickly and are difficult to apply.
C. They do not flow well and do not carry heat away well.
D. They do not protect against abrasive action of dirt, and they break down quickly.

5 Multiple-viscosity lubricants differ from single-viscosity lubricants because:

A. They have special additives that extend their effective temperature range.
B. They are best within a very narrow temperature range.

C. They will never degrade under high temperatures.
D. They last longer.

6 One advantage of multiple-viscosity lubricants is that:

A. They flow better at medium range temperatures than at either extreme.
B. They have a high bearing capacity.
C. They have a broad working temperature range.
D. They do not break down in the presence of water.

7 Which of the following is NOT a factor affecting the selection of a lubricant?

A. Machine speed
B. Environmental humidity
C. Operating temperature
D. Environmental temperatures

8 When choosing a lubricant, you want:

A. The lubricant to stay thin at high temperatures.
B. The lubricant to thicken at low temperatures.
C. The lubricant to thin at low temperatures.
D. The lubricant to maintain effective viscosity at its highest rated temperatures.

9 An oil cooler is used to:

A. Add heat to the oil to enable it to flow better at low temperatures.
B. Add heat to the oil to keep it from thinning at high temperatures.
C. Remove heat from the oil to prevent it from thinning at high temperatures.
D. Remove heat from the oil to prevent it from thickening at low temperatures.

10 What function do detergent additives in lubricants perform?

A. Keep metal surfaces clean
B. Keep the lubricant clean
C. Minimize the amount of foaming
D. All of the above

11 An anti-oxidation additive in a lubricant:

A. Controls the level of dirt.
B. Controls the amount of mixing with air.

C. Controls the level of foaming.
D. Prevents the lubricant from mixing with metal particles.

12 As a mechanic, you observe that a machine bearing is extremely hot and becoming discolored as it operates. Your conclusion is that the:

A. Lubricant is contaminated by water.
B. Bearing is about to seize.
C. Lubricant is causing acid corrosion on the bearing.
D. Bearing is not compatible with the lubricant.

13 When cooling an overheated bearing, what should you do first?

A. Wrap the bearing housing in hot, wet rags.
B. Spray cool water on the bearing.
C. Inject cool oil in the bearing.
D. Wrap the bearing housing in cool, wet rags.

14 Oil returning to the sump is visually cloudy and foaming. You conclude that the oil is:

A. Contaminated with soot.
B. Contaminated with water.
C. Contaminated with metal particles.
D. In need of detergent additives.

15 Undesired oil misting can be reduced by:

A. Increasing the temperature of the oil.
B. Increasing the speed of the machine.
C. Increasing the viscosity of the oil.
D. Reducing the viscosity of the oil.

16 A grease cup is defined as a:

A. Cup filled with grease that screws onto a fitting.
B. Timed lubrication system controlled by a rotating cam.
C. Gravity system that forces lubricant onto or into the area needing lubrication.
D. Fitting that applies oil in droplet form.

17 A lubricating system used in low-speed applications in which a needle valve meters a steady rate of lubricant to a machine without recycling the lubricant is a(n):

A. Dip lubricator.
B. Shot lubricator.

C. Drip lubricator.
D. Oil sump.

18 A lubrication system in which the component needing lubrication rotates through an enclosed housing containing oil and carries the oil to other components is called a(n):

A. Dip lubricator.
B. Shot lubricator.
C. Drip lubricator.
D. Oil sump.

19 In a force-feed lubrication system, lubricant is moved to the component needing lubrication by a:

A. Cooler.
B. Pump.
C. Filter.
D. Bearing.

20 What is the most undesirable by-product of oil misting?

A. Bearing failure
B. Shaft damage
C. Explosion potential
D. Oil breakdown

Knowledge Area: Bearings

1 The two basic categories of bearings are:

A. Plain and antifriction.
B. Ball and roller.
C. Journal and ball.
D. Pillow-block and roller.

2 Bearings:

A. Are found in machines with moving parts.
B. Function as guides.
C. Help reduce the friction between moving parts.
D. All of the above.

3 Thrust bearings:

A. Support axial loads on rotating members.
B. Support radial loads on rotating members.
C. Both A and B.
D. None of the above.

4 Antifriction bearings:

A. Contain balls.
B. Contain rollers.
C. Will run hot if they are overlubricated.
D. All of the above.

5 Bearing lubrication systems include:

A. Lubrication by hand.
B. Central grease systems.
C. Pressure-feed oil systems.
D. All of the above.

6 Plain bearings operate by:

A. Separating the races with balls or rollers.
B. Using an air gap.
C. Hydraulics.
D. Running on a film of lubricant.

7 Antifriction bearings operate by:

A. Separating the races with balls or rollers.
B. Using an air gap.
C. Hydraulics.
D. Running on a film of lubricant.

8 Roller bearings are used over ball bearings for which of the following situations?

A. High-speed applications
B. High-load applications
C. Wet environments
D. Mobile equipment engines

9 Bearing clearance can be described as:

A. The space between the rolling elements and the races.
B. The allowed difference between the shaft size and the bearing inner race.
C. The allowed differences between bearing inner and outer race.
D. None of the above.

10 Shaft tolerance can be defined as:

A. The allowed difference between the shaft size and the bearing inner race.
B. The force applied during installation.
C. The space between the rolling elements and the races.
D. None of the above.

11 The preferred method for installing an antifriction bearing is:

A. With a sledge hammer.
B. To sand down the shaft until the bearing slides on.
C. With a bearing heater.
D. Both B and C.

12 When tightening the lock-nut on a spherical roller bearing, the preferred tool is:

A. A spanner wrench.
B. A bearing heater.
C. A hammer and punch.
D. None of the above.

13 The bearing best suited for both radial and thrust loads is a ________ bearing.

A. tapered-sleeve
B. linear-motion
C. needle
D. tapered-roller

14 A bearing lubricated with oil is capable of ________ speeds than the same bearing lubricated with grease.

A. lower
B. higher
C. the same
D. different

15 As you tighten the nut on a spherical-roller bearing, the space between the race and the rolling element:

A. Increases.
B. Decreases.
C. Remains the same.
D. Develops cracks.

16 On a metric bearing with the number 7307, the ID of the bearing is:

A. 35 mm.
B. 7 mm.
C. .035".
D. .007".

17 To convert the metric shaft size of a bearing to inches, you multiply the millimeters by:

A. 5.
B. 39.
C. .03937.
D. .05.

18 A failed bearing that has a cracked inner race probably failed because:

A. the shaft was too large.
B. of a lack of lubricant.
C. the operator failed to do the proper inspection.
D. of overlubrication.

19 An antifriction bearing can run hot because:

A. of overlubrication.
B. it is about to fail.
C. of excessive load.
D. all of the above.

20 A 20% increase in bearing load, can result in a ________% decrease in bearing life.

A. 20
B. 100
C. 50
D. 10

Knowledge Area: Chain Drives

1 What is the maximum elongation that a roller chain can tolerate before it should be replaced?

A. 10%
B. 3 inches

C. 3%
D. $\frac{3}{16}$ inch per ft. of chain

2 What causes roller fatigue?

A. Lack of lubrication
B. Contamination
C. Improper chain tension
D. All of the above

3 What problems will chain misalignment cause?

A. Side plate wear
B. Sprocket wear
C. Pin wear
D. Chain break

4 What are the two most important factors in the life of a silent chain drive?

A. Lubrication and alignment
B. Lubrication and tension
C. Tension and alignment
D. All of the above

5 What type of machine guard limits the operator's access to the danger zone?

A. Safety chain
B. E-stop
C. A barrier guard
D. None of the above

6 What type of machine guard prevents access to the danger zone altogether?

A. An enclosure guard
B. Safety chain
C. Barrier guard
D. None of the above

7 Name the two main groups of mechanical couplings:

A. Rigid and flexible.
B. Chain and gear.
C. Grid and Lovejoy.
D. Spring and centrifugal.

8 Large fans are typically powered in what way?

A. Motor and belts
B. Motor and chain drive
C. Motor direct drive
D. Motor, gearbox, and belts

9 Which type of conductivity cells are usually equipped with a safety chain to prevent the cell from blowing out?

A. Immersions
B. Insertion
C. Flow
D. Screw-in

10 Chain-linked fences should be topped with:

A. Three strands of electrically energized wire.
B. Barbed wire.
C. A skirt of the same material that kicks back 45 degrees and extends 18 inches.
D. None of the above.

11 What are the two links that make up a standard roller chain?

A. Roller, rigid
B. Roller, pin
C. Master, half
D. Offset, half

12 On a roller chain with a designation of 35, what do the 3 and the 5 designate?

A. 3 is the size in inches; 5 is the duty
B. 3 is the duty; 5 is the size in thousandths of an inch
C. 3 is the size in eighths of an inch; 5 means a rollerless chain
D. 3 is the size; 5 means heavy duty

13 What is the maximum elongation that a roller chain can tolerate before it should be replaced?

A. 10%
B. 3 inches
C. 3%
D. $\frac{3}{16}$ inch per ft. of chain

14 What causes roller fatigue?

A. Lack of lubrication
B. Contamination
C. Improper chain tension
D. All of the above

15 What problems will chain misalignment cause?

A. Side plate wear
B. Pin wear
C. Chain break
D. All of the above

16 What are the two most important factors in the life of a silent chain drive?

A. Lubrication and alignment
B. Lubrication and tension
C. Tension and alignment
D. Tension and speed

17 Name the two main groups of mechanical couplings.

A. Rigid and flexible
B. Chain and gear
C. Grid and Lovejoy
D. Spring and centrifugal

18 What type of machine guard limits the operator's access to the danger zone?

A. Safety chain
B. E-stop
C. A barrier guard
D. None of the above

19 What type of machine guard prevents access to the danger zone altogether?

A. An enclosure guard
B. Safety chain
C. Barrier guard
D. None of the above

20 What is the tension or slack a roller chain is recommended to have?

A. 10% slack between centers
B. $\frac{1}{2}$ inch slack
C. 5% slack between centers
D. 2% slack between centers

Knowledge Area: Belt Drives

1 Of the following, what are the four main types of belt drives?

A. V-belt; flat belt; timing belt; ribbed belt
B. Straight belt; V-belt; gear belt; ribbed belt
C. Rubber; vitron; buna-n; teflon
D. Round belt; dual face belt; V-belt; multibelt

2 Identify routine maintenance performed on flat belts.

A. Tensioning
B. Cleaning
C. Dressing
D. Both A and B

3 Which of the following is an acceptable V-belt installation process?

A. Reduce center-to-center distance of the pulleys.
B. Use a pry bar to slide the belt over the pulleys.
C. Tension the belt until your hand bounces off the belt.
D. All of the above.

4 Which of the following contribute to rapid belt wear?

A. Oil, dirt, heat, and alignment
B. Water, dirt, sunlight, and tension
C. Guarding, shielding, reflection, and direction
D. Sunlight, dry-rot, mold, and water

5 Why do timing or gear belts not require high tension?

A. They do not require tensioning.
B. The material they are made of reduces slippage.
C. They provide slip-proof engagement.
D. There is not load fluctuation.

6 Different speeds are required for various materials when cutting with a band saw. How is saw speed adjusted?

A. Set the speed indicator to the desired rpm and start the machine.
B. Adjust the belt tensioner and start the machine.
C. Start the machine and adjust the speed indicator.
D. Dial in the correct job selector and start the machine.

7 Belt drives are designed to transmit power between a drive unit and a:

A. Conveyor belt.
B. Motor.

C. Driven unit.
D. Roller.

8 A drive belt with the designation "XL" would indicate which of the following types?

A. Cogged, fractional horsepower
B. Conventional V-belt
C. Narrow V-belt
D. Positive drive belt

9 A drive belt with the designation "C" would indicate which of the following types?

A. Power band belt
B. Conventional V-belt
C. Narrow V-belt
D. Fractional horsepower

10 A drive belt with the designation "V" would indicate which of the following types?

A. Positive drive belt
B. Conventional V-belt
C. Narrow V-belt
D. Fractional horsepower

11 A "C" belt with a nominal length of 60 inches is $\frac{4}{10}$" long. What identification should be marked on the belt?

A. C410
B. C60/4
C. C64
D. C56

12 When replacing one of a set of belts, the mechanic should:

A. Replace the worn belt only.
B. Replace the worn belt and the belt closest to the drive motor.
C. Replace all of the belts.
D. Replace the worn belt and the belt farthest from the drive motor.

13 What criteria should be used to determine if a belt needs to be replaced?

A. The belt protrudes above the top of the sheave no more than $\frac{1}{16}$".
B. The belt is flush with the top of the sheave groove.
C. The belt is recessed more than $\frac{1}{16}$" into the groove sheave.
D. The belt is recessed more than $\frac{1}{8}$" into the groove sheave.

14 The length of belt is determined by its:

A. Pitch line.
B. V-line.
C. Standard line.
D. None of the above.

15 When "timing" is a critical consideration with a belt, which of the following is typically used?

A. Joined belt
B. V-belt
C. Cogged belt
D. Positive drive belt

16 Regarding sheave grove angle, which of the following statements is correct?

A. Smaller diameter sheaves have less groove angle than larger diameter grooves.
B. Larger diameter sheaves have less groove angle than smaller diameter sheaves.
C. Groove angles are not affected by sheave diameter.
D. Can be compensated for by adjusting belt tension.

17 Which of the following conditions can cause excessive belt wear?

A. Dirty operating conditions
B. Improper sheave alignment
C. Improper tensioning
D. All of the above

Knowledge Area: Hydraulics

1 What function does a reservoir serve in a hydraulic system?

A. Stores surplus of oil
B. Cools the oil
C. Cleans the oil
D. A, B, and C

2 Why is there a baffle in a hydraulic reservoir?

A. To prevent baffling
B. To assist in cleaning and cooling of the oil
C. To keep the oil level the same
D. To prevent overpressurization of the reservoir

3 What should the fluid pass through before it reaches the pump?

A. Prepump precipitator
B. Metal diffuser
C. Heat exchanger
D. Strainer

4 What are the three basic types of positive displacement pumps?

A. Centrifugal, gear, piston
B. Air pump, piston, centrifugal
C. Piston, gear, vane

5 What function does a relief valve provide in a hydraulic system?

A. Protects the system from overpressurization
B. Dumps oil back to the tank
C. Cools the oil
D. A and B

6 What function does an actuator serve in a hydraulic system?

A. To hold pressure in a system
B. To assist the pump in flow
C. To allow work to be performed
D. A and B

7 What are the four basic types of center conditions (envelopes) in the directional control valves of a hydraulic system?

A. Open, close, tandem, return
B. Bidirectional, unidirectional, terminal, return
C. Tandem, float, closed, open
D. The center position is the same in all valves

8 What function do flow control valves serve in the hydraulic system?

A. Control pressure to a point in the system.
B. Control the speed of an actuator.
C. Reduce the speed of a pump.
D. Eliminate return in one direction.

9 What are the two basic types of flow control arrangements in a hydraulic system?

A. Meter in, meter out
B. Flow control, flow decontrol

C. Meter in, direct flow
D. Control, directional control

10 What are the two basic types of actuators in a hydraulic system?

A. Bladder, piston
B. Cylinder, motor
C. Double cylinder, single cylinder
D. Pump, actuator

11 Which of the following are means by which a directional control valve may be actuated?

A. Manual
B. Solenoid operated
C. Pilot pressure
D. All of the above

12 The variable volume pump used most often in high-pressure systems is what type of pump?

A. Vane
B. Piston
C. Gear
D. Centrifugal

13 Basic hydraulics theory tells us that pressure:

A. On a confined liquid is transmitted undiminished in every direction.
B. Acts with equal force on equal areas.
C. Acts at right angles only on the container side walls.
D. Acts with equal force on reduced areas.

14 Hydraulics is the science of transmitting force or motion through a medium of a:

A. Confined liquid.
B. Nonconfined liquid.
C. A and B.
D. None of the above.

15 Component parts connected to each other by fluid lines in the hydraulic system form a:

A. Closed center system.
B. Bypass.
C. Circuit.
D. Motor.

16 A needle valve in a hydraulic line provides a:

A. Flow control orifice.
B. Pressure control orifice.
C. Relief valve.
D. Master valve.

Knowledge Area: Couplings

1 A major advantage of flexible couplings is that they:

A. Compensate for thermal expansion.
B. Provide greater torsional transfer than rigid couplings.
C. Require only minimal motor alignment.
D. Operate at higher rpm than rigid couplings.

2 Name the two main groups of mechanical couplings:

A. Rigid and flexible.
B. Chain and gear.
C. Grid and Lovejoy.
D. Spring and centrifugal.

3 Which of the following couplings requires lubrication?

A. Slider
B. Steelflex
C. Torus rubber
D. Sure-Flex rubber

4 Measurements taken at the top and bottom of a coupling are taken to adjust:

A. Vertical alignment.
B. Horizontal alignment.
C. Angular alignment.
D. Parallel alignment.

5 When performing coupling alignment, the first activity that must be conducted is to:

A. Check runout.
B. Verify angular misalignment.
C. Verify parallel misalignment.
D. Check coupling spacing.

6 Which of the following measuring tools is most useful for the straight-edge method of coupling alignment?

A. Dial indicator
B. Dial caliper
C. Taper gauge
D. Vernier dial indicator

7 Cooling tower fan shafts are supported by:

A. Hubs.
B. Bearings.
C. Couplings.
D. Back plate.

8 How are cooling tower fan blades attached?

A. Brazed
B. Welded
C. With U-bolts
D. Coupling

9 When a high-speed gear mesh coupling is removed from a compressor, the coupling is typically match marked. What is the purpose of this?

A. Ensures that it can be reassembled correctly.
B. Ensures that it is disassembled correctly.
C. Ensures that one coupling half is aligned to mesh.

10 Some centrifugal compressors use a high-speed gear-type coupling. How is this type of gear attached to the compressor shaft?

A. Bolted
B. Welded
C. Keyed
D. Interference fit

11 With high-speed gear couplings, gear teeth showing a large polished area indicate which of the following?

A. Proper gear mesh
B. Unit operation beyond recommended capacity
C. Gear wear
D. Galling

12 High-speed gear-type coupling for motor-to-centrifugal compressor connections are typically lubricated by:

A. Grease.
B. Forced air.
C. Process fluid.
D. Oil.

13 Hose connections that are frequently made and "broken" would use which of the following types of couplings?

A. Flange
B. Push-on
C. Quick-connect
D. Reusable

14 KAMAG generators may use any number of drivers. Which of the following coupling types are NOT recommended to tie the two together?

A. Flexible
B. Gear
C. Solid
D. Both B and C

15 A common return signal wire shared by two or more signal wires is known as:

A. Impedance coupling.
B. Magnetic coupling.
C. Electrostatic coupling.
D. Dielectric coupling.

16 Which of the following comes in two pieces?

A. Coupling
B. Union
C. Bushing
D. Reducer

17 What type of fitting is used to connect tubing through a control panel or junction box?

A. Coupling
B. Bulkhead
C. Compression
D. Cross

Benchmark your mechanical knowledge: Go to Appendix A for the correct answers and grade as follows:

Safety: total right/20 = __________

Lubrication: total right/20 = __________

Bearings: total right/20 = __________

Chain Drives: total right/20 = __________

Belt Drives: total right/17 = __________

Hydraulics: total right/16 = __________

Couplings: total right/17 = __________

Final Score: total of the scores/130 = __________

69% or below = Needs extensive mechanical training

70% –79% = Needs moderate mechanical training

80% –95% = Needs to review mechanical subject areas

95% –100% = Needs no training

4 Safety First, Safety Always

Introduction

As an industrial mechanic, you should always make safety your primary concern in the workplace. No job should be performed that would endanger your life or the lives of others. Always remember, "Safety first, safety last, safety always."

Each year in the United States more people are injured or killed in accidents at home, at work, school, at play, or while traveling than were injured or killed in either the Korean or Vietnam wars.

Efforts to lessen or to eliminate the conditions that cause accidents are known as safety measures. Safety is a growing concern around the world, and safety skills are being taken more seriously today than ever before. People have come to realize that safety can be learned. Safety experts agree that it is possible to predict and prevent the majority of accidents. Few accidents simply "happen." *Most accidents are caused by ignorance, carelessness, neglect, or lack of skill.*

Since the introduction of automatic devices to move and handle materials, the exposure of workers to mechanical and handling hazards has been greatly reduced; however, the basic principles of safeguarding continue to be of great importance. Injuries result when loose clothing or hair is caught in rotating mechanisms; when fingers and hands are crushed in rollers, meshing gear teeth, belts, and chain drives; and when moving parts supply cutting, shearing, or crushing forces. Machine guards are designed to prevent such injuries. They may be fixed guards or automatic interlocking guards that prevent the operation of a machine unless a guard is in position at the danger point. Other types of guards prevent the operator from coming in contact with the dangerous part through a barrier, through devices that push the hands away, or through the use of sensor devices that stop the machine when hands are put into the danger zone.

Making sure that the machinery, tools, and furniture associated with a job fit the workers who do that job is known as ergonomics, or human engineering. A properly designed workplace can reduce worker fatigue and increase safety on the job.

Although many aspects of materials handling have been taken over by machine, a great deal of manual lifting and carrying must still be done in industry. In such cases, it is important to avoid unsafe work practices, such as improper lifting, carrying too heavy a load, or incorrect gripping. Workers must take responsibility for their own safety.

The Risk: Performing a Risk Assessment (the Preventive Management Tool)

The identification and the analysis of risk, and making rational decisions based on the known risk, are the best preventive management tools a maintenance workforce can use. The risk assessment will allow a company to maximize its safety program and thus reduce accidents. The risk assessment process should be formalized in order to reduce accidents that could cause injuries, death, machine damage, and longer equipment stoppage. One must learn in maintenance that one must reduce risk at all times. In any task there are risk and thus the reduction or elimination of risk will make all tasks more successful and safe.

In order to perform a risk assessment one must know the risk and then identify countermeasures to reduce or eliminate risk. Once one has reduced or eliminated the risk, a maintenance person can then perform the task with known acceptable risk. The Risk Assessment Worksheet shown below will assist in the explanation of this process.

Risk Assessment Worksheet

The Risk Assessment Worksheet should be used for all maintenance tasks. What is known in maintenance is that a maintenance organization, in order to be successful, must have defined processes, and thus this worksheet is one of the important processes one must utilize in order to be successful. What makes this process so important is that lives are at risk if we do not follow it.

Note: The score noted on the far right has three numbers to score each risk after countermeasures are in place. Once the assessment is completed then all scores are then added and matched to the Task Risk Scale at the bottom of the chart.

WARNING: Sample Only—Table 4.1: This table is designed to provide a maintenance person with the understanding of the risk assessment process and should not be used for any determination of risk or safety criteria.

Table 4.1 *Risk Assessment Worksheet*

Risk Assessment Worksheet
Task: Replace 100hp Electric Motor
Date: 10/19/02
Name of Person Performing Risk Assessment: Ricky Smith

Risk	**Potential safety hazard**	**Countermeasures**	**Score**
Lifting Chain Break—single chain lift	Death, injury, equipment damage, loss of equipment run time	Inspect chain and safety latches to insure they do not have damage (replace if damaged; do not use) and are marked and identified to lift the direct lifting load you are about to lift (note lifting chains must have identified lifting capabilities marked by the chain manufacturer).	1
Lifting shackle does not have center pin and thus a grade 8 bolt will be used	Death, injury, equipment damage, loss of equipment run time	None needed.	3
One maintenance person and one unskilled person changing the motor	Death, injury, equipment damage, loss of equipment run time	Identify another person skilled and trained in rigging and hoisting.	1
Boom truck (crane truck)—Has not been load tested within the last 12 months	Death, injury, equipment damage, loss of equipment run time	None needed.	3
			Score total = 8

Risk = High/Danger

Note: Task risk scale: The score for a risk is only determined once the countermeasures are followed; 1 = no risk; 2 = moderate risk; 3 = high risk (other countermeasures must be found). ANY 3 must be addressed immediately before the rigging and hoisting can begin.

Total task risk: add all scores from each risk and match to the final score scale.

Final score scale = minimal risk: 4–5; moderate risk (review with higher management before task begins): 6–7; high risk/danger (task will not begin until reduction of risk to level acceptable by management): 8 and above.

Risk Assessment Conclusion

As one can establish from the chart above, if one can identify the known risk in a specific task then an organization can reduce or eliminate accidents. One can use the risk assessment process for not only large, complex tasks, but for all tasks. A simple laminated risk assessment card can be use by maintenance supervisors and maintenance personnel for emergency and just-in-time tasks. A maintenance planner should use the risk assessment process for all jobs planned. A copy of the Risk Assessment Worksheet should be a part of the job plan package given to maintenance technicians.

Lockout/Tagout/Tryout

All machinery or equipment capable of movement must be de-energized or disengaged and blocked or locked out during cleaning, servicing, adjusting, or setting up operations, whenever required. The locking out of the control circuits in lieu of locking out main power disconnects is prohibited.

All equipment control valve handles must be provided with a means for locking out. The lockout procedure requires that stored energy (i.e., mechanical, hydraulic, air) be released or blocked before equipment is locked out for repairs.

Appropriate employees are provided with individually keyed personal safety locks. Employees are required to keep personal control of their key(s) while they have safety locks in use. Employees must check the safety of the lockout by attempting a startup after making sure no one is exposed.

Where the power disconnect does not also disconnect the electrical control circuit, the appropriate electrical enclosures must be identified. The control circuit can also be disconnected and locked out.

Manual Lifting Rules

Manual lifting and handling of material must be done by methods that ensure the safety of both the employee and the material. It is important that employees whose work assignments require heavy lifting be trained and physically qualified by medical examination if necessary.

The following are rules for manual lifting: Inspect for sharp edges, slivers, and wet or greasy spots. Wear gloves when lifting or handling objects with sharp or splintered edges. These gloves must be free of oil, grease, or other agents that may cause a poor grip.

Inspect the route over which the load is to be carried. It should be free of obstructions or spillage that could cause tripping or slipping.

Consider the distance the load is to be carried. Recognize the fact that your gripping power may weaken over long distances. Size up the load and make a preliminary "lift" to be sure the load is easily within your lifting capacity. If not, get help. If team lifting is required, personnel should be similar in size and physique. One person should act as leader and give the commands to lift, lower, etc. Two persons carrying a long piece of pipe or lumber should carry it on the same shoulder and walk in step.

Shoulder pads may be used to prevent cutting shoulders and help reduce fatigue. To lift an object off the ground, the following are manual lifting steps: Make sure of good footing and set your feet about 10 to 15 inches apart. It may help to set one foot forward. Assume a knee-bend or squatting position, keeping your back straight and upright. Get a firm grip and lift the object by straightening your knees—not your back. Carry the load close to your body (not on extended arms). To turn or change your position, shift your feet—don't twist your back. The steps for setting an object on the ground are the same as above, but reversed.

Power-Actuated Tools

Employees using power-actuated tools must be properly trained. All power-actuated tools must be left disconnected until they are actually ready to be used. Each day before using, each power-actuated tool must be inspected for obstructions or defects.

The power-actuated tool operators must have and use appropriate personal protective equipment such as hard hats, safety goggles, safety shoes, and ear protectors whenever they are using the equipment.

Machine Guarding

Before operating any machine, the employee should have completed a training program on safe methods of operation. All machinery and equipment must be kept clean and properly maintained.

There must be sufficient clearance provided around and between machines to allow for safe operations, setup, servicing, material handling, and waste removal. All equipment and machinery should be securely placed, and anchored when necessary, to prevent tipping or other movement that could result in damage or personal injury.

Most machinery should be bolted to the floor to prevent falling during an earthquake. Also, the electrical cord should be fixed to a breaker or other shutoff device to stop power in case of machine movement. There should be a power shutoff switch within reach of the operator's position. Electrical power to each machine must be capable of being locked out for maintenance, repair, or security.

The noncurrent-carrying metal parts of electrically operated machines must be bonded and grounded. Foot-operated switches should be guarded and/or arranged to prevent accidental actuation by personnel or falling objects.

All manually operated valves and switches controlling the operation of equipment and machines must be clearly identified and readily accessible. All EMERGENCY stop buttons should be colored RED. All the sheaves and belts that are within 7 feet of the floor or working level should be properly guarded.

All moving chains and gears must be properly guarded. All splash guards mounted on machines that use coolant must be positioned to prevent coolant from splashing the employees. The machinery guards must be secure and arranged so they do not present a hazard. All special hand tools used for placing and removing material must protect the operator's hands. All revolving drums, barrels, and containers should be guarded by an enclosure that is interlocked with the drive mechanisms, so that revolution cannot occur unless the guard enclosure is in place. All arbors and mandrels must have firm and secure bearings and be free of play. A protective mechanism should be installed to prevent machines from automatically starting when power is restored after a power failure or shutdown.

Machines should be constructed so as to be free from excessive vibration when under full load or mounted and running at full speed. If the machinery is cleaned with compressed air, the air must be pressure controlled, and personal protective equipment or other safeguards must be used to protect operators and other workers from eye and bodily injury.

All fan blades should be protected by a guard having openings no larger than $\frac{1}{2}$ inch when operating within 7 feet of the floor. Saws used for ripping equipment must be installed with antikickback devices and spreaders. All radial arm saws must be arranged so that the cutting head will gently return to the back of the table when released.

5 Rotor Balancing

Mechanical imbalance is one of the most common causes of machinery vibration and is present to some degree on nearly all machines that have rotating parts or rotors. Static, or standing, imbalance is the condition when there is more weight on one side of a centerline than the other. However, a rotor may be in perfect static balance and not be in a balanced state when rotating at high speed.

If the rotor is a thin disk, careful static balancing may be accurate enough for high speeds. However, if the rotating part is long in proportion to its diameter, and the unbalanced portions are at opposite ends or in different planes, the balancing must counteract the centrifugal force of these heavy parts when they are rotating rapidly.

This section provides information needed to understand and solve the majority of balancing problems using a vibration/balance analyzer, a portable device that detects the level of imbalance, misalignment, etc., in a rotating part based on the measurement of vibration signals.

Sources of Vibration due to Mechanical Imbalance

Two major sources of vibration due to mechanical imbalance in equipment with rotating parts or rotors are: (1) assembly errors and (2) incorrect key length guesses during balancing.

Assembly Errors

Even when parts are precision balanced to extremely close tolerances, vibration due to mechanical imbalance can be much greater than necessary due to assembly errors. Potential errors include relative placement of each part's center of rotation, location of the shaft relative to the bore, and cocked rotors.

Center of Rotation

Assembly errors are not simply the additive effects of tolerances, but also include the relative placement of each part's center of rotation. For example, a "perfectly" balanced blower rotor can be assembled to a "perfectly" balanced shaft, and yet the resultant imbalance can be high. This can happen if the rotor is balanced on a balancing shaft that fits the rotor bore within 0.5 mils (0.5 thousandths of an inch) and then is mounted on a standard cold-rolled steel shaft allowing a clearance of over 2 mils.

Shifting any rotor from the rotational center on which it was balanced to the piece of machinery on which it is intended to operate can cause an assembly imbalance four to five times greater than that resulting simply from tolerances. For this reason, all rotors should be balanced on a shaft having a diameter as nearly the same as the shaft on which they will be assembled.

For best results, balance the rotor *on its own shaft* rather than on a balancing shaft. This may require some rotors to be balanced in an overhung position, a procedure the balancing shop often wishes to avoid. However, it is better to use this technique rather than being forced to make too many balancing shafts. The extra precision balance attained by using this procedure is well worth the effort.

Method of Locating Position of Shaft Relative to Bore

Imbalance often results with rotors that do not incorporate setscrews to locate the shaft relative to the bore (e.g., rotors that are end-clamped). In this case, the balancing shaft is usually horizontal. When the operator slides the rotor on the shaft, gravity causes the rotor's bore to make contact at the 12 o'clock position on the top surface of the shaft. In this position, the rotor is end-clamped in place and then balanced.

If the operator removes the rotor from the balancing shaft without marking the point of bore and shaft contact, it may not be in the same position when reassembled. This often shifts the rotor by several mils as compared to the axis on which it was balanced, thus causing an imbalance to be introduced. The vibration that results is usually enough to spoil what should have been a precision balance and produce a barely acceptable vibration level. In addition, if the resultant vibration is resonant with some part of the machine or structure, a more serious vibration could result.

To prevent this type of error, the balancer operators and those who do final assembly should follow the following procedure. The balancer operator should permanently mark the location of the contact point between the bore and the shaft during balancing. When the equipment is reassembled in the plant or the shop, the assembler should also use this mark. For end-clamped rotors, the assembler should slide the bore on the horizontal shaft, rotating both until the mark is at the 12 o'clock position, and then clamp it in place.

Cocked Rotor

If a rotor is cocked on a shaft in a position different from the one in which it was originally balanced, an imbalanced assembly will result. If, for example, a pulley has a wide face that requires more than one setscrew, it could be mounted on-center, but be cocked in a different position than during balancing. This can happen by reversing the order in which the setscrews are tightened against a straight key during final mounting as compared to the order in which the setscrews were tightened on the balancing arbor. This can introduce a pure couple imbalance, which adds to the small couple imbalance already existing in the rotor and causes unnecessary vibration.

For very narrow rotors (i.e., disk-shaped pump impellers or pulleys), the distance between the centrifugal forces of each half may be very small. Nevertheless, a very high centrifugal force, which is mostly counterbalanced statically by its counterpart in the other half of the rotor, can result. If the rotor is slightly cocked, the small axial distance between the two very large centrifugal forces causes an appreciable couple imbalance, which is often several times the allowable tolerance. This is due to the fact that the centrifugal force is proportional to half the rotor weight (at any one time, half of the rotor is pulling against the other half) times the radial distance from the axis of rotation to the center of gravity of that half.

To prevent this, the assembler should tighten each setscrew gradually—first one, then the other, and back again—so that the rotor is aligned evenly. On flange-mounted rotors such as flywheels, it is important to clean the mating surfaces and the bolt holes. Clean bolt holes are important because high couple imbalance can result from the assembly bolt pushing a small amount of dirt between the surfaces, cocking the rotor. Burrs on bolt holes also can produce the same problem.

Other

There are other assembly errors that can cause vibration. Variances in bolt weights when one bolt is replaced by one of a different length or material can cause vibration. For setscrews that are 90 degrees apart, the tightening sequence may not be the same at final assembly as during balancing. To prevent this, the balancer operator should mark which was tightened first.

Key Length

With a keyed-shaft rotor, the balancing process can introduce machine vibration if the assumed key length is different from the length of the one used during operation. Such an imbalance usually results in a mediocre or "good" running machine as opposed to a very smooth running machine.

For example, a "good" vibration level that can be obtained without following the precautions described in this section is amplitude of 0.12 inches/second (3.0 mm/sec.). By following the precautions, the orbit can be reduced to about 0.04 in./sec. (1 mm/sec.). This smaller orbit results in longer bearing or seal life, which is worth the effort required to make sure that the proper key length is used.

When balancing a keyed-shaft rotor, one half of the key's weight is assumed to be part of the shaft's male portion. The other half is considered to be part of the female portion that is coupled to it. However, when the two rotor parts are sent to a balancing shop for rebalancing, the actual key is rarely included. As a result, the balance operator usually guesses at the key's length, makes up a half key, and then balances the part. (Note: A "half key" is of full-key length, but only half-key depth.)

In order to prevent an imbalance from occurring, **do not allow the balance operator to guess the key length**. It is strongly suggested that the actual key length be recorded on a tag that is attached to the rotor to be balanced. The tag should be attached in such a way that another device (such as a coupling half, pulley, fan, etc.) cannot be attached until the balance operator removes the tag.

Theory of Imbalance

Imbalance is the condition in which there is more weight on one side of a centerline than the other. This condition results in unnecessary vibration, which generally can be corrected by the addition of counterweights. There are four types of imbalance: (1) static, (2) dynamic, (3) coupled, and (4) dynamic imbalance combinations of static and couple.

Static

Static imbalance is single-plane imbalance acting through the center of gravity of the rotor, perpendicular to the shaft axis. The imbalance also can be separated into two separate single-plane imbalances, each acting in-phase or at the same angular relationship to each other (i.e., 0 degrees apart). However, the net effect is as if one force is acting through the center of gravity. For a uniform straight cylinder such as a simple paper machine roll or a multigrooved sheave, the forces of static imbalance measured at each end of the rotor are equal in magnitude (i.e., the ounce-inches or gram-centimeters in one plane are equal to the ounce-inches or gram-centimeters in the other).

In static imbalance, the only force involved is weight. For example, assume that a rotor is perfectly balanced and, therefore, will not vibrate regardless of the speed of rotation. Also assume that this rotor is placed on frictionless rollers or "knife edges." If a weight is applied on the rim at the center of gravity line between two ends, the weighted portion immediately rolls to the 6 o'clock position due to the gravitational force.

When rotation occurs, static imbalance translates into a centrifugal force. As a result, this type of imbalance is sometimes referred to as "force imbalance," and some balancing machine manufacturers use the word "force" instead of "static" on their machines. However, when the term "force imbalance" was just starting to be accepted as the proper term, an American standardization committee on balancing terminology standardized the term "static" instead of "force." The rationale was that the role of the standardization committee was not to determine and/or correct right or wrong practices, but to standardize those currently in use by industry. As a result, the term "static imbalance" is now widely accepted as the international standard and, therefore, is the term used here.

Dynamic

Dynamic imbalance is any imbalance resolved to at least two correction planes (i.e., planes in which a balancing correction is made by adding or removing weight). The imbalance in each of these two planes may be the result of many imbalances in many planes, but the final effects can be limited to only two planes in almost all situations.

An example of a case where more than two planes are required is flexible rotors (i.e., long rotors running at high speeds). High speeds are considered to be revolutions per minute (rpm) higher than about 80% of the rotor's first critical speed. However, in over 95% of all run-of-the-mill rotors (e.g., pump impellers, armatures, generators, fans, couplings, pulleys, etc.), two-plane dynamic balance is sufficient. Therefore, flexible rotors are not covered in this document because of the low number in operation and the fact that specially trained people at the manufacturer's plant almost always perform balancing operations.

In dynamic imbalance, the two imbalances do not have to be equal in magnitude to each other, nor do they have to have any particular angular reference to each other. For example, they could be 0 (in-phase), 10, 80, or 180 degrees from each other.

Although the definition of dynamic imbalance covers all two-plane situations, an understanding of the components of dynamic imbalance is needed so that its causes can be understood. Also, an understanding of the components makes it easier to understand why certain types of balancing do not always work with many older balancing machines for overhung rotors and very narrow rotors. The primary components of dynamic imbalance include: number of points of imbalance, amount of imbalance, phase relationships, and rotor speed.

Points of Imbalance

The first consideration of dynamic balancing is the number of imbalance points on the rotor, as there can be more than one point of imbalance within a rotor assembly. This is especially true in rotor assemblies with more than one rotating element, such as a three-rotor fan or multistage pump.

Amount of Imbalance

The amplitude of each point of imbalance must be known to resolve dynamic balance problems. Most dynamic balancing machines or *in situ* balancing instruments are able to isolate and define the specific amount of imbalance at each point on the rotor.

Phase Relationship

The phase relationship of each point of imbalance is the third factor that must be known. Balancing instruments isolate each point of imbalance and determine their phase relationship. Plotting each point of imbalance on a polar plot does this. In simple terms, a polar plot is a circular display of the shaft end. Each point of imbalance is located on the polar plot as a specific radial, ranging from 0 to 360 degrees.

Rotor Speed

Rotor speed is the final factor that must be considered. Most rotating elements are balanced at their normal running speed or over their normal speed range. As a result, they may be out of balance at some speeds that are not included in the balancing solution. As an example, the wheel and tires on your car are dynamically balanced for speeds ranging from zero to the maximum expected speed (i.e., eighty miles per hour). At speeds above eighty miles per hour, they may be out of balance.

Coupled

Coupled imbalance is caused by two equal noncollinear imbalance forces that oppose each other angularly (i.e., 180 degrees apart). Assume that a rotor with pure coupled imbalance is placed on frictionless rollers. Because the imbalance weights or forces are 180 degrees apart and equal, the rotor is statically balanced. However, a pure coupled imbalance occurs if this same rotor is revolved at an appreciable speed.

Each weight causes a centrifugal force, which results in a rocking motion or rotor wobble. This condition can be simulated by placing a pencil on a table, then at one end pushing the side of the pencil with one finger. At the same time, push in the opposite direction at the other end. The pencil will tend to rotate end-over-end. This end-over-end action causes two imbalance "orbits," both 180 degrees out of phase, resulting in a "wobble" motion.

Dynamic Imbalance Combinations of Static and Coupled

Visualize a rotor that has only one imbalance in a single plane. Also visualize that the plane is *not* at the rotor's center of gravity, but is off to one side. Although there is no other source of couple, this force to one side of the rotor not only causes translation (parallel motion due to pure static imbalance), but also causes the rotor to rotate or wobble end-over-end as from a couple. In other words, such a force would create a combination of both static and couple imbalance. This again is dynamic imbalance.

In addition, a rotor may have two imbalance forces exactly 180 degrees opposite to each other. However, if the forces are not equal in magnitude, the rotor has a static imbalance in combination with its pure couple. This combination is also dynamic imbalance.

Another way of looking at it is to visualize the usual rendition of dynamic imbalance—imbalance in two separate planes at an angle and magnitude relative to each other not necessarily that of pure static or pure couple.

For example, assume that the angular relationship is 80 degrees and the magnitudes are 8 units in one plane and 3 units in the other. Normally, you would simply balance this rotor on an ordinary two-plane dynamic balancer and that would be satisfactory. But for further understanding of balancing, imagine that this same rotor is placed on static balancing rollers, whereby gravity brings the static imbalance components of this dynamically out-of-balance rotor to the 6 o'clock position.

The static imbalance can be removed by adding counter-balancing weights at the 12 o'clock position. Although statically balanced, however, the two remaining forces result in a pure coupled imbalance. With the entire static imbalance removed, these two forces are equal in magnitude and exactly 180 degrees apart. The coupled imbalance can be removed, as with any other coupled imbalance, by using a two-plane dynamic balancer and adding counterweights.

Note that whenever you hear the word "imbalance," you should mentally add the word "dynamic" to it. Then when you hear "dynamic imbalance," mentally visualize "combination of static and coupled imbalance." This will be of much help not only in balancing, but in understanding phase and coupling misalignment as well.

Balancing

Imbalance is one of the most common sources of major vibration in machinery. It is the main source in about 40% of the excessive vibration situations. The vibration frequency of imbalance is equal to one times the rpm (1 × rpm) of the imbalanced rotating part.

Before a part can be balanced using the vibration analyzer, certain conditions must be met:

- The vibration must be due to mechanical imbalance;
- Weight corrections can be made on the rotating component.

In order to calculate imbalance units, simply multiply the amount of imbalance by the radius at which it is acting. In other words, one ounce of imbalance at a one-inch radius will result in one oz.-in. of imbalance. Five ounces at one-half inch radius results in $2\frac{1}{2}$ oz.-in. of imbalance. (Dynamic imbalance units are measured in ounce-inches [oz.-in.] or gram-millimeters [g.-mm.].) Although this refers to a single plane, dynamic balancing is performed in at least two separate planes. Therefore, the tolerance is usually given in single-plane units for each plane of correction.

Important balancing techniques and concepts to be discussed in the sections to follow include: in-place balancing, single-plane versus two-plane balancing, precision balancing, techniques that make use of a phase shift, and balancing standards.

In-Place Balancing

In most cases, weight corrections can be made with the rotor mounted in its normal housing. The process of balancing a part without taking it out of the machine is called *in-place balancing*. This technique eliminates costly and time consuming disassembly. It also prevents the possibility of damage to the rotor, which can occur during removal, transportation to and from the balancing machine, and reinstallation in the machine.

Single-Plane versus Two-Plane Balancing

The most common rule of thumb is that a disk-shaped rotating part usually can be balanced in one correction plane only, whereas parts that have appreciable width require two-plane balancing. Precision tolerances, which become more meaningful for higher performance (even on relatively narrow face width), suggest two-plane balancing. However, the width should be the guide, not the diameter-to-width ratio.

For example, a 20" wide rotor could have a large enough couple imbalance component in its dynamic imbalance to require two-plane balancing. (Note: The couple component makes two-plane balancing important.) Yet, if the 20" width is on a rotor of large diameter that qualifies as a "disk-shaped rotor," even some of the balance manufacturers erroneously would call for a single-plane balance.

It is true that the narrower the rotor, the less the chance for a large couple component and, therefore, the greater the possibility of getting by with a single-plane balance. For rotors over 4" to 5" in width, it is best to check for real dynamic imbalance (or for couple imbalance). Unfortunately, you cannot always get by with a static- and couple-type balance, even for very narrow flywheels used in automobiles. Although most of the flywheels are only 1" to $1\frac{1}{2}$" wide, more than half have enough couple imbalance to cause excessive vibration. This obviously is not due to a large distance between the planes (width), but due to the fact that the flywheel's mounting surface can cause it to be slightly cocked or tilted. Instead of the flywheel being 90 degrees to the shaft axis, it may be perhaps 85 to 95 degrees, causing a large couple despite its narrow width.

This situation is very common with narrow and disc-shaped industrial rotors such as single-stage turbine wheels, narrow fans, and pump impellers. The original manufacturer often accepts the guidelines supplied by others and performs a single-plane balance only. By obtaining separate readings for static and couple, the manufacturer could and should easily remove the remaining couple.

An important point to remember is that static imbalance is always removed first. In static and couple balancing, remove the static imbalance first, and then remove the couple.

Precision Balancing

Most original-equipment manufacturers balance to commercial tolerances, a practice that has become acceptable to most buyers. However, due to frequent customer demands, some of the equipment manufacturers now provide precision balancing. Part of the driving force for providing this service is that many large mills and refineries have started doing their own precision balancing to tolerances considerably closer than those used by the original-equipment manufacturer. For example, the International Standards Organization (ISO) for process plant machinery calls for a G6.3 level of balancing in its balancing guide. This was calculated based on a rotor running free in space with a restraint vibration of 6.3 mm/sec. (0.25 in./sec.) vibration velocity.

Precision balancing requires a G2.5 guide number, which is based on 2.5 mm/sec. (0.1 in./sec.) vibration velocity. As can be seen from this, 6.3 mm/sec. (0.25 in./sec.) balanced rotors will vibrate more than the 2.5 mm/sec. (0.1 in./sec.) precision balanced rotors. Many vibration guidelines now consider 2.5 mm/sec. (0.1 in./sec.) "good," creating the demand for precision balancing. Precision balancing tolerances can produce velocities of 0.01 in./sec. (0.3 mm/sec.) and lower.

It is true that the extra weight of nonrotating parts (i.e., frame and foundation) reduces the vibration somewhat from the free-in-space amplitude. However, it is possible to reach precision balancing levels in only two or three additional runs, providing the smoothest running rotor. The extra effort to the balance operator is minimal because he already has the "feel" of the rotor and has the proper setup and tools in hand. In addition, there is a large financial payoff for this minimal extra effort due to decreased bearing and seal wear.

Techniques Using Phase Shift

If we assume that there is no other source of vibration other than imbalance (i.e., we have perfect alignment, a perfectly straight shaft, etc.), it is readily seen that pure static imbalance gives in-phase vibrations, and pure coupled imbalance gives various phase relationships. Compare the *vertical* reading of a bearing at one end of the rotor with the *vertical* reading at the other end of the rotor to determine how that part is shaking vertically. Then compare the *horizontal* reading at one end with the *horizontal* reading at the other end to determine how the part is shaking horizontally.

If there is no resonant condition to modify the resultant vibration phase, then the phase for both vertical and horizontal readings is essentially the same even though the vertical and horizontal amplitudes do not necessarily correspond. In actual practice, this may be slightly off due to other vibration sources such as misalignment. In performing the analysis, what counts is that when the source of the vibration is *primarily from imbalance*, then the vertical reading phase differences between one end of the rotor and the other will be very similar to the phase differences when measured horizontally. For example, vibrations 60 degrees out of phase vertically would show 60 degrees out of phase horizontally within 20%.

However, the horizontal reading on one bearing will not show the same phase relationship as the vertical reading on the same bearing. This is due to the pickup axis being oriented in a different angular position, as well as the phase adjustment due to possible resonance. For example, the horizontal vibration frequency may be below the horizontal resonance of various major portions of machinery, whereas the vertical vibration frequency may be above the natural frequency of the floor supporting the machine.

First, determine how the rotor is vibrating vertically by comparing "vertical only" readings with each other. Then, determine how the rotor is vibrating horizontally. If, the rotor is shaking horizontally and vertically and the phase differences are relatively similar, then the source of vibration is likely to be *imbalance*. However, before coming to a final conclusion, be sure that other 1 × rpm sources (e.g., bent shaft, eccentric armature, misaligned coupling) are not at fault.

Balancing Standards

The ISO has published standards for acceptable limits for residual imbalance in various classifications of rotor assemblies. Balancing standards are given in oz-in. or lb-in. per pound of rotor weight or the equivalent in metric units (g-mm/kg). The oz-in. are for each correction plane for which the imbalance is measured and corrected.

Caution must be exercised when using balancing standards. The recommended levels are for residual imbalance, which is defined as imbalance of any kind that remains *after* balancing.

Figure 5.1 and Table 5.1 are the norms established for most rotating equipment. Additional information can be obtained from ISO 5406 and 5343.

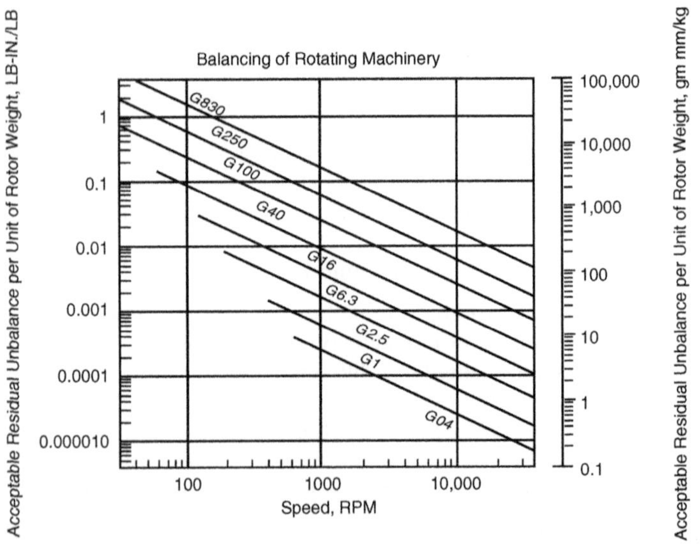

Figure 5.1 *Balancing standards: residual imbalance per unit rotor weight*

Table 5.1 *Balance quality grades for various groups of rigid rotors*

Balance quality grade	Type of rotor
G4,000	Crankshaft drives of rigidly mounted slow marine diesel engines with uneven number of cylinders.
G1,600	Crankshaft drives of rigidly mounted large two-cycle engines.
G630	Crankshaft drives of rigidly mounted large four-cycle engines; crankshaft drives of elastically mounted marine diesel engines.
G250	Crankshaft drives of rigidly mounted fast four-cylinder diesel engines.
G100	Crankshaft drives of fast diesel engines with six or more cylinders; complete engines (gasoline or diesel) for cars and trucks.
G40	Car wheels, wheel rims, wheel sets, drive shafts; crankshaft drives of elastically mounted fast four-cycle engines (gasoline and diesel) with six or more cylinders; crankshaft drives for engines of cars and trucks.
G16	Parts of agricultural machinery; individual components of engines (gasoline or diesel) for cars and trucks.
G6.3	Parts or process plant machines; marine main-turbine gears; centrifuge drums; fans; assembled aircraft gas-turbine rotors; flywheels; pump impellers; machine-tool and general machinery parts; electrical armatures.
G2.5	Gas and steam turbines; rigid turbo-generator rotors; rotors; turbo-compressors; machine-tool drives; small electrical armatures; turbine-driven pumps.
G1	Tape recorder and phonograph drives; grinding-machine drives.
G0.4	Spindles, disks, and armatures of precision grinders; gyroscopes.

Similar standards are available from the American National Standards Institute (ANSI) in their publication ANSI S2.43-1984.

So far, there has been no consideration of the angular positions of the usual two points of imbalance relative to each other or the distance between the two correction planes. For example, if the residual imbalances in each of the two planes were in-phase, they would add to each other to create more static imbalance.

Most balancing standards are based on a *residual* imbalance and do not include multiplane imbalance. If they are approximately 180 degrees to each other, they form a couple. If the distance between the planes is small, the resulting couple is small; if the distance is large, the couple is large. A couple creates considerably more vibration than when the two residual imbalances are in-phase. Unfortunately, there is nothing in the balancing standards that takes this into consideration.

There is another problem that could also result in excessive imbalance-related vibration even though the ISO standards were met. The ISO standards call for a balancing grade of G6.3 for components such as pump impellers, normal electric armatures, and parts of process plant machines. This results in an operating speed vibration velocity of 6.3 mm/sec. (0.25 in./sec.) vibration velocity. However, practice has shown that an acceptable vibration velocity is 0.1 in./sec. and the ISO standard of G2.5 is really required. As a result of these discrepancies, changes in the recommended balancing grade are expected in the future.

6 Bearings

A bearing is a machine element that supports a part, such as a shaft, that rotates, slides, or oscillates in or on it. There are two broad classifications of bearings: plain and rolling element (also called antifriction). Plain bearings are based on sliding motion made possible through the use of a lubricant. Antifriction bearings are based on rolling motion, which is made possible by balls or other types of rollers. In modern rotor systems operating at relatively high speeds and loads, the proper selection and design of the bearings and bearing-support structure are key factors affecting system life.

Types of Movement

The type of bearing used in a particular application is determined by the nature of the relative movement and other application constraints. Movement can be grouped into the following categories: rotation about a point, rotation about a line, translation along a line, rotation in a plane, and translation in a plane. These movements can be either continuous or oscillating.

Although many bearings perform more than one function, they can generally be classified based on types of movement. There are three major classifications of both plain and rolling element bearings: radial, thrust, and guide. *Radial* bearings support loads that act radially and at right angles to the shaft center line. These loads may be visualized as radiating into or away from a center point like the spokes on a bicycle wheel. *Thrust* bearings support or resist loads that act axially. These may be described as endwise loads that act parallel to the center line toward the ends of the shaft. This type of bearing prevents lengthwise or axial motion of a rotating shaft. *Guide* bearings support and align members having sliding or reciprocating motion. This type of bearing guides a machine element in its lengthwise motion, usually without rotation of the element.

Table 6.1 gives examples of bearings that are suitable for continuous movement; Table 6.2 shows bearings that are appropriate for oscillatory movement only. For the bearings that allow movements in addition to the one listed, the effect on machine design is described in the column "Effect

Constraint applied to the movement	Examples of arrangements which allow movement only within this constraint	Examples of arrangements which allow movement but also have other degrees of freedom	Effect of the other degrees of freedom
About a point	Gimbals 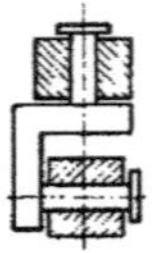	Ball on a recessed plate 	Ball must be forced into contact with the plate
About a line	Journal bearing with double thrust location 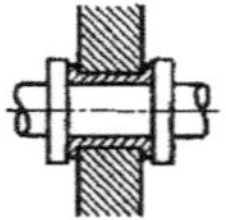	Journal bearing 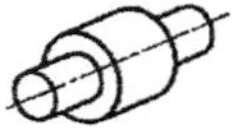	Simple journal bearing allows free axial movement as well
	Double conical bearing 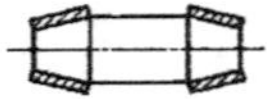	Screw and nut 	Gives some related axial movement as well
		Ball joint or spherical roller bearing 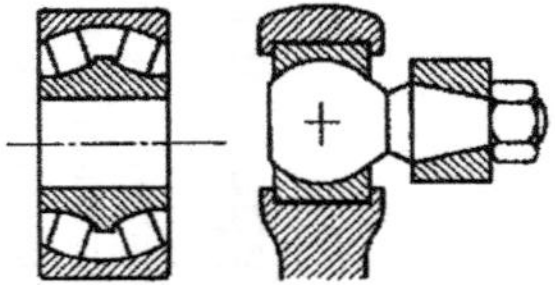	Allows some angular freedom to the line of rotation

Table 6.1 *continued*

About a line	Crane wheels restrained between two rails 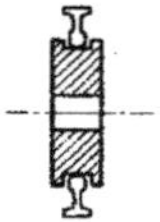	Railway or crane wheel on a track 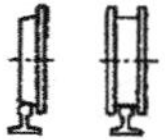 Pulley wheel on a cable 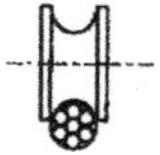 Hovercraft or hoverpad on a track 	These arrangements need to be loaded into contact. This is usually done by gravity. Wheels on a single rail or cable need restraint to prevent rotation about the track member
In a plane (rotation)	Double thrust bearing 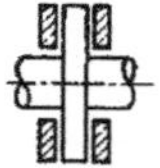	Single thrust bearing 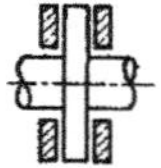	Single thrust bearing must be loaded into contact
In a plane (translation)		Hovercraft or hoverpad 	Needs to be loaded into contact usually by gravity

Source: M.J. Neale, Society of Automotive Engineers Inc. *Bearings—A Tribology Handbook.* Oxford: Butterworth–Heinemann, 1993.

Table 6.2 Bearing selection guide (oscillatory movement)

Constraint applied to the movement	Examples of arrangements which allow movement only within this constraint	Examples of arrangements which allow this movement but also have other degrees of freedom	Effect of the other degrees of freedom
About a point	Hookes joint 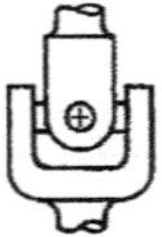	Cable connection between components 	Cable needs to be kept in tension
About a line	Crossed strip flexure pivot 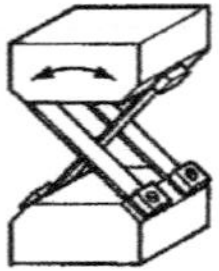	Torsion suspension 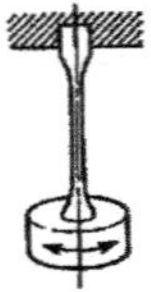	A single torsion suspension gives no lateral location
		Knife-edge pivot 	Must be loaded into contact
		Rubber bush 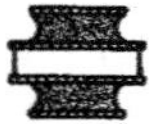	Gives some axial and lateral flexibility as well
		Rocker pad 	Gives some related translation as well. Must be loaded into contact

Table 6.2 *continued*

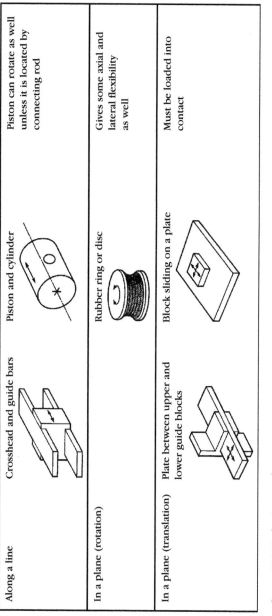

Along a line	Crosshead and guide bars 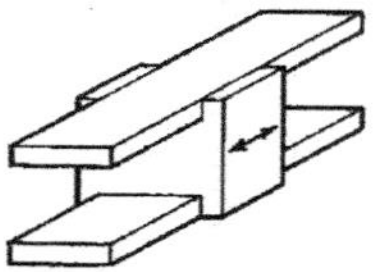	Piston and cylinder 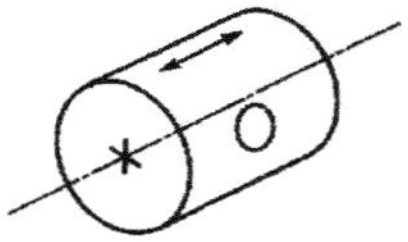	Piston can rotate as well unless it is located by connecting rod
In a plane (rotation)		Rubber ring or disc 	Gives some axial and lateral flexibility as well
In a plane (translation)	Plate between upper and lower guide blocks 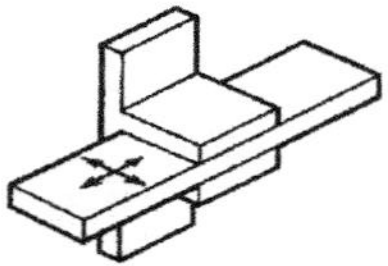	Block sliding on a plate 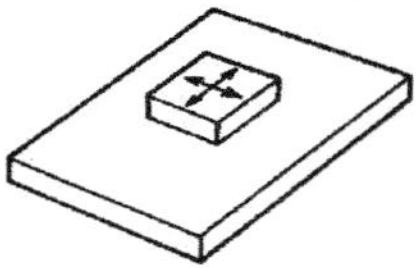	Must be loaded into contact

Source: M.J. Neale, Society of Automotive Engineers Inc. *Bearings—A Tribology Handbook.* Oxford: Butterworth–Heinemann, 1993.

Table 6.3 *Comparison of plain and rolling element bearings*

Rolling element	Plain
Assembly on crankshaft is virtually impossible, except with very short or built-up crankshafts	Assembly on crankshaft is no problem as split bearings can be used
Cost relatively high	Cost relatively low
Hardness of shaft unimportant	Hardness of shaft important with harder bearings
Heavier than plain bearings	Lighter than rolling element bearings
Housing requirement not critical	Rigidity and clamping most important housing requirement
Less rigid than plain bearings	More rigid than rolling element bearings
Life limited by material fatigue	Life not generally limited by material fatigue
Lower friction results in lower power consumption	Higher friction causes more power consumption
Lubrication easy to accomplish; the required flow is low except at high speed	Lubrication pressure feed critically important; required flow is large, susceptible to damage by contaminants and interrupted lubricant flow
Noisy operation	Quiet operation
Poor tolerance of shaft deflection	Moderate tolerance of shaft deflection
Poor tolerance of hard dirt particles	Moderate tolerance of dirt particles, depending on hardness of bearing
Requires more overall space: Length: Smaller than plain Diameter: Larger than plain	Requires less overall space: Length: Larger than rolling element Diameter: Smaller than rolling element
Running Friction: Very low at low speeds May be high at high speeds	Running Friction: Higher at low speeds Moderate at usual crank speeds
Smaller radial clearance than plain	Larger radial clearance than rolling element

Source: Integrated Systems Inc.

of the other degrees of freedom." Table 6.3 compares the characteristics, advantages, and disadvantages of plain and rolling element bearings.

About a Point (Rotational)

Continuous movement about a point is rotation, a motion that requires repeated use of accurate surfaces. If the motion is oscillatory rather than continuous, some additional arrangements must be made in which the geometric layout prevents continuous rotation.

About a Line (Rotational)

Continuous movement about a line is also referred to as rotation, and the same comments apply as for movement about a point.

Along a Line (Translational)

Movement along a line is referred to as translation. One surface is generally long and continuous, and the moving component is usually supported on a fluid film or rolling contact in order to achieve an acceptable wear rate. If the translational movement is reciprocation, the application makes repeated use of accurate surfaces, and a variety of economical bearing mechanisms are available.

In a Plane (Rotational/Translational)

If the movement in a plane is rotational or both rotational and oscillatory, the same comments apply as for movement about a point. If the movement in a plane is translational or both translational and oscillatory, the same comments apply as for movement along a line.

Commonly Used Bearing Types

As mentioned before, the major bearing classifications are plain and rolling element. These types of bearings are discussed in the sections to follow. Table 6.4 is a bearings characteristics summary. Table 6.5 is a selection guide for bearings operating with continuous rotation and special environmental conditions. Table 6.6 is a selection guide for bearings operating with continuous rotation and special performance requirements. Table 6.7 is a selection

Table 6.4 *Bearings characteristic summary*

Bearing type	Description
Plain	See Table 6.3.
Lobed	See Radial, elliptical.
Radial or journal	
Cylindrical	Gas lubricated, low-speed applications.
Elliptical	Oil lubricated, gear and turbine applications, stiffer and somewhat more stable bearing.
Four-axial grooved	Oil lubricated, higher-speed applications than cylindrical.
Partial arc	Not a bearing type, but a theoretical component of grooved and lobed bearing configurations.
Tilting pad	High-speed applications where hydrodynamic instability and misalignment are common problems.
Thrust	Semifluid lubrication state, relatively high friction, lower service pressures with multicollar version, used at low speeds.
Rolling element	See Table 6.3. Radial and axial loads, moderate- to high-speed applications.
Ball	Higher speed and lighter load applications than roller bearings.
Single-row	
Radial nonfilling slot	Also referred to as Conrad or deep-groove bearing. Sustains combined radial and thrust loads, or thrust loads alone, in either direction, even at high speeds. Not self-aligning.
Radial filling slot	Handles heavier loads than nonfilling slot.
Angular contact radial thrust	Radial loads combined with thrust loads, or heavy thrust loads alone. Axial deflection must be limited.
Ball-thrust	Very high thrust loads in one direction only, no radial loading, cannot be operated at high speeds.
Double-row	Heavy radial with minimal bearing deflection and light thrust loads.
Double-roll, self-aligning	Moderate radial and limited thrust loads.
Roller	Handles heavier loads and shock better than ball bearings, but are more limited in speed than ball bearings.

Continued

Table 6.4 *continued*

Bearing type	Description
Cylindrical	Heavy radial loads, fairly high speeds, can allow free axial shaft movement.
Needle-type cylindrical or barrel	Does not normally support thrust loads, used in space-limited applications, angular mounting of rolls in double-row version tolerates combined axial and thrust loads.
Spherical	High radial and moderate-to-heavy thrust loads, usually comes in double-row mounting that is inherently self-aligning.
Tapered	Heavy radial and thrust loads. Can be preloaded for maximum system rigidity.

Source: Integrated Systems, Inc.

guide for oscillating movement and special environment or performance requirements.

Plain Bearings

All plain bearings also are referred to as fluid-film bearings. In addition, radial plain bearings also are commonly referred to as journal bearings. Plain bearings are available in a wide variety of types or styles and may be self-contained units or built into a machine assembly. Table 6.8 is a selection guide for radial and thrust plain bearings.

Plain bearings are dependent on maintaining an adequate lubricant film to prevent the bearing and shaft surfaces from coming into contact, which is necessary to prevent premature bearing failure. However, this is difficult to achieve, and some contact usually occurs during operation. Material selection plays a critical role in the amount of friction and the resulting seizure and wear that occurs with surface contact. Note that fluid-film bearings do not have the ability to carry the full load of the rotor assembly at any speed and must have turning gear to support the rotor's weight at low speeds.

Thrust or Fixed

Thrust plain bearings consist of fixed shaft shoulders or collars that rest against flat bearing rings. The lubrication state may be semifluid, and friction

Table 6.5 *Bearings selection guide for special environmental conditions (continuous rotation)*

Bearing type	High temp.	Low temp.	Vacuum	Wet/humid	Dirt/dust	External vibration
Plain, externally pressurized	1 (With gas lubrication)	2	No (affected by lubricant feed)	2	2 (1 when gas lubricated)	1
Plain, porous metal (oil impregnated)	4 (Lubricant oxidizes)	3 (May have high starting torque)	Possible with special lubricant	2	Seals essential	2
Plain, rubbing(non-metallic)	2 (Up to temp. limit of material)	2	1	2 (Shaft must not corrode)	2 (Seals help)	2
Plain, fluid film	2 (Up to temp. limit of lubricant)	2 (May have high starting torque)	Possible with special lubricant	2	2 (With seals and filtration)	2
Rolling	Consult manufacturer above 150°C	2	3 (With special lubricant)	3 (With seals)	Sealing essential	3 (Consult manufac-turers)
Things to watch with all bearings	Effect of thermal expansion on fits	Effect of thermal expansion on fits		Corrosion		Fretting
Rating: 1 - Excellent, 2 - Good, 3 - Fair, 4 - Poor						

Source: Adapted by Integrated Systems, Inc. from *Bearings—A Tribology Handbook*, M.J. Neale, Society of AutomotiveEngineers, Inc., Butterworth–Heinemann Ltd., Oxford, Great Britain, 1993.

Table 6.6 *Bearing selection guide for particular performance requirements (continuous rotation)*

Bearing type	Accurate radial location	Axial load capacity as well	Low starting torque	Silent running	Standard parts available	Simple lubrication
Plain, externally pressurized	1	No (need separate thrust bearing)	1	1	No	4 (Need special system)
Plain, fluid film	3	No (need separate thrust bearing)	2	1	Some	2 (Usually requires circulation system)
Plain, porous metal (oil impregnated)	2	Some	2	1	Yes	1
Plain, rubbing (nonmetallic)	4	Some in most instances	4	3	Some	1
Rolling	2	Yes in most instances	1	Usually satis-factory	Yes	2 (When grease lubricated)
Rating: 1 - Excellent, 2 - Good, 3 - Fair, 4 - Poor						

Source: Adapted by Integrated Systems Inc. from M. J. Neale, Society of Automotive Engineers Inc. *Bearings—A Tribology Handbook*. Oxford: Butterworth–Heinemann, 1993.

Table 6.7 *Bearing selection guide for special environments or performance (oscillating movement)*

Bearing type	High temp.	Low temp.	Low friction	Wet/humid	Dirt/dust	External vibration
Knife edge pivots	2	2	1	2 (Watch corrosion)	2	4
Plain, porous metal (oil impregnated)	4 (Lubricant oxidizes)	3 (Friction can be high)	2	2	Sealing essential	2
Plain, rubbing	2 (Up to temp. limit of material)	1	2 (With PTFE)	2 (Shaft must not corrode)	2 (Sealing helps)	1
Rolling	Consult manufacturer above 150°C	2	1	2 (With seals)	Sealing essential	4
Rubber bushes	4	4	Elastically stiff	1	1	1
Strip flexures	2	1	1	2 (Watch corrosion)	1	1
Rating: 1 - Excellent, 2 - Good, 3 - Fair, 4 - Poor						

Source: Adapted by Integrated Systems Inc. from M.J. Neale, Society of Automotive Engineers Inc. *Bearings—A Tribology Handbook*. Oxford: Butterworth–Heinemann, 1993.

Table 6.8 *Plain bearing selection guide*

Journal bearings

Characteristics	Direct lined	Insert liners
Accuracy	Dependent upon facilities and skill available	Precision components
Quality (consistency)	Doubtful	Consistent
Cost	Initial cost may be lower	Initial cost may be higher
Ease of Repair	Difficult and costly	Easily done by replacement
Condition upon extensive use	Likely to be weak in fatigue	Ability to sustain higher peak loads
Materials used	Limited to white metals	Extensive range available

Thrust bearings

Characteristics	Flanged journal bearings	Separate thrust washer
Cost	Costly to manufacture	Much lower initial cost
Replacement	Involves whole journal/thrust component	Easily replaced without moving journal bearing
Materials used	Thrust face materials limited in larger sizes	Extensive range available
Benefits	Aids assembly on a production line	Aligns itself with the housing

Source: Adapted by Integrated Systems Inc. from M.J. Neale, Society of Automotive Engineers Inc. *Bearings—A Tribology Handbook.* Oxford: Butterworth–Heinemann, 1993.

is relatively high. In multicollar thrust bearings, allowable service pressures are considerably lower because of the difficulty in distributing the load evenly between several collars. However, thrust ring performance can be improved by introducing tapered grooves. Figure 6.1 shows a mounting half-section for a vertical thrust bearing.

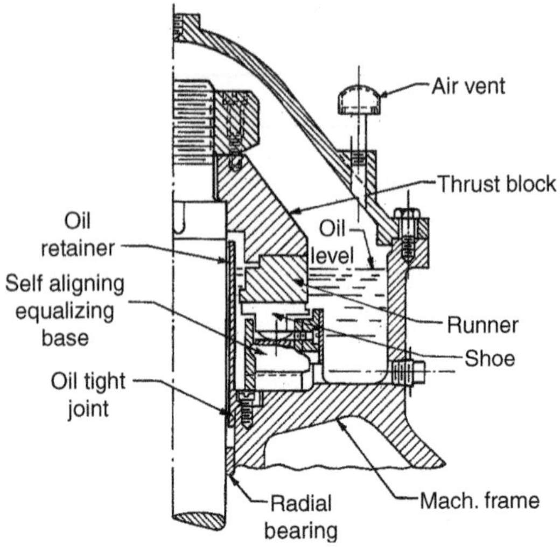

Figure 6.1 *Half section of mounting for vertical thrust bearing*

Radial or Journal

Plain radial, or journal, bearings also are referred to as sleeve or Babbit bearings. The most common type is the full journal bearing, which has 360-degree contact with its mating journal. The partial journal bearing has less than 180-degree contact and is used when the load direction is constant. The sections to follow describe the major types of fluid-film journal bearings: plain cylindrical, four-axial groove, elliptical, partial-arc, and tilting-pad.

Plain Cylindrical

The plain cylindrical journal bearing (Figure 6.2) is the simplest of all journal bearing types. The performance characteristics of cylindrical bearings are well established, and extensive design information is available. Practically, use of the unmodified cylindrical bearing is generally limited to gas-lubricated bearings and low-speed machinery.

Four-Axial Groove Bearing

To make the plain cylindrical bearing practical for oil or other liquid lubricants, it is necessary to modify it by the addition of grooves or holes through which the lubricant can be introduced. Sometimes, a single circumferential

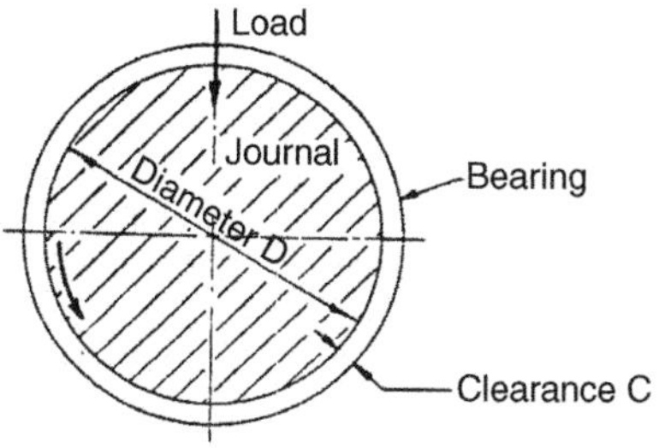

Figure 6.2 *Plain cylindrical bearing*

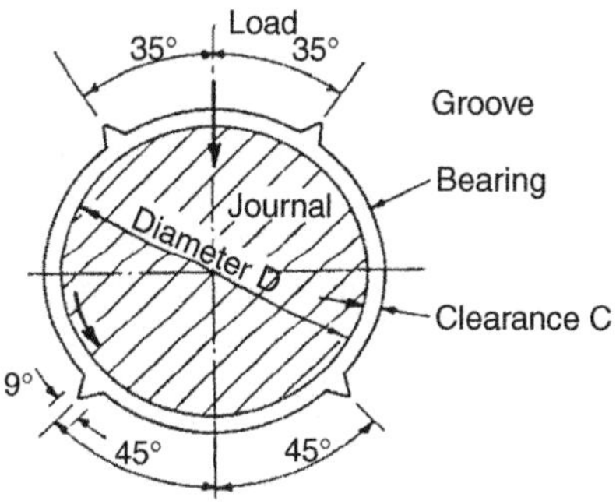

Figure 6.3 *Four-axial groove bearing*

groove in the middle of the bearing is used. In other cases, one or more axial grooves are provided.

The four-axial groove bearing is the most commonly used oil-lubricated sleeve bearing. The oil is supplied at a nominal gauge pressure that ensures an adequate oil flow and some cooling capability. Figure 6.3 illustrates this type of bearing.

Elliptical Bearing

The elliptical bearing is oil-lubricated and typically is used in gear and turbine applications. It is classified as a lobed bearing in contrast to a grooved bearing. Where the grooved bearing consists of a number of partial arcs with a common center, the lobed bearing is made up of partial arcs whose centers

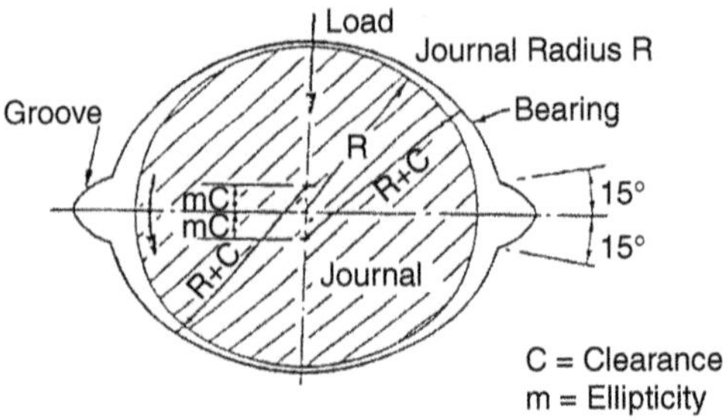

Figure 6.4 *Elliptical bearing*

do not coincide. The elliptical bearing consists of two partial arcs where the bottom arc has its center a distance above the bearing center. This arrangement has the effect of preloading the bearing, where the journal center eccentricity with respect to the loaded arc is increased and never becomes zero. This results in the bearing being stiffened, somewhat improving its stability. An elliptical bearing is shown in Figure 6.4.

Partial-Arc Bearings

A partial-arc bearing is not a separate type of bearing. Instead, it refers to a variation of previously discussed bearings (e.g., grooved and lobed bearings) that incorporates partial arcs. It is necessary to use partial-arc bearing data to incorporate partial arcs in a variety of grooved and lobed bearing configurations. In all cases, the lubricant is a liquid and the bearing film is laminar. Figure 6.5 illustrates a typical partial-arc bearing.

Tilting-Pad Bearings

Tilting-pad bearings are widely used in high-speed applications where hydrodynamic instability and misalignment are common problems. This bearing consists of a number of shoes mounted on pivots, with each shoe being a partial-arc bearing. The shoes adjust and follow the motions of the journal, ensuring inherent stability if the inertia of the shoes does not interfere with the adjustment ability of the bearing. The load direction may either pass between the two bottom shoes or it may pass through the pivot of the bottom shoe. The lubricant is incompressible (i.e., liquid), and the lubricant film is laminar. Figure 6.6 illustrates a tilting-pad bearing.

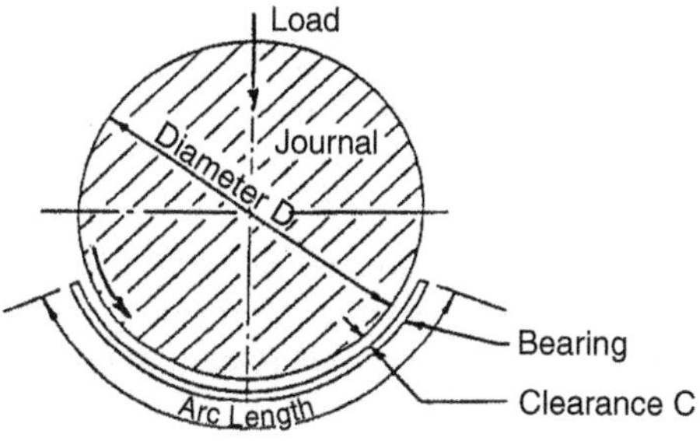

Figure 6.5 *Partial-arc bearing*

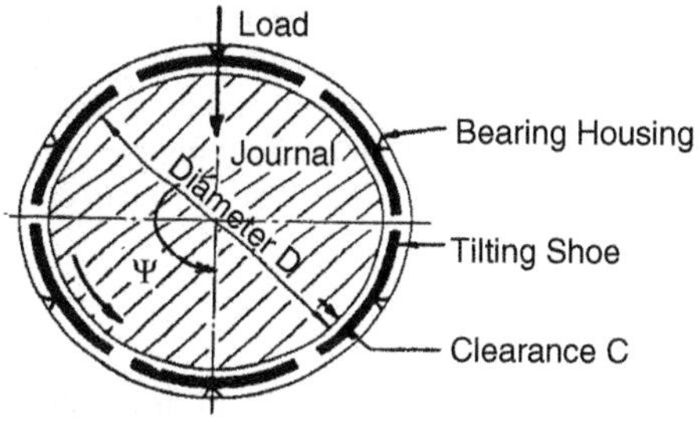

Figure 6.6 *Tilting-pad bearing*

Rolling Element or Antifriction

Rolling element antifriction bearings are one of the most common types used in machinery. Antifriction bearings are based on rolling motion as opposed to the sliding motion of plain bearings. The use of rolling elements between rotating and stationary surfaces reduces the friction to a fraction of that resulting with the use of plain bearings. Use of rolling element bearings is determined by many factors, including load, speed, misalignment sensitivity, space limitations, and desire for precise shaft positioning. They support both radial and axial loads and are generally used in moderate- to high-speed applications.

Unlike fluid-film plain bearings, rolling element bearings have the added ability to carry the full load of the rotor assembly at any speed. Where fluid-film bearings must have turning gear to support the rotor's weight at low speeds, rolling element bearings can maintain the proper shaft centerline through the entire speed range of the machine.

Grade Classifications

Rolling element bearings are available in either commercial- or precision-grade classifications. Most *commercial-grade bearings* are made to non-specific standards and are not manufactured to the same precise standards as precision-grade bearings. This limits the speeds at which they can operate efficiently, and given the brand of bearings may or may not be interchangeable.

Precision bearings are used extensively on many machines such as pumps, air compressors, gear drives, electric motors, and gas turbines. The shape of the rolling elements determines the use of the bearing in machinery. Because of standardization in bearing envelope dimensions, precision bearings were once considered to be interchangeable, even if manufactured by different companies. *It has been discovered, however, that interchanging bearings is a major cause of machinery failure and should be done with extreme caution.*

Rolling Element Types

There are two major classifications of rolling elements: ball and roller. Ball bearings function on point contact and are suited for higher speeds and lighter loads than roller bearings. Roller element bearings function on line contact and generally are more expensive than ball bearings, except for the larger sizes. Roller bearings carry heavy loads and handle shock more satisfactorily than ball bearings, but are more limited in speed. Figure 6.7 provides general guidelines to determine if a ball or roller bearing should be selected. This figure is based on a rated life of 30,000 hours.

Although there are many types of rolling elements, each bearing design is based on a series of hardened rolling elements sandwiched between hardened inner and outer rings. The rings provide continuous tracks or races for the rollers or balls to roll in. Each ball or roller is separated from its neighbor by a separator cage or retainer, which properly spaces the rolling elements around the track and guides them through the load zone. Bearing size is usually given in terms of boundary dimensions: outside diameter, bore, and width.

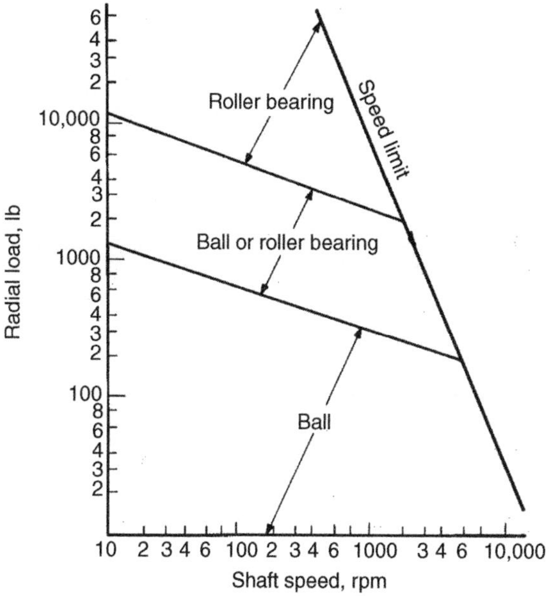

Figure 6.7 *Guide to selecting ball or roller bearings*

Ball Bearings

Common functional groupings of ball bearings are radial, thrust, and angular-contact bearings. Radial bearings carry a load in a direction perpendicular to the axis of rotation. Thrust bearings carry only thrust loads, a force parallel to the axis of rotation tending to cause endwise motion of the shaft. Angular-contact bearings support combined radial and thrust loads. These loads are illustrated in Figure 6.8. Another common classification of ball bearings is single-row (also referred to as Conrad or deep-groove bearing) and double-row.

Single-Row

Types of single-row ball bearings are radial nonfilling slot bearings, radial filling slot bearings, angular contact bearings, and ball thrust bearings.

Radial, Nonfilling Slot Bearings

This ball bearing is often referred to as the Conrad-type or deep-groove bearing and is the most widely used of all ball bearings (and probably of

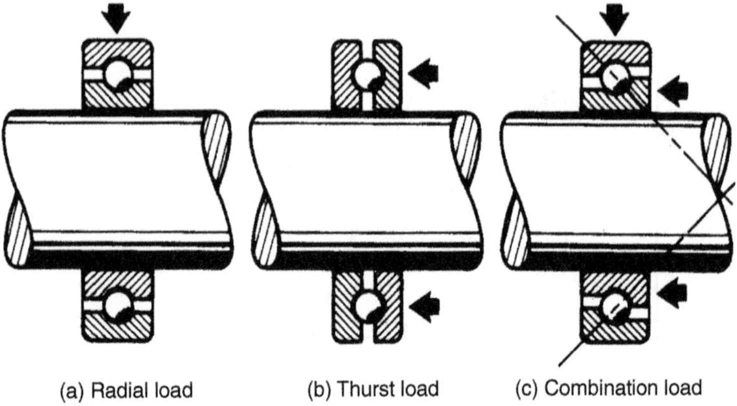

Figure 6.8 *Three principal types of ball bearing loads*

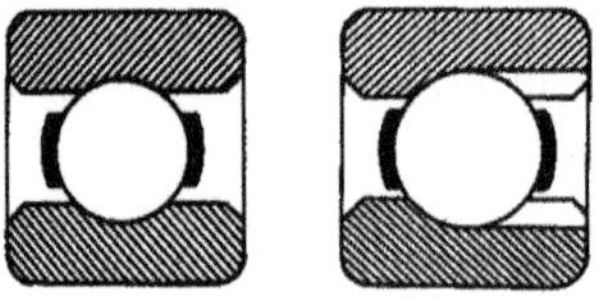

Figure 6.9 *Single-row radial, nonfilling slot bearing*

all antifriction bearings). It is available in many variations, with single or double shields or seals. They sustain combined radial and thrust loads, or thrust loads alone, in either direction—even at extremely high speeds. This bearing is not designed to be self-aligning; therefore, it is imperative that the shaft and the housing bore be accurately aligned (Figure 6.9).

Figure 6.10 labels the parts of the Conrad antifriction ball bearing. This design is widely used and is versatile because the deep-grooved raceways permit the rotating balls to rapidly adjust to radial and thrust loadings, or a combination of these loadings.

Radial, Filling Slot Bearing

The geometry of this ball bearing is similar to the Conrad bearing, except for the filling slot. This slot allows more balls in the complement and thus can

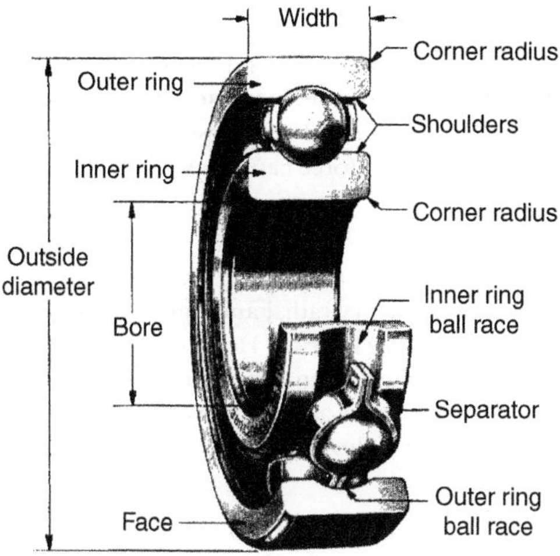

Figure 6.10 *Conrad antifriction ball bearing parts*

carry heavier radial loads. The bearing is assembled with as many balls that fit in the gap created by eccentrically displacing the inner ring. The balls are evenly spaced by a slight spreading of the rings and heat expansion of the outer ring. However, because of the filling slot, the thrust capacity in both directions is reduced. In combination with radial loads, this bearing design accomodates thrust of less than 60% of the radial load.

Angular Contact Radial Thrust

This ball bearing is designed to support radial loads combined with thrust loads, or heavy thrust loads (depending on the contact-angle magnitude). The outer ring is designed with one shoulder higher than the other, which allows it to accommodate thrust loads. The shoulder on the other side of the ring is just high enough to prevent the bearing from separating. This type of bearing is used for pure thrust load in one direction and is applied either in opposed pairs (duplex), or one at each end of the shaft. They can be mounted either face to face or back to back and in tandem for constant thrust in one direction. This bearing is designed for combination loads where the thrust component is greater than the capacity of single-row, deep-groove bearings. Axial deflection must be confined to very close tolerances.

Ball-Thrust Bearing

The ball-thrust bearing supports very high thrust loads in one direction only, but supports **no** radial loading. To operate successfully, this type of bearing must be at least moderately thrust-loaded at all times. It should not be operated at high speeds, since centrifugal force causes excessive loading of the outer edges of the races.

Double-Row

Double-row ball bearings accommodate heavy radial and light thrust loads without increasing the outer diameter of the bearing. However, the double-row bearing is approximately 60 to 80% wider than a comparable single-row bearing. The double-row bearing incorporates a filling slot, which requires the thrust load to be light. Figure 6.11 shows a double-row type ball bearing.

This unit is, in effect, two single-row angular contact bearings built as a unit with the internal fit between balls and raceway fixed during assembly. As a result, fit and internal stiffness are not dependent upon mounting methods. These bearings usually have a known amount of internal preload, or compression, built in for maximum resistance to deflection under combined

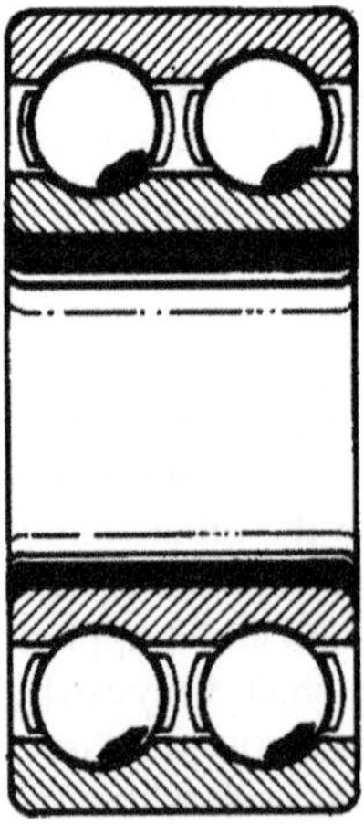

Figure 6.11 *Double row-type ball bearing*

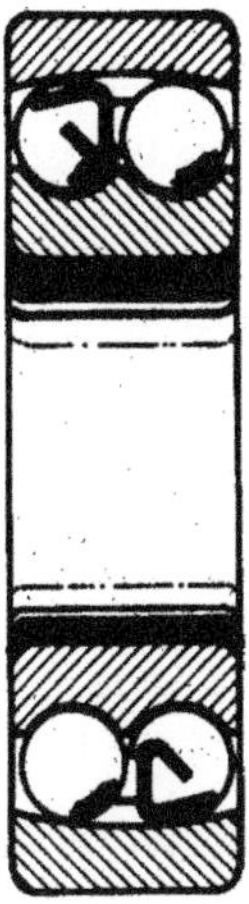

Figure 6.12 *Double-row internal self-aligning bearing*

loads with thrust from either direction. As a result of this compression prior to external loading, the bearings are very effective for radial loads where bearing deflection must be minimized.

Another double-row ball bearing is the internal self-aligning type, which is shown in Figure 6.12. It compensates for angular misalignment, which can be caused by errors in mounting, shaft deflection, misalignment, etc. This bearing supports moderate radial loads and limited thrust loads.

Roller

As with plain and ball bearings, roller bearings also may be classified by their ability to support radial, thrust, and combination loads. Note that combination load-supporting roller bearings *are not* called angular-contact bearings as they are with ball bearings. For example, the taper-roller bearing is a combination load-carrying bearing by virtue of the shape of its rollers.

Figure 6.13 shows the different types of roller elements used in these bearings. Roller elements are classified as cylindrical, barrel, spherical, and tapered. Note that barrel rollers are called needle rollers when they are less than $\frac{1}{4}$" in diameter and have a relatively high ratio of length to diameter.

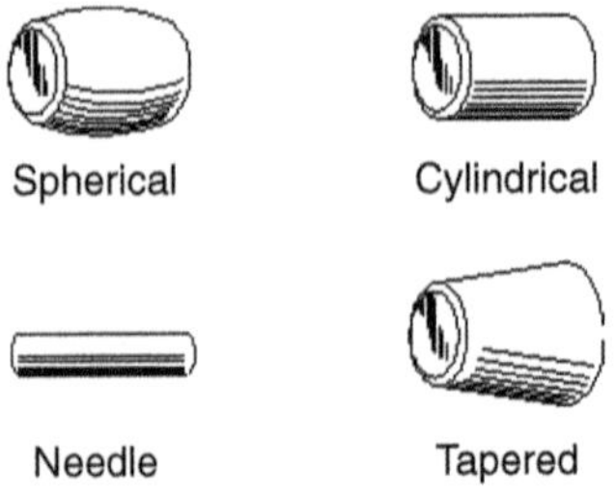

Figure 6.13 *Types of roller elements*

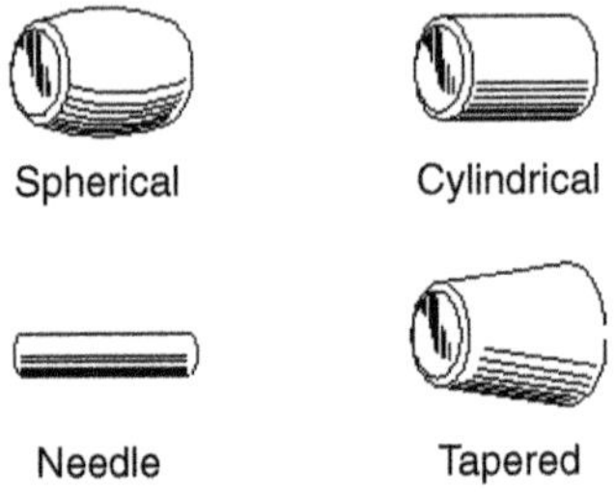

Figure 6.14 *Cylindrical roller bearing*

Cylindrical

Cylindrical bearings have solid or helically wound hollow cylindrically shaped rollers, which have an approximate length-diameter ratio ranging from 1:1 to 1:3. They normally are used for heavy radial loads beyond the capacities of comparably sized radial ball bearings.

Cylindrical bearings are especially useful for free axial movement of the shaft. The free ring may have a restraining flange to provide some restraint to endwise movement in one direction. Another configuration comes without a flange, which allows the bearing rings to be displaced axially.

Either the rolls or the roller path on the races may be slightly crowned to prevent edge loading under slight shaft misalignment. Low friction makes this bearing type suitable for fairly high speeds. Figure 6.14 shows a typical cylindrical roller bearing.

Figure 6.15 shows separable inner-ring cylindrical roller bearings. Figure 6.16 shows separable inner-ring cylindrical roller bearings with a different inner ring.

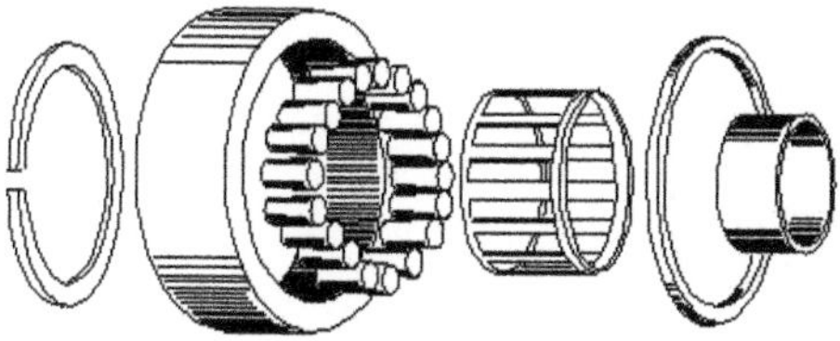

Figure 6.15 *Separable inner ring-type cylindrical roller bearings*

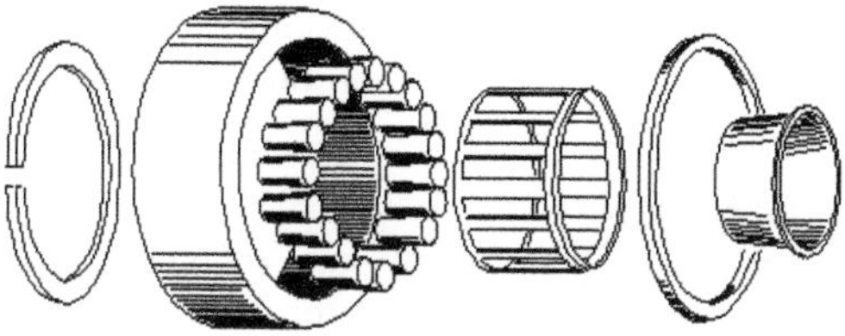

Figure 6.16 *Separable inner ring-type roller bearings with different inner ring*

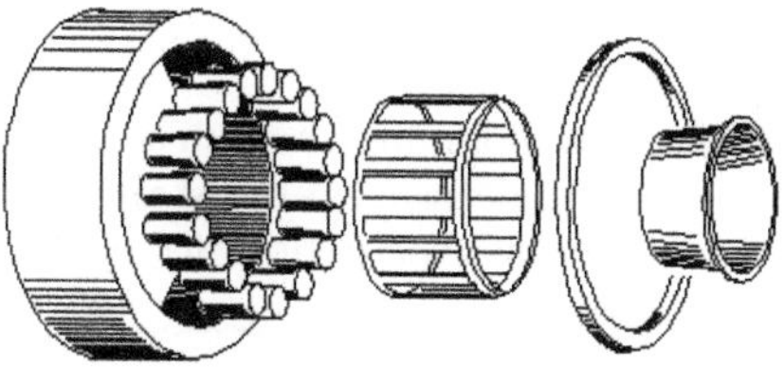

Figure 6.17 *Separable inner ring-type cylindrical roller bearings with elimination of a retainer ring on one side*

The roller assembly in Figure 6.15 is located in the outer ring with retaining rings. The inner ring can be omitted and the roller operated on hardened ground shaft surfaces.

The style in Figure 6.16 is similar to the one in Figure 6.15, except the rib on the inner ring is different. This prohibits the outer ring from moving in a direction toward the rib.

Figure 6.17 shows separable inner ring-type cylindrical roller bearings with elimination of a retainer ring on one side.

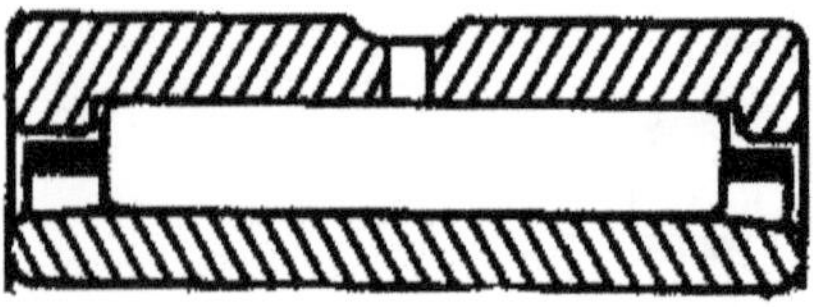

Figure 6.18 *Needle bearings*

The style shown in Figure 6.17 is similar to the two previous styles except for the elimination of a retainer ring on one side. It can carry small thrust loads in only **one** direction.

Needle-Type Cylindrical or Barrel

Needle-type cylindrical bearings (Figure 6.18) incorporate rollers that are symmetrical, with a length at least four times their diameter. They are sometimes referred to as barrel rollers. These bearings are most useful where space is limited and thrust-load support is not required. They are available with or without an inner race. If a shaft takes the place of an inner race, it must be hardened and ground. The full-complement type is used for high loads and oscillating or slow speeds. The cage type should be used for rotational motion.

They come in both single-row and double-row mountings. As with all cylindrical roller bearings, the single-row mounting type has a low thrust capacity, but angular mounting of rolls in the double-row type permits its use for combined axial and thrust loads.

Spherical

Spherical bearings are usually furnished in a double-row mounting that is inherently self-aligning. Both rows of rollers have a common spherical outer raceway. The rollers are barrel-shaped with one end smaller to provide a small thrust to keep the rollers in contact with the center guide flange.

This type of roller bearing has a high radial and moderate-to-heavy thrust load-carrying capacity. It maintains this capability with some degree of shaft and bearing housing misalignment. While their internal self-aligning feature is useful, care should be taken in specifying this type of bearing to compensate for misalignment. Figure 6.19 shows a typical spherical roller bearing assembly. Figure 6.20 shows a series of spherical roller bearings for a given shaft size.

Figure 6.19 *Spherical roller bearing assembly*

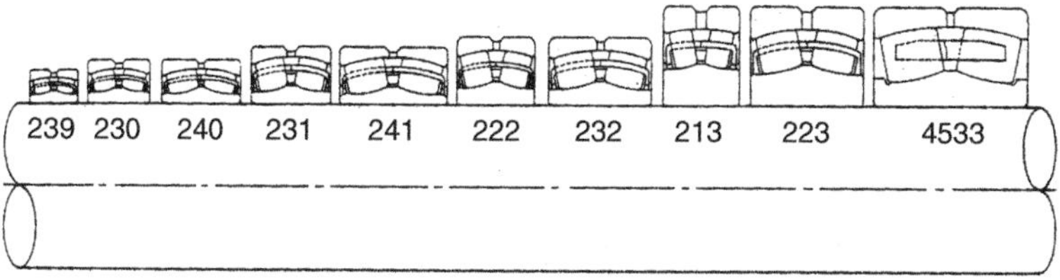

Figure 6.20 *Series of spherical roller bearings for a given shaft size (available in several series)*

Tapered

Tapered bearings are used for heavy radial and thrust loads. They have straight tapered rollers, which are held in accurate alignment by means of a guide flange on the inner ring. Figure 6.21 shows a typical tapered-roller bearing. Figure 6.22 shows necessary information to identify a taper-roller bearing. Figure 6.23 shows various types of tapered roller bearings.

True rolling occurs because they are designed so all elements in the rolling surface and the raceways intersect at a common point on the axis. The basic characteristic of these bearings is that if the apexes of the tapered working surfaces of both rollers and races were extended, they would coincide on

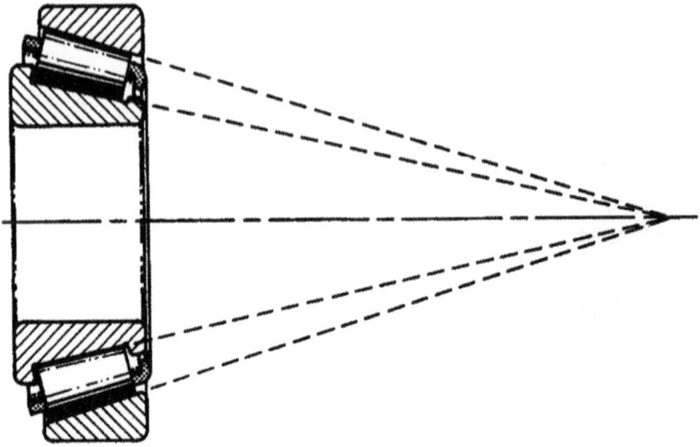

Figure 6.21 *Tapered-roller bearing*

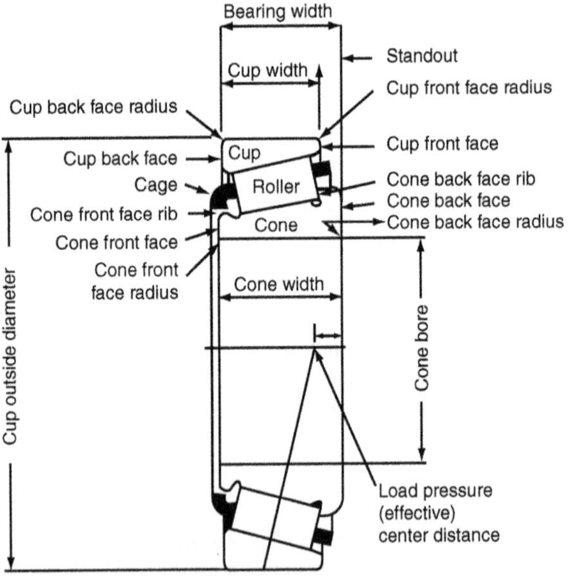

Figure 6.22 *Information needed to identify a taper-roller bearing*

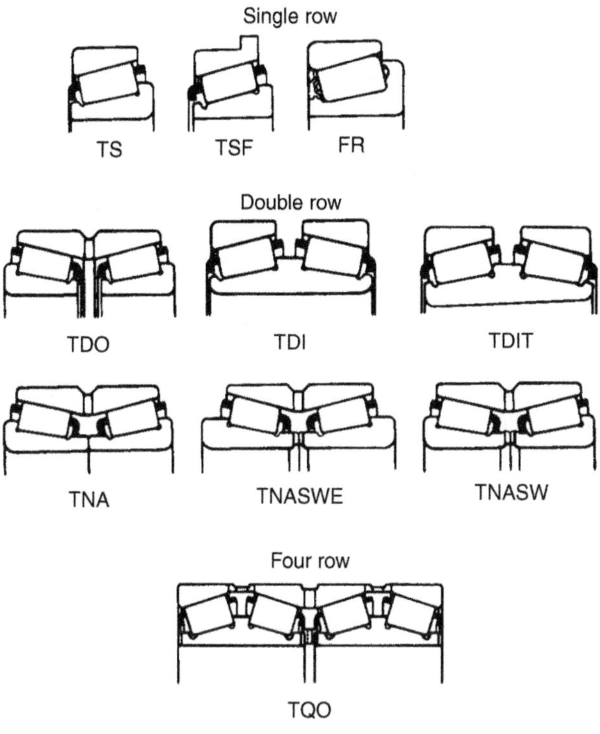

Figure 6.23 *Various types of tapered roller bearings*

the bearing axis. Where maximum system rigidity is required, they can be adjusted for a preload. These bearings are separable.

Bearing Materials

Because two contacting metal surfaces are in motion in bearing applications, material selection plays a crucial role in their life. Properties of the materials used in bearing construction determine the amount of sliding friction that occurs, a key factor affecting bearing life. When two similar metals are in contact without the presence of adequate lubrication, friction is generally high, and the surfaces will seize (i.e., weld) at relatively low pressures or surface loads. However, certain combinations of materials support substantial loads without seizing or welding as a result of their low frictional qualities.

In most machinery, shafts are made of steel. Bearings are generally made of softer materials that have low frictional as well as sacrificial qualities when in contact with steel. A softer, sacrificial material is used for bearings because it is easier and cheaper to replace a worn bearing as opposed to a worn shaft. Common bearing materials are cast iron, bronze, and babbitt. Other less commonly used materials include wood, plastics, and other synthetics.

There are several important characteristics to consider when specifying bearing materials, including: (1) strength or the ability to withstand loads without plastic deformation; (2) ability to permit embedding of grit or dirt particles that are present in the lubricant; (3) ability to elastically deform in order to permit load distribution over the full bearing surface; (4) ability to dissipate heat and prevent hot spots that might seize; and (5) corrosion resistance.

Plain

As indicated above, dissimilar metals with low frictional characteristics are most suitable for plain bearing applications. With steel shafts, plain bearings made of bronze or babbitt are commonly used. Bronze is one of the harder bearing materials and is generally used for low speeds and heavy loads.

A plain bearing may sometimes be made of a combination of materials. The outer portion may be constructed of bronze, steel, or iron to provide the strength needed to provide a load-carrying capability. The bearing may be lined with a softer material such as babbitt to provide the sacrificial capability needed to protect the shaft.

Rolling Element

A specially developed steel alloy is used for an estimated 98% of all rolling element bearing uses. In certain special applications, however, materials such as glass, plastic, and other substances are sometimes used in rolling element construction.

Bearing steel is a high-carbon chrome alloy with high hardenability and good toughness characteristics in the hardened and drawn state. All load-carrying members of most rolling contact bearings are made with this steel.

Controlled procedures and practices are necessary to ensure specification of the proper alloy, maintain material cleanliness, and ensure freedom from defects—all of which affect bearing reliability. Alloying practices that conform to rigid specifications are required to reduce anomalies and inclusions that adversely affect a bearing's useful life. Magnaflux inspections ensure that rolling elements are free from material defects and cracks. Light etching is used between rough and finish grinding processes to stop burning during heavy machining operations.

Lubrication

It is critical to consider lubrication requirements when specifying bearings. Factors affecting lubricants include relatively high speeds, difficulty in performing relubrication, nonhorizontal shafts, and applications where leakage cannot be tolerated. This section briefly discusses lubrication mechanisms and techniques for bearings.

Plain Bearings

In plain bearings, the lubricating fluid must be replenished to compensate for end leakage in order to maintain the bearings' load-carrying capacity. Pressure lubrication from a pump- or gravity-fed tank, or automatic lubricating devices such as oil rings or oil disks, are provided in self-contained bearings. Another means of lubrication is to submerge the bearing (in particular, thrust bearings for vertical shafts) in an oil bath.

Lubricating Fluids

Almost any process fluid may be used to lubricate plain bearings if parameters such as viscosity, corrosive action, toxicity, change in state (where a liquid is close to its boiling point), and in the case of a gaseous fluid, its compressibility, are appropriate for the application. Fluid-film journal and thrust bearings have run successfully, for example, on water, kerosene, gasoline, acid, liquid refrigerants, mercury, molten metals, and a wide variety of gases.

Gases, however, lack the cooling and boundary-lubrication capabilities of most liquid lubricants. Therefore, the operation of self-acting gas bearings is restricted by start/stop friction and wear. If start/stop is performed under load, then the design is limited to about seven pounds per square inch (lb/in^2) or 48 kilo-Newtons per square meter (kN/m^2) on the projected bearing area, depending upon the choice of materials. In general, the materials used for these bearings are those of dry rubbing bearings (e.g., either a hard/hard combination such as ceramics with or without a molecular layer of boundary lubricant, or a hard/soft combination with a plastic surface).

Externally pressurized gas journal bearings have the same principle of operation as hydrostatic liquid-lubricated bearings. Any clear gas can be used, but many of the design charts are based on air. There are three forms of external flow restrictors in use with these bearings: pocketed (simple) orifice, unpocketed (annular) orifice, and slot.

State of Lubrication

Fluid or complete lubrication, the condition where the surfaces are completely separated by a fluid film, provides the lowest friction losses and prevents wear.

The semifluid lubrication state exists between the journal and bearing when a load-carrying fluid film does not form to separate the surfaces. This occurs at comparatively low speed with intermittent or oscillating motion, heavy load, and insufficient oil supply to the bearing. Semifluid lubrication also may exist in thrust bearings with fixed parallel-thrust collars; guide bearings of machine tools; in bearings with plenty of lubrication that have a bent or misaligned shaft; or where the bearing surface has improperly arranged oil grooves. The coefficient of friction in such bearings may range from 0.02 to 0.08.

In situations where the bearing is well lubricated, but the speed of rotation is very slow or the bearing is barely greasy, boundary lubrication takes place. In this situation, which occurs in bearings when the shaft is starting from rest, the coefficient of friction may vary from 0.08 to 0.14.

A bearing may run completely dry in exceptional cases of design or with a complete failure of lubrication. Depending on the contacting surface materials, the coefficient of friction will be between 0.25 and 0.40.

Rolling Element Bearings

Rolling element bearings also need a lubricant to meet or exceed their rated life. In the absence of high temperatures, however, excellent performance can be obtained with a very small quantity of lubricant. Excess lubricant causes excessive heating, which accelerates lubricant deterioration.

Table 6.9 *Ball-bearing grease relubrication intervals (hours of operation)*

Bearing bore, mm	Bearing speed, rpm				
	5,000	3,600	1,750	1,000	200
10	8,700	12,000	25,000	44,000	220,000
20	5,500	8,000	17,000	30,000	150,000
30	4,000	6,000	13,000	24,000	127,000
40	2,800	4,500	11,000	20,000	111,000
50		3,500	9,300	18,000	97,000
60		2,600	8,000	16,000	88,000
70			6,700	14,000	81,000
80			5,700	12,000	75,000
90			4,800	11,000	70,000
100			4,000	10,000	66,000

Source: Theodore Baumeister, ed. *Marks' Standard Handbook for Mechanical Engineers,* Eighth Edition. New York: McGraw-Hill, 1978.

The most popular type of lubrication is the sealed grease ball-bearing cartridge. Grease is commonly used for lubrication because of its convenience and minimum maintenance requirements. A high-quality lithium-based NLGI 2 grease is commonly used for temperatures up to 180°F (82°C). Grease must be replenished and relubrication intervals in hours of operation are dependent on temperature, speed, and bearing size. Table 6.9 is a general guide to the time after which it is advisable to add a small amount of grease.

Some applications, however, cannot use the cartridge design, for example, when the operating environment is too hot for the seals. Another example is when minute leaks or the accumulation of traces of dirt at the lip seals cannot be tolerated (e.g., food processing machines). In these cases, bearings with specialized sealing and lubrication systems must be used.

In applications involving high speed, oil lubrication is typically required. Table 6.10 is a general guide in selecting oil of the proper viscosity for these bearings. For applications involving high-speed shafts, bearing selection must take into account the inherent speed limitations of certain bearing designs, cooling needs, and lubrication issues such as churning and aeration suppression. A typical case is the effect of cage design and roller-end thrust-flange contact on the lubrication requirements in taper roller bearings. These design elements limit the speed and the thrust load that these

Table 6.10 *Oil lubrication viscosity (ISO identification numbers)*

	Bearing speed, rpm				
Bearing bore, mm	10,000	3,600	1,800	600	50
4–7	68	150	220		
10–20	32	68	150	220	460
25–45	10	32	68	150	320
50–70	7	22	68	150	320
75–90	3	10	22	68	220
100	3	7	22	68	220

Source: Theodore Baumeister, ed. *Marks' Standard Handbook for Mechanical Engineers*, Eighth Edition. New York: McGraw-Hill, 1978.

bearings can endure. As a result, it is important always to refer to the bearing manufacturer's instructions on load-carrying design and lubrication specifications.

Installation and General Handling Precautions

Proper handling and installation practices are crucial to optimal bearing performance and life. In addition to standard handling and installation practices, the issue of emergency bearing substitutions is an area of critical importance. If substitute bearings are used as an emergency means of getting a machine back into production quickly, the substitution should *be entered into the historical records for that machine.* This documents the temporary change and avoids the possibility of the substitute bearing becoming a *permanent* replacement. This error can be extremely costly, particularly if the incorrectly specified bearing continually fails prematurely. It is important that an inferior substitute be removed as soon as possible and replaced with the originally specified bearing.

Plain Bearing Installation

It is important to keep plain bearings from shifting sideways during installation and to ensure an axial position that does not interfere with shaft fillets. Both of these can be accomplished with a locating lug at the parting line.

Less frequently used is a dowel in the housing, which protrudes partially into a mating hole in the bearing.

The distance across the outside parting edges of a plain bearing are manufactured slightly greater than the housing bore diameter. During installation, a light force is necessary to snap it into place and, once installed, the bearing stays in place because of the pressure against the housing bore.

It is necessary to prevent a bearing from spinning during operation, which can cause a catastrophic failure. Spinning is prevented by what is referred to as "crush." Bearings are slightly longer circumferentially than their mating housings and upon installation this excess length is elastically deformed or "crushed." This sets up a high radial contact pressure between the bearing and housing, which ensures good back contact for heat conduction and, in combination with the bore-to-bearing friction, prevents spinning. It is important that *under no circumstances* should the bearing parting lines be filed or otherwise altered to remove the crush.

Roller Bearing Installation

A basic rule of rolling element bearing installation is that one ring must be mounted on its mating shaft or in its housing with an interference fit to prevent rotation. This is necessary because it is virtually impossible to prevent rotation by clamping the ring axially.

Mounting Hardware

Bearings come as separate parts that require mounting hardware or as pre-mounted units that are supplied with their own housings, adapters, and seals.

Bearing Mountings

Typical bearing mountings, which are shown in Figure 6.24, locate and hold the shaft axially and allow for thermal expansion and/or contraction of the shaft. Locating and holding the shaft axially is generally accomplished by clamping one of the bearings on the shaft so that all machine parts remain in proper relationship dimensionally. The inner ring is locked axially relative to the shaft by locating it between a shaft shoulder and some type of removable locking device once the inner ring has a tight fit. Typical removable locking devices are specially designed nuts, which are used for a through shaft, and clamp plates, which are commonly used when the bearing is mounted on the end of the shaft. For the locating or held bearing, the outer

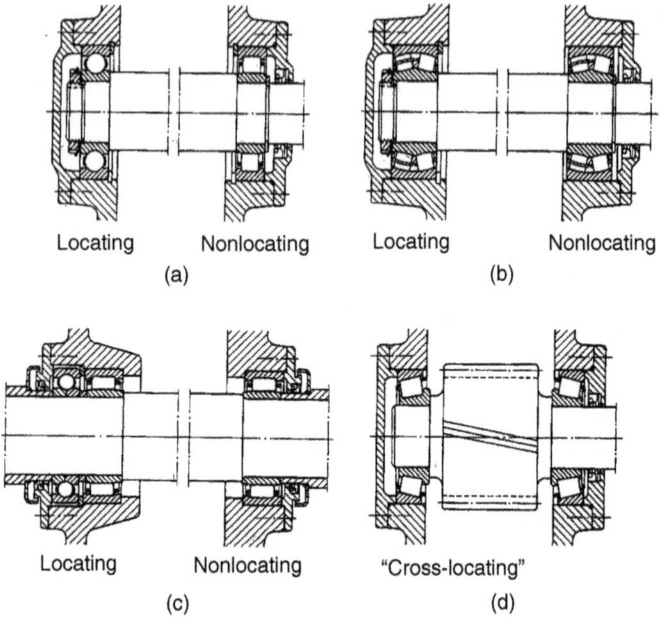

Figure 6.24 *Typical bearing mounting*

ring is clamped axially, usually between housing shoulders or end-cap pilots.

With general types of cylindrical roller bearings, shaft expansion is absorbed internally simply by allowing one ring to move relative to the other [Figure 6.24(a) and 6.24(c), nonlocating positions]. The advantage of this type of mounting is that both inner and outer rings may have a tight fit, which is desirable or even mandatory if significant vibration and/or imbalance exists in addition to the applied load.

Premounted Bearing

Premounted bearings, referred to as pillow-block and flanged-housing mountings, are of considerable importance to millwrights. They are particularly adaptable to "line-shafting" applications, which are a series of ball and roller bearings supplied with their own housings, adapters, and seals. Premounted bearings come with a wide variety of flange mountings, which permit them to be located on faces parallel or perpendicular to the shaft

Figure 6.25 *Typical pillow block*

Figure 6.26 *Flanged bearing unit*

axis. Figure 6.25 shows a typical pillow block. Figure 6.26 shows a flanged bearing unit.

Inner races can be mounted directly on ground shafts or can be adapter-mounted to "drill-rod" or to commercial-shafting. For installations sensitive to imbalance and vibration, the use of accurately ground shaft seats is recommended.

Most pillow-block designs incorporate self-aligning bearing types and do not require the precision mountings utilized with other bearing installations.

Mounting Techniques

When mounting or dismounting a roller bearing, the most important thing to remember is to apply the mounting or dismounting force to the side face of the ring with the interference fit. This force should not pass from one ring to the other through the ball or roller set, as internal damage can easily occur.

Mounting tapered-bore bearings can be accomplished simply by tightening the locknut or clamping plate. This locates it on the shaft until the bearing is forced the proper distance up the taper. This technique requires a significant amount of force, particularly for large bearings.

Cold Mounting

Cold mounting, or force fitting a bearing onto a shaft or into a housing, is appropriate for all small bearings (i.e., 4" bore and smaller). The force, however, must be applied as uniformly as possible around the side face of the bearing and to the ring to be press-fitted. Mounting fixtures, such as a simple piece of tubing of appropriate size and a flat plate, should be used. It is not appropriate to use a drift and hammer to force the bearing on, which will cause the bearing to cock. It is possible to apply force by striking the plate with a hammer or by an arbor press. However, before forcing the bearing on the shaft, a coat of light oil should be applied to the bearing seat on the shaft and the bearing bores. All sealed and shielded ball bearings should be cold mounted in this manner.

Temperature Mounting

The simplest way to mount any open straight-bore bearing regardless of its size is temperature mounting, which entails heating the entire bearing, pushing it on its seat, and holding it in place until it cools enough to grip the shaft. The housing may be heated if practical for tight outside-diameter fits; however, temperatures should not exceed 250°F. If heating of the housing is not practical, the bearing may be cooled with dry ice. The risk of cooling is that if the ambient conditions are humid, moisture is introduced and there is a potential for corrosion in the future. Acceptable ways of heating bearings are by hot plate, temperature-controlled oven, induction heaters, and hot-oil bath.

With the hot plate method, the bearing is simply laid on the plate until it reaches the approved temperature, using a pyrometer or Tempilstik to make certain it is not overheated. Difficulty in controlling the temperature is the major disadvantage of this method.

When using a temperature-controlled oven, the bearings should be left in the oven long enough to heat thoroughly, but they should never be left overnight.

The use of induction heaters is a quick method of heating bearings. However, some method of measuring the ring temperature (e.g., pyrometer or a Tempilstik) must be used or damage to the bearing may occur. Note that bearings must be demagnetized after the use of this method.

The use of a hot-oil bath is the most practical means of heating larger bearings. Disadvantages are that the temperature of the oil is hard to control and may ignite or overheat the bearing. The use of a soluble oil and water mixture (10 to 15% oil) can eliminate these problems and still attain a boiling temperature of 210°F. The bearing should be kept off the bottom of the container by a grate or screen located several inches off the bottom. This is important to allow contaminants to sink to the bottom of the container and away from the bearing.

Dismounting

Commercially available bearing pullers allow rolling element bearings to be dismounted from their seats without damage. When removing a bearing, force should be applied to the ring with the tight fit, although sometimes it is necessary to use supplementary plates or fixtures. An arbor press is equally effective at removing smaller bearings as well as mounting them.

Ball Installation

Figure 6.27 shows the ball installation procedure for roller bearings. The designed load carrying capacity of Conrad-type bearings is determined by the number of balls that can be installed between the rings. Ball installation is accomplished by the following procedure:

- Slip the inner ring slightly to one side;
- Insert balls into the gap, which centers the inner ring as the balls are positioned between the rings;
- Place stamped retainer rings on either side of the balls before riveting together. This positions the balls equidistant around the bearing.

General Roller-Element Bearing Handling Precautions

In order for rolling element bearings to achieve their design life and perform with no abnormal noise, temperature rise, or shaft excursions, the following precautions should be taken:

- Always select the best bearing design for the application and not the cheapest. The cost of the original bearing is usually small by comparison

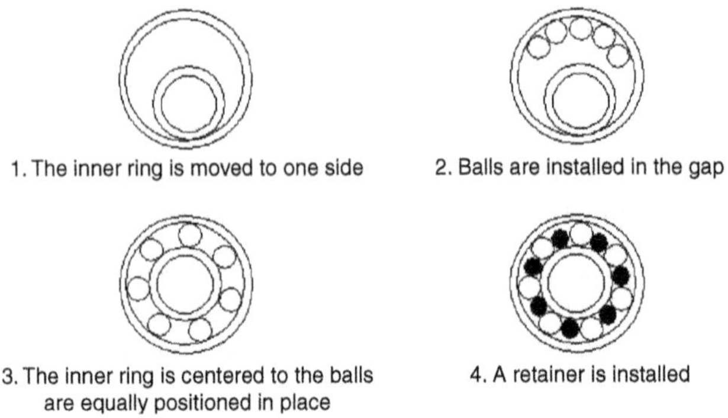

1. The inner ring is moved to one side
2. Balls are installed in the gap
3. The inner ring is centered to the balls are equally positioned in place
4. A retainer is installed

Figure 6.27 *Ball installation procedures*

to the costs of replacement components and the downtime in production when premature bearing failure occurs because an inappropriate bearing was used.

- If in doubt about bearings and their uses, consult the manufacturer's representative and the product literature.
- Bearings should always be handled with great care. Never ignore the handling and installation instructions from the manufacturer.
- Always work with clean hands, clean tools, and the cleanest environment available.
- Never wash or wipe bearings prior to installation unless the instructions specifically state that this should be done. Exceptions to this rule are when oil-mist lubrication is to be used and the slushing compound has hardened in storage or is blocking lubrication holes in the bearing rings. In this situation, it is best to clean the bearing with kerosene or other appropriate petroleum-based solvent. The other exception is if the slushing compound has been contaminated with dirt or foreign matter before mounting.
- Keep new bearings in their greased paper wrappings until they are ready to install. Place unwrapped bearings on clean paper or lint-free cloth if they cannot be kept in their original containers. Wrap bearings in clean, oil-proof paper when not in use.
- Never use wooden mallets, brittle or chipped tools, or dirty fixtures and tools when bearings are being installed.

- Do not spin bearings (particularly dirty ones) with compressed service air.
- Avoid scratching or nicking bearing surfaces. Care must be taken when polishing bearings with emery cloth to avoid scratching.
- Never strike or press on race flanges.
- Always use adapters for mounting that ensure uniform steady pressure rather than hammering on a drift or sleeve. *Never* use brass or bronze drifts to install bearings as these materials chip very easily into minute particles that will quickly damage a bearing.
- Avoid cocking bearings onto shafts during installation.
- Always inspect the mounting surface on the shaft and housing to insure that there are no burrs or defects.
- When bearings are being removed, clean housings and shafts before exposing the bearings.
- Dirt is abrasive and detrimental to the designed life span of bearings.
- Always treat used bearings as if they are new, especially if they are to be reused.
- Protect dismantled bearings from moisture and dirt.
- Use clean filtered, water-free Stoddard's solvent or flushing oil to clean bearings.
- When heating is used to mount bearings onto shafts, follow the manufacturer's instructions.
- When assembling and mounting bearings onto shafts, *never* strike the outer race or press on it to force the inner race. Apply the pressure on the inner race only. When dismantling, follow the same procedure.
- Never press, strike, or otherwise force the seal or shield on factory-sealed bearings.

Bearing Failures, Deficiencies, and Their Causes

The general classifications of failures and deficiencies requiring bearing removal are overheating, vibration, turning on the shaft, binding of the shaft, noise during operation, and lubricant leakage. Table 6.11 is a troubleshooting guide that lists the common causes for each of these failures and deficiencies. As indicated by the causes of failure listed, bearing failures are rarely caused by the bearing itself.

Many abnormal vibrations generated by actual bearing problems are the result of improper sizing of the bearing liner or improper lubrication. However, numerous machine and process-related problems generate abnormal vibration spectra in bearing data. The primary contributors to abnormal bearing signatures are: (1) imbalance, (2) misalignment, (3) rotor instability, (4) excessive or abnormal loads, and (5) mechanical looseness.

Defective bearings that leave the manufacturer are very rare, and it is estimated that defective bearings contribute to only 2% of total failures. The failure is invariably linked to symptoms of misalignment, imbalance, resonance, and lubrication—or the lack of it. Most of the problems that occur result from the following reasons: dirt, shipping damage, storage and handling, poor fit resulting in installation damage, wrong type of bearing design, overloading, improper lubrication practices, misalignment, bent shaft, imbalance, resonance, and soft foot. Anyone of these conditions will eventually destroy a bearing—two or more of these problems can result in disaster!

Although most industrial machine designers provide adequate bearings for their equipment, there are some cases in which bearings are improperly designed, manufactured, or installed at the factory. Usually, however, the trouble is caused by one or more of the following reasons: (1) improper on-site bearing selection and/or installation, (2) incorrect grooving, (3) unsuitable surface finish, (4) insufficient clearance, (5) faulty relining practices, (6) operating conditions, (7) excessive operating temperature, (8) contaminated oil supply, and (9) oil-film instability.

Improper Bearing Selection and/or Installation

There are several things to consider when selecting and installing bearings, including the issue of interchangeability, materials of construction, and damage that might have occurred during shipping, storage, and handling.

Interchangeability

Because of the standardization in envelope dimensions, precision bearings were once regarded as interchangeable among manufacturers.

Table 6.11 *Troubleshooting guide*

Overheating	Vibration	Turning on the shaft	Binding of the shaft	Noisy bearing	Lubricant leakage
Inadequate or insufficient lubrication	Dirt or chips in bearing	Growth of race due to overheating	Lubricant breakdown	Lubrication breakdown	Overfilling of lubricant
Excessive lubrication	Fatigued race or rolling elements	Fretting wear	Contamination by abrasive or corrosive materials	Inadequate lubrication	Grease churning due to too soft a consistency
Grease liquifaction or aeration	Rotor unbalance	Improper initial fit	Housing distortion or out-of-round pinching bearing	Pinched bearing	Grease deterioration due to excessive operating temperature
Oil foaming	Out-of-round shaft	Excessive shaft deflection	Uneven shimming of housing with loss of clearance	Contamination	Operating beyond grease life
Abrasion or corrosion due to contaminants	Race misalignment	Initial coarse finish on shaft	Tight rubbing seals	Seal rubbing	Seal wear
Housing distortion due to warping or out-of-round	Housing resonance	Seal rub on inner race	Preloaded bearings	Bearing slipping on shaft or in housing	Wrong shaft attitude (bearing seals designed for horizontal mounting only)

Continued

Table 6.11 *continued*

Overheating	Vibration	Turning on the shaft	Binding of the shaft	Noisy bearing	Lubricant leakage
Seal rubbing or failure	Cage wear		Cocked races	Flatted roller or ball	Seal failure
Inadequate or blocked scavenge oil passages	Flats on races or rolling elements		Loss of clearance due to excessive adapter tightening	Brinelling due to assembly abuse, handling, or shock loads	Clogged breather
Inadequate bearing clearance or bearing preload	Race turning		Thermal shaft expansion	Variation in size of rolling elements	Oil foaming due to churning or air flow through housing
Race turning	Excessive clearance			Out-of-round or lobular shaft	Gasket (O-ring) failure or misapplication
Cage wear	Corrosion			Housing bore waviness	Porous housing or closure
	False brinelling or indentation of races			Chips or scores under bearing seat	Lubricator set at the wrong flow rate
	Electrical arcing				
	Mixed rolling element diameters				
	Out-of-square rolling paths in races				

Source: Integrated Systems Inc.

This interchangeability has since been considered a major cause of failures in machinery, and the practice should be used with extreme caution.

Most of the problems with interchangeability stem from selecting and replacing bearings based only on bore size and outside diameters. Often, very little consideration is paid to the number of rolling elements contained in the bearings. This can seriously affect the operational frequency vibrations of the bearing and may generate destructive resonance in the host machine or adjacent machines.

More bearings are destroyed during their installation than fail in operation. Installation with a heavy hammer is the usual method in many plants. Heating the bearing with an oxy-acetylene burner is another classical method. However, the bearing does not stand a chance of reaching its life expectancy when either of these installation practices are used. The bearing manufacturer's installation instructions should always be followed.

Shipping Damage

Bearings and the machinery containing them should be properly packaged to avoid damage during shipping. However, many installed bearings are exposed to vibrations, bending, and massive shock loadings through bad handling practices during shipping. It has been estimated that approximately 40% of newly received machines have "bad" bearings.

Because of this, all new machinery should be thoroughly inspected for defects before installation. Acceptance criteria should include guidelines that clearly define acceptable design/operational specifications. This practice pays big dividends by increasing productivity and decreasing unscheduled downtime.

Storage and Handling

Storeroom and other appropriate personnel must be made aware of the potential havoc they can cause by their mishandling of bearings. Bearing failure often starts in the storeroom rather than the machinery. Premature opening of packages containing bearings should be avoided whenever possible. If packages must be opened for inspection, they should be protected from exposure to harmful dirt sources and then resealed in the original wrappings. The bearing should never be dropped or bumped as this can cause shock loading on the bearing surface.

Incorrect Placement of Oil Grooves

Incorrectly placed oil grooves can cause bearing failure. Locating the grooves in high-pressure areas causes them to act as pressure-relief passages. This interferes with the formation of the hydrodynamic film, resulting in reduced load-carrying capability.

Unsuitable Surface Finish

Smooth surface finishes on both the shaft and the bearing are important to prevent surface variations from penetrating the oil film. Rough surfaces can cause scoring, overheating, and bearing failure. The smoother the finishes, the closer the shaft may approach the bearing without danger of surface contact. Although important in all bearing applications, surface finish is critical with the use of harder bearing materials such as bronze.

Insufficient Clearance

There must be sufficient clearance between the journal and bearing in order to allow an oil film to form. An average diametral clearance of 0.001 inches per inch of shaft diameter is often used. This value may be adjusted depending on the type of bearing material, the load, speed, and the accuracy of the shaft position desired.

Faulty Relining

Faulty relining occurs primarily with babitted bearings rather than precision machine-made inserts. Babbitted bearings are fabricated by a pouring process that should be performed under carefully controlled conditions. Some reasons for faulty relining are: (1) improper preparation of the bonding surface, (2) poor pouring technique, (3) contamination of babbitt, and (4) pouring bearing to size with journal in place.

Operating Conditions

Abnormal operating conditions or neglect of necessary maintenance precautions cause most bearing failures. Bearings may experience premature and/or catastrophic failure on machines that are operated heavily loaded, speeded up, or being used for a purpose not appropriate for the system design. Improper use of lubricants can also result in bearing failure. Some typical causes of premature failure include: (1) excessive operating temperatures, (2) foreign material in the lubricant supply, (3) corrosion, (4) material fatigue, and (5) use of unsuitable lubricants.

Excessive Temperatures

Excessive temperatures affect the strength, hardness, and life of bearing materials. Lower temperatures are required for thick babbitt liners than for thin precision babbitt inserts. Not only do high temperatures affect bearing materials, they also reduce the viscosity of the lubricant and affect the thickness of the film, which affects the bearing's load-carrying capacity. In addition, high temperatures result in more rapid oxidation of the lubricating oil, which can result in unsatisfactory performance.

Dirt and Contamination in Oil Supply

Dirt is one of the biggest culprits in the demise of bearings. Dirt makes its appearance in bearings in many subtle ways, and it can be introduced by bad work habits. It also can be introduced through lubricants that have been exposed to dirt, a problem that is responsible for approximately half of bearing failures throughout the industry.

To combat this problem, soft materials such as babbit are used when it is known that a bearing will be exposed to abrasive materials. Babbitt metal embeds hard particles, which protects the shaft against abrasion. When harder materials are used in the presence of abrasives, scoring and galling occurs as a result of abrasives caught between the journal and bearing.

In addition to the use of softer bearing materials for applications where abrasives may potentially be present, it is important to properly maintain filters and breathers, which should regularly be examined. In order to avoid oil supply contamination, foreign material that collects at the bottom of the bearing sump should be removed on a regular basis.

Oil-Film Instability

The primary vibration frequency components associated with fluid-film bearings problems are in fact displays of turbulent or nonuniform oil film. Such instability problems are classified as either *oil whirl* or *oil whip* depending on the severity of the instability.

Machine-trains that use sleeve bearings are designed based on the assumption that rotating elements and shafts operate in a balanced and, therefore, centered position. Under this assumption, the machine-train shaft will operate with an even, concentric oil film between the shaft and sleeve bearing.

For a normal machine, this assumption is valid after the rotating element has achieved equilibrium. When the forces associated with rotation are

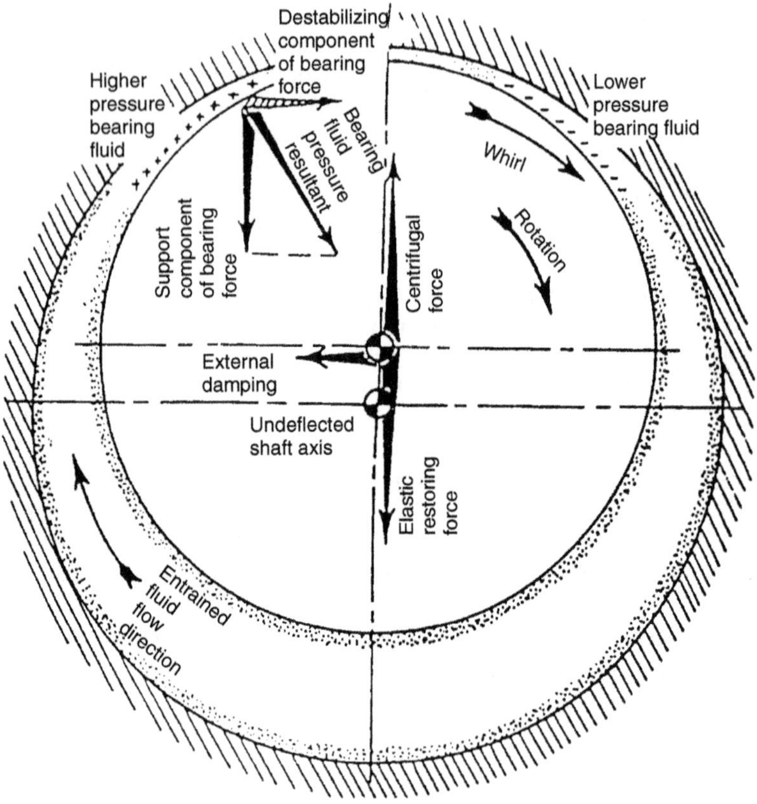

Figure 6.28 *Oil whirl, oil whip*

in balance, the rotating element will center the shaft within the bearing. However, several problems directly affect this self-centering operation. First, the machine-train must be at designed operating speed and load to achieve equilibrium. Second, any imbalance or abnormal operation limits the machine-train's ability to center itself within the bearing.

A typical example is a steam turbine. A turbine must be supported by auxiliary running gear during startup or shutdown to prevent damage to the sleeve bearings. The lower speeds during the startup and shutdown phase of operation prevent the self-centering ability of the rotating element. Once the turbine has achieved full speed and load, the rotating element and shaft should operate without assistance in the center of the sleeve bearings.

Oil Whirl

In an abnormal mode of operation, the rotating shaft may not hold the centerline of the sleeve bearing. When this happens, an instability called oil whirl occurs. Oil whirl is an imbalance in the hydraulic forces within a sleeve bearing. Under normal operation, the hydraulic forces such as velocity and pressure are balanced. If the rotating shaft is offset from the true centerline of the bearing, instability occurs.

As Figure 6.28 illustrates, a restriction is created by the offset. This restriction creates a high pressure and another force vector in the direction of rotation. Oil whirl accelerates the wear and failure of the bearing and bearing support structure.

Oil Whip

The most severe damage results if the oil whirl is allowed to degrade into oil whip. Oil whip occurs when the clearance between the rotating shaft and sleeve bearing is allowed to close to a point approaching actual metal-to-metal contact. When the clearance between the shaft and bearing approaches contact, the oil film is no longer free to flow between the shaft and bearing. As a result, the oil film is forced to change directions. When this occurs, the high-pressure area created in the region behind the shaft is greatly increased. This vortex of oil increases the abnormal force vector created by the offset and rotational force to the point that metal-to-metal contact between the shaft and bearing occurs. In almost all instances where oil whip is allowed, severe damage to the sleeve bearing occurs.

7 Chain Drives

"Only Permanent Repairs Made Here"

Introduction

Chain drives are an important part of a conveyor system. They are used to transmit needed power from the drive unit to a portion of the conveyor system. This chapter will cover:

1 Various types of chains that are used to transmit power in a conveyor system.

2 The advantages and disadvantages of using chain drives.

3 The correct installation procedure for chain drives.

4 How to maintain chain drives.

5 How to calculate speeds and ratios that will enable you to make corrections or adjustments to conveyor speeds.

6 How to determine chain length and sprocket sizes when making speed adjustments.

Chain Drives

Chain drives are used to transmit power between a drive unit and a driven unit. For example, if we have a gearbox and a contact roll on a conveyor, we need a way to transmit the power from the gearbox to the roll. This can be done easily and efficiently with a chain drive unit.

Chain drives can consist of one or multiple strand chains, depending on the load that the unit must transmit. The chains need to be the matched with the sprocket type, and they must be tight enough to prevent slippage.

Chain is sized by the pitch or the center-to-center distance between the pins. This is done in $\frac{1}{8}$" increments, and the pitch number is found on the side bars. Examples of the different chain and sprocket sizes can be seen in Figures 7.1 and 7.2.

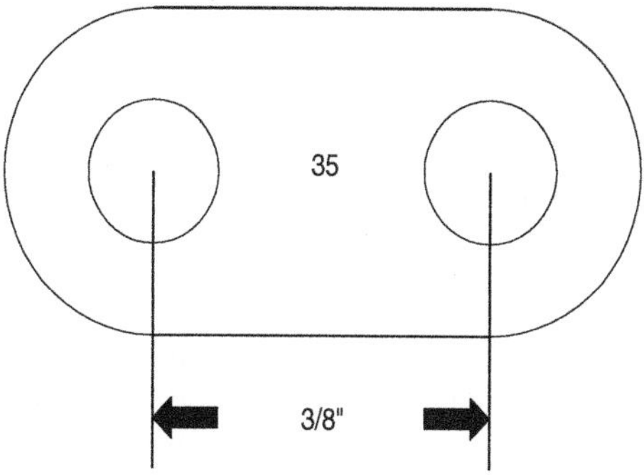

Figure 7.1 *Chain size*

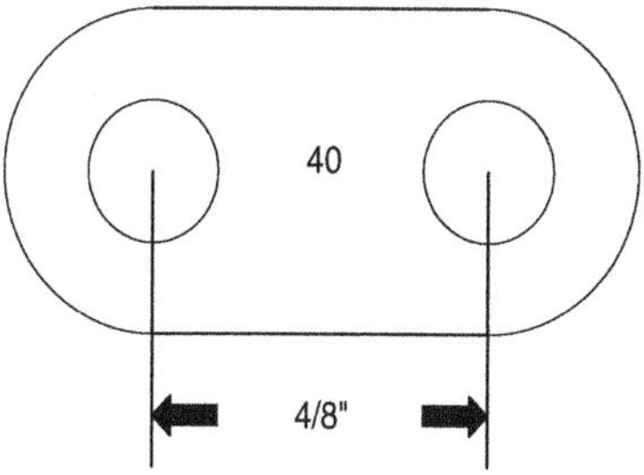

Figure 7.2 *Chain size*

Sometimes chains are linked to form two multistrand chains. The number designation for this chain would have the same pitch number as standard chain, but the pitch would be followed by the number of strands (80-4).

Chain Selection

Plain or Detachable-Link Chain

Plain chains are usually used in slow speed applications like conveyors. They are rugged, designed to carry heavy loads, and when properly maintained can offer years of reliable service. They are made up of a series of detachable links that do not have rollers.

The problem is that if the direction of the chain is reversed, the chain can come apart. When replacing a motor, the rotation of the coupling must be the same before you connect the coupling to the driven unit.

Roller Chain

Roller chains are made up of roller links that are joined with pin links. The links are made up of two side bars, two rollers, and two bushings. The roller reduces the friction between the chain and the sprocket, thereby increasing the life of the unit.

Roller chains can operate at faster speeds than plain chains, and properly maintained, they will offer years of reliable service.

Some roller chains come with a double pitch, meaning that the pitch is double that of a standard chain, but the width and roller size remains the same. Double-pitch chain can be used on standard sprockets, but double-pitch sprockets are also available. The main advantage to the double-pitch chain is that it is cheaper than the standard pitch chain. So, they are often used for applications that require slow speeds, as in for lifting pieces of equipment in a hot press application.

Sprockets

Sprockets are fabricated from a variety of materials; this would depend upon the application of the drive. Large fabricated steel sprockets are manufactured with holes to reduce the weight of the sprocket on the equipment. Because roller chain drives sometimes have restricted spaces for their installation or mounting, the hubs are made in several different styles. See Figure 7.3.

Type A sprockets are flat and have no hub at all. They are usually mounted on flanges or hubs of the device that they are driving. This is accomplished through a series of holes that are either plain or tapered.

Type B sprocket hubs are flush on one side and extend slightly on the other side. The hub is extended to one side to allow the sprocket to be fitted close to the machinery that it is being mounted on. This eliminates a large overhung load on the bearings of the equipment.

Type C sprockets are extended on both sides of the plate surface. They are usually used on the driven sprocket where the pitch diameter is larger and

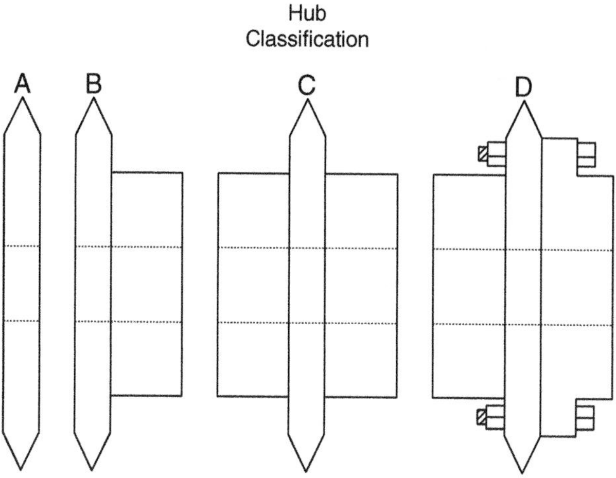

Figure 7.3 *Types of sprocket hubs*

where there is more weight to support on the shaft. Remember this the larger the load is, the larger the hub should be.

Type D sprockets use an A sprocket mounted on a solid or split hub. The type A sprocket is split and bolted to the hub. This is done for ease of removal and not practicality. It allows the speed ratio to be changed easily by simply unbolting the sprocket and changing it without having the remove bearings or other equipment.

Chain Installation

When the proper procedures are followed for installing chains, they will yield years of trouble-free service. Use the following procedure to perform this task:

1 The shafts must be parallel, or the life of the chain will be shortened. The first step is to level the shafts. This is done by placing a level on each of the shafts, then shimming the low side until the shaft is level. See Figure 7.4.

2 The next step is to make sure that the shafts are parallel. This is done by measuring at different points on the shaft and adjusting the shafts until

they are an equal distance apart. Make sure that the shafts are pulled in as close as possible before performing this procedure. The jacking bolts can be used to move the shafts apart evenly after the chain is installed. See Figure 7.5.

3 Before installing a set of used sprockets, verify the size and condition of the sprockets.

4 Install the sprockets on the shafts following the manufacturer's recommendations. Locate and install the first sprocket, then use a straightedge or a string to line the other one up with the one previously installed.

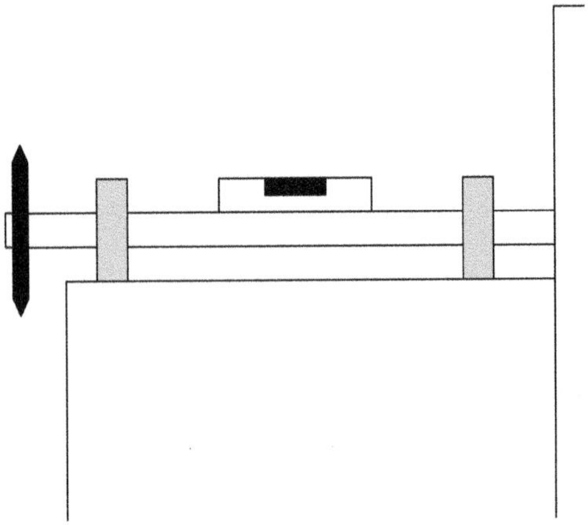

Figure 7.4 *Alignment*

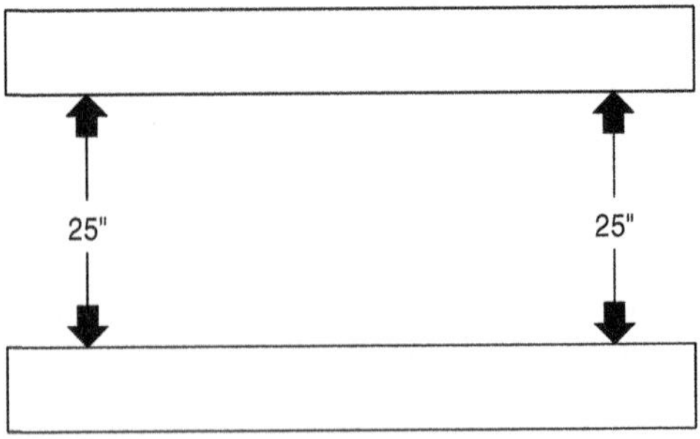

Figure 7.5 *Alignment*

5 Install the chain on the sprockets, then begin increasing the distance between the sprockets by turning the jacking bolts; do this until the chain is snug but not tight. To set the proper chain sag, deflect the chain $\frac{1}{4}$" per foot of span between the shafts. Use a string or straightedge and place it across the top of the chain. Then push down on the chain just enough to remove the slack. Use a tape measure to measure the amount of sag. See Figure 7.6.

6 Do a final check for parallel alignment. Remember: the closer the alignment, the longer the chain will run. See Figure 7.7.

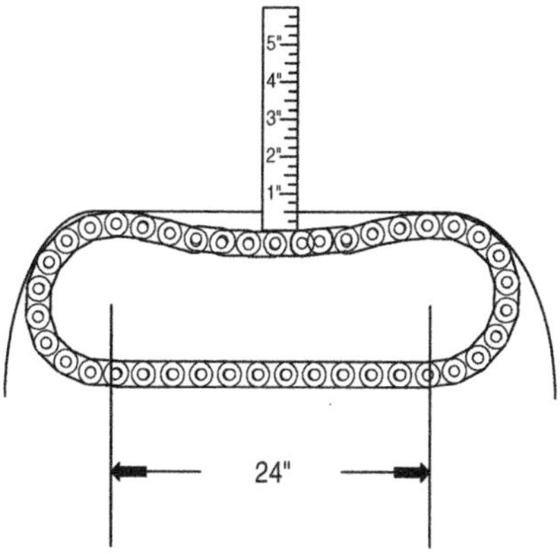

Figure 7.6 *Tensioning*

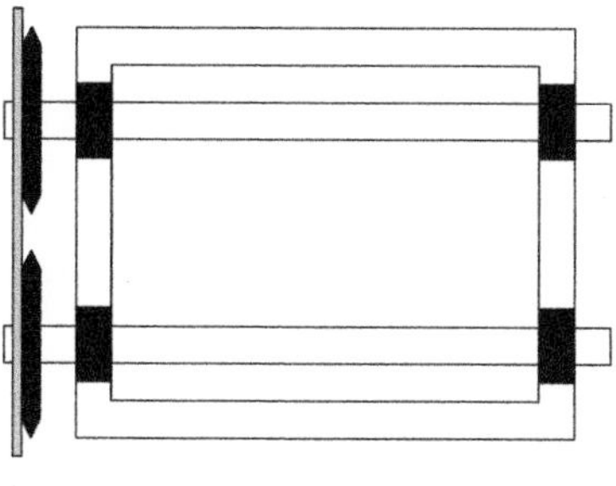

Figure 7.7 *Final alignment*

Power Train Formulas

Shaft Speed

The size of the sprockets in a chain drive system determines the speed relationship between the drive and driven sprockets. For example, if the drive sprocket has the same size sprocket as the driven, then the speed will be equal. See Figure 7.8.

If we change the size of driven sprocket, then the speed of the shaft will also change. If we know what the speed of the electric motor is, and the size of the sprockets, we can calculate the speed of the driven shaft by using the following formula (see Figure 7.9):

$$\text{Driven shaft rpm} = \frac{\text{Drive sprocket \# teeth} \times \text{drive shaft rpm}}{\text{Driven sprocket \# teeth}}$$

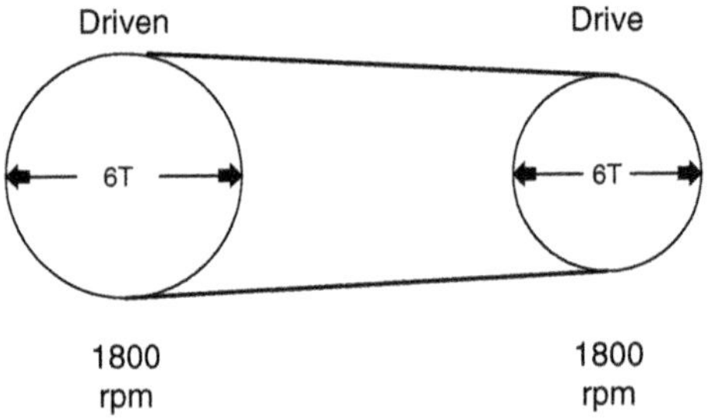

Figure 7.8 *Ratio*

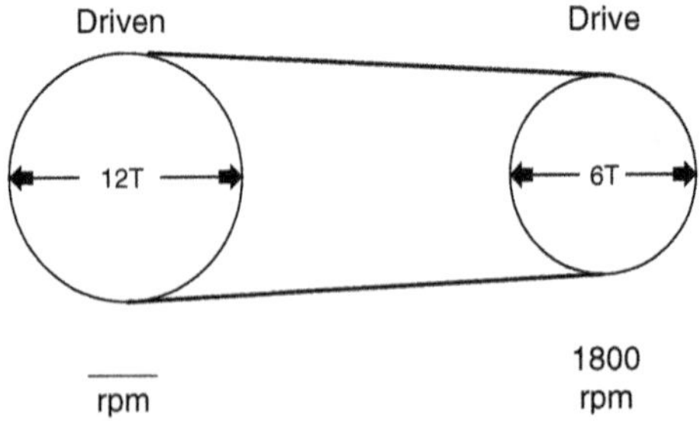

Figure 7.9 *Speed ratio example*

$$\text{Driven shaft rpm} = \frac{6 \times 1800}{12}$$

$$900 = \frac{6 \times 1800}{12}$$

Now we understand how changing the size of a sprocket will also change the shaft speed. Knowing this, we could also assume that to change the shaft rpm we must change the sprocket size.

The problem is how do we know the exact size sprocket that we need to reach the desired speed? Use the same formula that was used to calculate shaft speed, only switch the location of the driven shaft speed and the driven sprocket size:

$$\text{Driven shaft rpm} = \frac{\text{Drive sprocket \# teeth} \times \text{drive shaft rpm}}{\text{Driven sprocket \# teeth}}$$

Let's change the problem to look like this:

$$\text{Driven sprocket teeth} = \frac{\text{Drive sprocket \# teeth} \times \text{drive shaft rpm}}{\text{Driven shaft rpm}}$$

Let's say that we have a problem similar to the ones that we just did, but we want to change the shaft speed of the driven unit. If we know the speed we are looking for, we can use the formula above to calculate the sprocket size required.

Let's change the speed of the driven shaft to 900 rpm (see Figure 7.10):

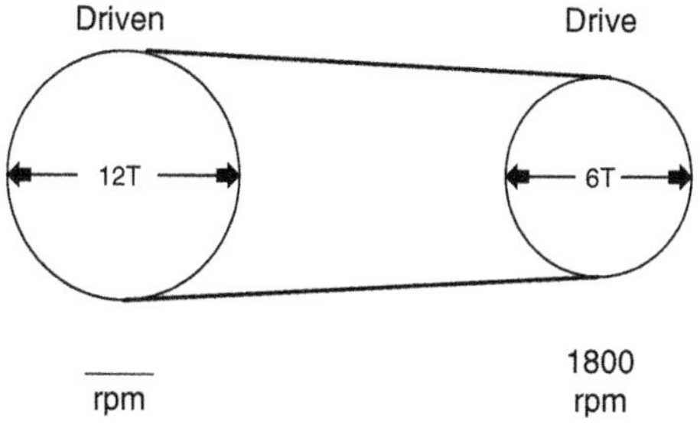

Figure 7.10 *Sprocket calculations*

$$\text{Driven shaft rpm} = \frac{6 \times 1800}{900}$$

$$12 = \frac{6 \times 1800}{900}$$

Chain Length

Many times when a mechanic has to change out chains there is no way of knowing how long the chain should be. One way is to lay the new chain down beside the old chain, but remember that the old chain has been stretched.

Or, maybe you are installing a new drive and you want to have the chain made up before you install it. So what do you do? One method is to take a tape measure and wrap it around the sprockets to get the chain length.

However, this is not a very accurate way to determine the length. Instead, let's take a couple of measurements, then use a simple formula to calculate the actual length that is needed.

First, move the sprockets together until they are as close as the adjustments will allow. Then move the motor or drive out $\frac{1}{4}$ of its travel. Now we are ready to take our measurements. The following information is needed for an equation to find the chain length:

1 Number of teeth on the drive sprocket.

2 Center-to-center distance between the shafts.

3 The chain pitch in inches.

Now use the following formula to solve the equation (see Figure 7.11):

$$\text{Chain length} = \frac{\#\text{ teeth drive} \times \text{pitch}}{2} + \frac{\#\text{ teeth driven} \times \text{pitch}}{2} + \text{center to center} \times 2$$

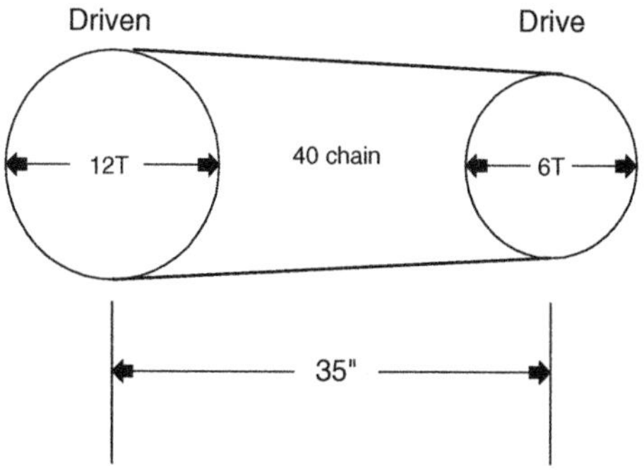

Figure 7.11 *Chain size calculations*

Use the formula above to find the chain length.

$$\text{Chain length} = \frac{6\times.5}{2} + \frac{12\times.5}{2} + 35'' \times 2$$

$$74.5'' = \frac{6\times.5}{2} + \frac{12\times.5}{2} + 35'' \times 2$$

Multiple Sprockets

When calculating multiple sprocket systems, think of each set of sprockets as a two-sprocket system.

Chain Speed

In order to calculate the speed of a chain in feet per minute (FPM), we need the following information:

1 The number of teeth on the sprocket.

2 The shaft rpm of the sprocket.

3 The pitch of the chain in inches.

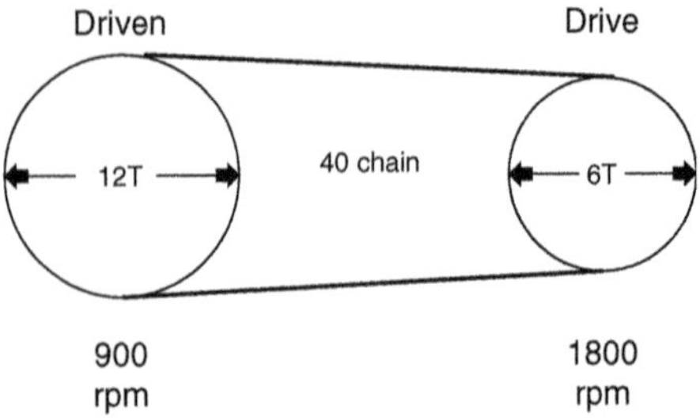

Figure 7.12 *Speed calculations*

With this information we can use the following formula:

$$\text{FPM} = \frac{\#\ \text{teeth} \times \text{pitch} \times \text{rpm}}{12}$$

Use this formula to find the speed of the following chain (see Figure 7.12):

$$\text{FPM} = \frac{\#\ \text{teeth} \times \text{pitch} \times \text{rpm}}{12}$$

$$450 = \frac{6\text{T} \times .5'' \times 1800}{12}$$

Chain drives are used to transmit power between a drive unit and a driven unit. For example, if we have a gearbox and a contact roll on a conveyor, we need a way to transmit the power from the gearbox to the roll. This can be done easily and efficiently with a chain drive unit.

Chain drives can consist of one or multiple strand chains, depending on the load that the unit must transmit. The chains need to be matched with the sprocket type, and they must be tight enough to prevent slippage.

An effective preventive maintenance program will provide extended life to a chain drive system, and through proper corrective maintenance procedures, we can prevent premature failures.

Preventive Maintenance Procedures

Inspection (risk of failures for not following the procedures below is noted along with a rating): LOW: minimal risk/low chance of failure; MEDIUM: failure is possible but equipment not operation to specification is highly probable; HIGH: failure will happen prematurely.

- Inspect a chain for wear by inspecting the links for worn bushings. If worn bushings are noted, write a corrective maintenance work order so that the replacement can be planned and scheduled at a later time.

Risk if the procedure is not followed: HIGH. Chain breakage will occur.

- Lubricate chain with lightweight oil recommended by chain manufacturer. (Ask your chain supplier to visit your site and make recommendations based on documentation they can present to you.)

Risk if the procedure is not followed: HIGH. Chain breakage will occur.

- Check chain sag. Measure the chain sag using a straight edge or string and measure the specifications noted on this PM task. (The chain sag specification can be provided by your chain supplier, or you can use the procedure noted earlier in this chapter.) **WARNING**: The specification must be noted on the PM procedure.
- Set tension, and make a note at the bottom of the PM work order, if a deficiency is noted.

Risk if the procedure is not followed: MEDIUM. Sprocket and chain wear will accelerate, thus causing equipment stoppage.

- Inspect sprockets for worn teeth and abnormal wear on the sides of the sprockets. (The question is: Can the sprockets and chain last for two more weeks without equipment stoppage?) If the sprockets and chain can last two weeks then write a corrective maintenance work order in order for this job to be planned and scheduled with the correct parts. If the sprocket cannot last two weeks, then change all sprockets and the chain. Set and check sheave and chain alignment and tension. **WARNING:** When changing a sprocket, all sprockets, and the chain, should be changed because the difference between a worn and new sprocket in pitch diameter can be extreme, thus causing premature failure of the sprockets and chain.

Risk if the procedure is not followed: High. Worn sprockets are an indication of the equipment being in a failure mode. Action must be taken.

8 Compressors

A compressor is a machine that is used to increase the pressure of a gas or vapor. They can be grouped into two major classifications: centrifugal and positive displacement. This section provides a general discussion of these types of compressors.

Centrifugal

In general, the centrifugal designation is used when the gas flow is radial and the energy transfer is predominantly due to a change in the centrifugal forces acting on the gas. The force utilized by the centrifugal compressor is the same as that utilized by centrifugal pumps.

In a centrifugal compressor, air or gas at atmospheric pressure enters the eye of the impeller. As the impeller rotates, the gas is accelerated by the rotating element within the confined space that is created by the volute of the compressor's casing. The gas is compressed as more gas is forced into the volute by the impeller blades. The pressure of the gas increases as it is pushed through the reduced free space within the volute.

As in centrifugal pumps, there may be several stages to a centrifugal air compressor. In these multistage units, a progressively higher pressure is produced by each stage of compression.

Configuration

The actual dynamics of centrifugal compressors are determined by their design. Common designs are: overhung or cantilever, centerline, and bullgear.

Overhung or Cantilever

The cantilever design is more susceptible to process instability than centerline centrifugal compressors. Figure 8.1 illustrates a typical cantilever design.

The overhung design of the rotor (i.e., no outboard bearing) increases the potential for radical shaft deflection. Any variation in laminar flow, volume,

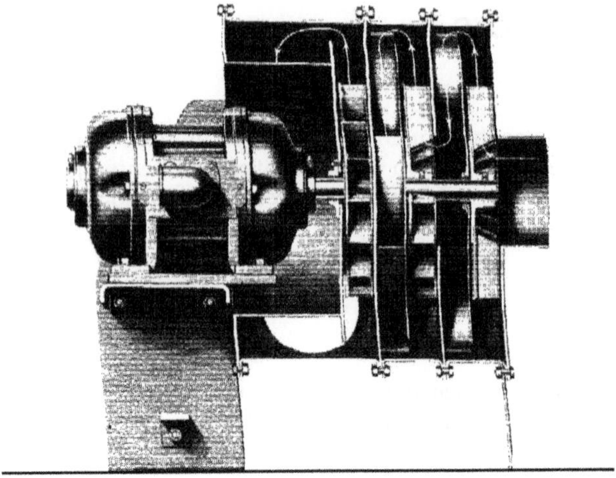

Figure 8.1 *Cantilever centrifugal compressor is susceptible to instability*

or load of the inlet or discharge gas forces the shaft to bend or deflect from its true centerline. As a result, the mode shape of the shaft must be monitored closely.

Centerline

Centerline designs, such as horizontal and vertical split-case, are more stable over a wider operating range, but should not be operated in a variable-demand system. Figure 8.2 illustrates the normal airflow pattern through a horizontal split-case compressor. Inlet air enters the first stage of the compressor, where pressure and velocity increases occur. The partially compressed air is routed to the second stage where the velocity and pressure are increased further. Adding additional stages until the desired final discharge pressure is achieved can continue this process.

Two factors are critical to the operation of these compressors: impeller configuration and laminar flow, which must be maintained through all of the stages.

The impeller configuration has a major impact on stability and operating envelope. There are two impeller configurations: in-line and back-to-back, or opposed. With the in-line design, all impellers face in the same direction. With the opposed design, impeller direction is reversed in adjacent stages.

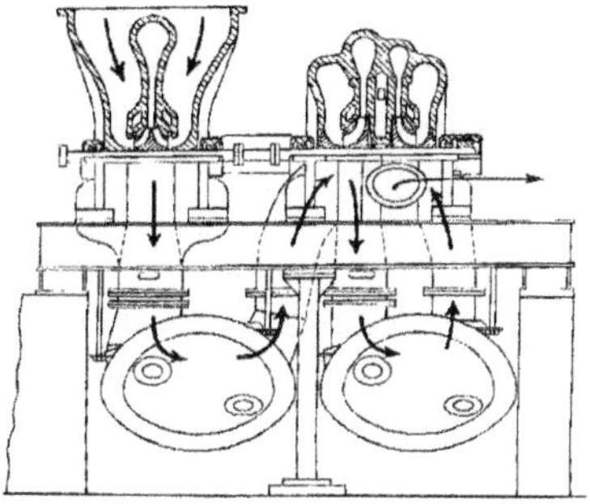

Figure 8.2 *Airflow through a centerline centrifugal compressor*

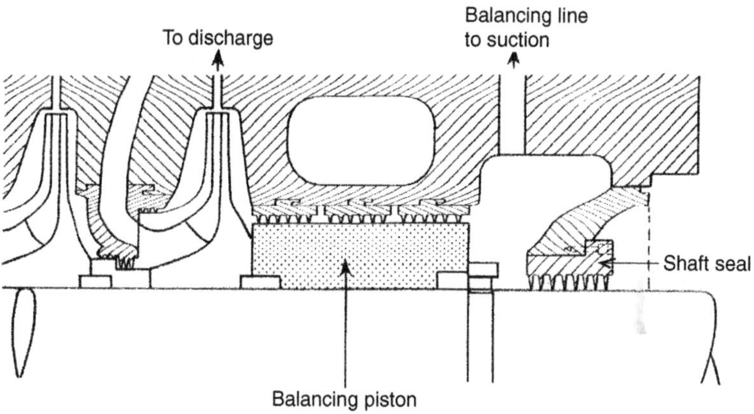

Figure 8.3 *Balancing piston resists axial thrust from the in-line impeller design of a centerline centrifugal compressor*

In-Line

A compressor with all impellers facing in the same direction generates substantial axial forces. The axial pressures generated by each impeller for all the stages are additive. As a result, massive axial loads are transmitted to the fixed bearing. Because of this load, most of these compressors use either a Kingsbury thrust bearing or a balancing piston to resist axial thrusting. Figure 8.3 illustrates a typical balancing piston.

All compressors that use in-line impellers must be monitored closely for axial thrusting. If the compressor is subjected to frequent or constant unloading, the axial clearance will increase due to this thrusting cycle. Ultimately, this frequent thrust loading will lead to catastrophic failure of the compressor.

Opposed

By reversing the direction of alternating impellers, the axial forces generated by each impeller or stage can be minimized. In effect, the opposed impellers tend to cancel the axial forces generated by the preceding stage. This design is more stable and should not generate measurable axial thrusting. This allows these units to contain a normal float and fixed rolling-element bearing.

Bullgear

The bullgear design uses a direct-driven helical gear to transmit power from the primary driver to a series of pinion-gear-driven impellers that are located around the circumference of the bullgear. Figure 8.4 illustrates a typical bullgear compressor layout.

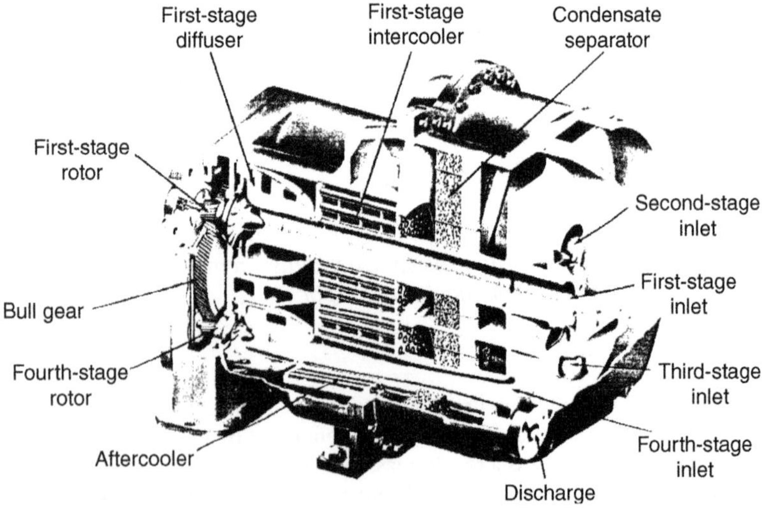

Figure 8.4 *Bullgear centrifugal compressor*

The pinion shafts are typically a cantilever-type design that has an enclosed impeller on one end and a tilting-pad bearing on the other. The pinion gear is between these two components. The number of impeller-pinions (i.e., stages) varies with the application and the original equipment vendor. However, all bullgear compressors contain multiple pinions that operate in series.

Atmospheric air or gas enters the first-stage pinion, where the pressure is increased by the centrifugal force created by the first-stage impeller. The partially compressed air leaves the first stage, passes through an intercooler, and enters the second-stage impeller. This process is repeated until the fully compressed air leaves through the final pinion-impeller, or stage.

Most bullgear compressors are designed to operate with a gear speed of 3,600 rpm. In a typical four-stage compressor, the pinions operate at progressively higher speeds. A typical range is between 12,000 rpm (first stage) and 70,000 rpm (fourth stage).

Because of their cantilever design and pinion rotating speeds, bullgear compressors are extremely sensitive to variations in demand or downstream pressure changes. Because of this sensitivity, their use should be limited to baseload applications.

Bullgear compressors are not designed for, nor will they tolerate, load-following applications. They should not be installed in the same discharge manifold with positive-displacement compressors, especially reciprocating compressors. The standing-wave pulses created by many positive-displacement compressors create enough variation in the discharge manifold to cause potentially serious instability.

In addition, the large helical gear used for the bullgear creates an axial oscillation or thrusting that contributes to instability within the compressor. This axial movement is transmitted throughout the machine-train.

Performance

The physical laws of thermodynamics, which define their efficiency and system dynamics, govern compressed-air systems and compressors. This section discusses both the first and second laws of thermodynamics, which apply to all compressors and compressed-air systems. Also applying to these systems are the Ideal Gas Law and the concepts of pressure and compression.

First Law of Thermodynamics

This law states that energy cannot be created or destroyed during a process, such as compression and delivery of air or gas, although it may change from one form of energy to another. In other words, whenever a quantity of one kind of energy disappears, an exactly equivalent total of other kinds of energy must be produced. This is expressed for a steady-flow open system such as a compressor by the following relationship:

Net energy added to system as heat and work	+	Stored energy of mass entering system	−	Stored energy of mass leaving system	= 0

Second Law of Thermodynamics

The second law of thermodynamics states that energy exists at various levels and is available for use only if it can move from a higher to a lower level. For example, it is impossible for any device to operate in a cycle and produce work while exchanging heat only with bodies at a single fixed temperature. In thermodynamics a measure of the unavailability of energy has been devised and is known as entropy. As a measure of unavailability, entropy increases as a system loses heat, but it remains constant when there is no gain or loss of heat as in an adiabatic process. It is defined by the following differential equation:

$$dS = \frac{dQ}{T}$$

where:

T = Temperature (Fahrenheit)

Q = Heat added (BTU)

Pressure/Volume/Temperature (PVT) Relationship

Pressure, temperature, and volume are properties of gases that are completely interrelated. Boyle's Law and Charles' Law may be combined into one equation that is referred to as the Ideal Gas Law. This equation is always true for ideal gases and is true for real gases under certain conditions.

$$\frac{P_1 V_1}{T_1} = \frac{P_2 V_2}{T_2}$$

For air at room temperature, the error in this equation is less than 1% for pressures as high as 400 psia. For air at one atmosphere of pressure, the error is less than 1% for temperatures as low as $-200°F$. These error factors will vary for different gases.

Pressure/Compression

In a compressor, pressure is generated by pumping quantities of gas into a tank or other pressure vessel. Progressively increasing the amount of gas in the confined or fixed-volume space increases the pressure. The effects of pressure exerted by a confined gas result from the force acting on the container walls. This force is caused by the rapid and repeated bombardment from the enormous number of molecules that are present in a given quantity of gas.

Compression occurs when the space is decreased between the molecules. Less volume means that each particle has a shorter distance to travel, thus proportionately more collisions occur in a given span of time, resulting in a higher pressure. Air compressors are designed to generate particular pressures to meet specific application requirements.

Other Performance Indicators

The same performance indicators as those for centrifugal pumps or fans govern centrifugal compressors.

Installation

Dynamic compressors seldom pose serious foundation problems. Since moments and shaking forces are not generated during compressor operation, there are no variable loads to be supported by the foundation. A foundation or mounting of sufficient area and mass to maintain compressor level and alignment and to assure safe soil loading is all that is required. The units may be supported on structural steel if necessary. The principles defined for centrifugal pumps also apply to centrifugal compressors.

It is necessary to install pressure-relief valves on most dynamic compressors to protect them due to restrictions placed on casing pressure, power input, and to keep out of the compressor's surge range. Always install a valve capable of bypassing the full-load capacity of the compressor between its discharge port and the first isolation valve.

Operating Methods

The acceptable operating envelope for centrifugal compressors is very limited. Therefore, care should be taken to minimize any variation in suction supply, backpressure caused by changes in demand, and frequency of unloading. The operating guidelines provided in the compressor vendor's O&M manual should be followed to prevent abnormal operating behavior or premature wear or failure of the system.

Centrifugal compressors are designed to be baseloaded and may exhibit abnormal behavior or chronic reliability problems when used in a load-following mode of operation. This is especially true of bullgear and cantilever compressors. For example, a one-psig change in discharge pressure may be enough to cause catastrophic failure of a bullgear compressor. Variations in demand or backpressure on a cantilever design can cause the entire rotating element and its shaft to flex. This not only affects the compressor's efficiency, but also accelerates wear and may lead to premature shaft or rotor failure.

All compressor types have moving parts, high noise levels, high pressures, and high-temperature cylinder and discharge-piping surfaces.

Positive Displacement

Positive-displacement compressors can be divided into two major classifications: rotary and reciprocating.

Rotary

The rotary compressor is adaptable to direct drive by the use of induction motors or multicylinder gasoline or diesel engines. These compressors are compact, relatively inexpensive, and require a minimum of operating attention and maintenance. They occupy a fraction of the space and weight of a reciprocating machine having equivalent capacity.

Configuration

Rotary compressors are classified into three general groups: sliding vane, helical lobe, and liquid-seal ring.

Sliding Vane

The basic element of the sliding-vane compressor is the cylindrical housing and the rotor assembly. This compressor, which is illustrated in Figure 8.5,

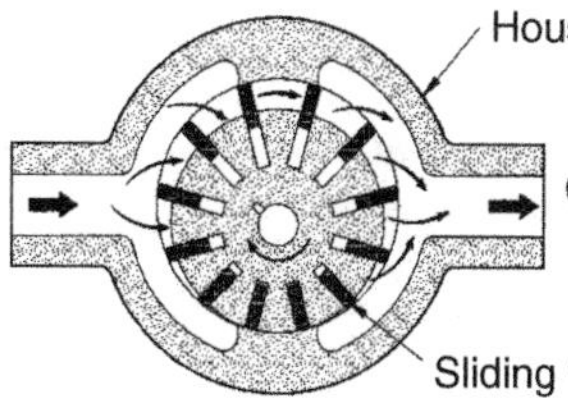

Figure 8.5 *Rotary sliding-vane compressor*

has longitudinal vanes that slide radially in a slotted rotor mounted eccentrically in a cylinder. The centrifugal force carries the sliding vanes against the cylindrical case with the vanes forming a number of individual longitudinal cells in the eccentric annulus between the case and rotor. The suction port is located where the longitudinal cells are largest. The size of each cell is reduced by the eccentricity of the rotor as the vanes approach the discharge port, thus compressing the gas.

Cyclical opening and closing of the inlet and discharge ports occurs by the rotor's vanes passing over them. The inlet port is normally a wide opening that is designed to admit gas in the pocket between two vanes. The port closes momentarily when the second vane of each air-containing pocket passes over the inlet port.

When running at design pressure, the theoretical operation curves are identical (see Figure 8.6) to those of a reciprocating compressor. However, there is one major difference between a sliding-vane and a reciprocating compressor. The reciprocating unit has spring-loaded valves that open automatically with small pressure differentials between the outside and inside cylinder. The sliding-vane compressor has no valves.

The fundamental design considerations of a sliding-vane compressor are the rotor assembly, cylinder housing, and the lubrication system.

Housing and Rotor Assembly

Cast iron is the standard material used to construct the cylindrical housing, but other materials may be used if corrosive conditions exist. The rotor is usually a continuous piece of steel that includes the shaft and is made from bar stock. Special materials can be selected for corrosive applications. Occasionally, the rotor may be a separate iron casting keyed to a shaft. On most standard air compressors, the rotor-shaft seals are semimetallic packing in a stuffing box. Commercial mechanical rotary seals can be supplied

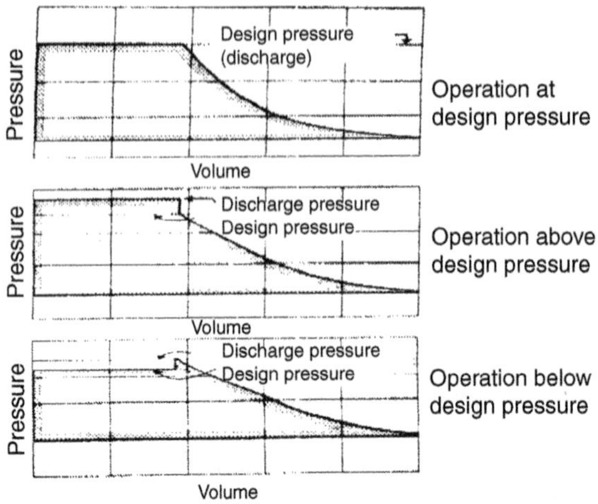

Figure 8.6 *Theoretical operation curves for rotary compressors with built-in porting*

when needed. Cylindrical roller bearings are generally used in these assemblies.

Vanes are usually asbestos or cotton cloth impregnated with a phenolic resin. Bronze or aluminum also may be used for vane construction. Each vane fits into a milled slot extending the full length of the rotor and slides radially in and out of this slot once per revolution. Vanes are the most maintenance-prone part in the compressor. There are from 8 to 20 vanes on each rotor, depending upon its diameter. A greater number of vanes increase compartmentalization, which reduces the pressure differential across each vane.

Lubrication System

A V-belt-driven, force-fed oil lubrication system is used on water-cooled compressors. Oil goes to both bearings and to several points in the cylinder. Ten times as much oil is recommended to lubricate the rotary cylinder as is required for the cylinder of a corresponding reciprocating compressor. The oil carried over with the gas to the line may be reduced 50% with an oil separator on the discharge. Use of an aftercooler ahead of the separator permits removal of 85 to 90% of the entrained oil.

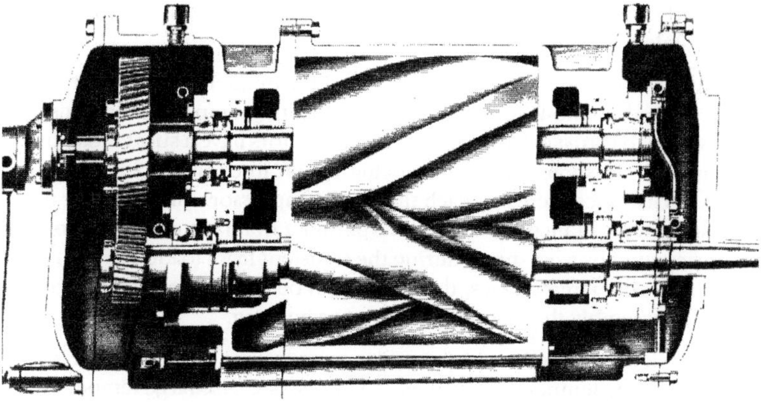

Figure 8.7 *Helical lobe, or screw, rotary air compressor*

Helical Lobe or Screw

The helical lobe, or screw, compressor is shown in Figure 8.7. It has two or more mating sets of lobe-type rotors mounted in a common housing. The male lobe, or rotor, is usually direct-driven by an electric motor. The female lobe, or mating rotor, is driven by a helical gear set that is mounted on the outboard end of the rotor shafts. The gears provide both motive power for the female rotor and absolute timing between the rotors.

The rotor set has extremely close mating clearance (i.e., about 0.5 mils) but no metal-to-metal contact. Most of these compressors are designed for oil-free operation. In other words, no oil is used to lubricate or seal the rotors. Instead, oil lubrication is limited to the timing gears and bearings that are outside the air chamber. Because of this, maintaining proper clearance between the two rotors is critical.

This type of compressor is classified as a constant volume, variable-pressure machine that is quite similar to the vane-type rotary in general characteristics. Both have a built-in compression ratio.

Helical-lobe compressors are best suited for base-load applications where they can provide a constant volume and pressure of discharge gas. The only recommended method of volume control is the use of variable-speed motors. With variable-speed drives, capacity variations can be obtained with a proportionate reduction in speed. A 50% speed reduction is the maximum permissible control range.

Helical-lobe compressors are not designed for frequent or constant cycles between load and no-load operation. Each time the compressor unloads, the rotors tend to thrust axially. Even though the rotors have a substantial thrust bearing and, in some cases, a balancing piston to counteract axial thrust, the axial clearance increases each time the compressor unloads. Over time, this clearance will increase enough to permit a dramatic rise in the impact energy created by axial thrust during the transient from loaded to unloaded conditions. In extreme cases, the energy can be enough to physically push the rotor assembly through the compressor housing.

Compression ratio and maximum inlet temperature determine the maximum discharge temperature of these compressors. Discharge temperatures must be limited to prevent excessive distortion between the inlet and discharge ends of the casing and rotor expansion. High-pressure units are water-jacketed in order to obtain uniform casing temperature. Rotors also may be cooled to permit a higher operating temperature.

If either casing distortion or rotor expansion occur, the clearance between the rotating parts will decrease, and metal-to-metal contact will occur. Since the rotors typically rotate at speeds between 3,600 and 10,000 rpm, metal-to-metal contact normally results in instantaneous, catastrophic compressor failure.

Changes in differential pressures can be caused by variations in either inlet or discharge conditions (i.e., temperature, volume, or pressure). Such changes can cause the rotors to become unstable and change the load zones in the shaft-support bearings. The result is premature wear and/or failure of the bearings.

Always install a relief valve that is capable of bypassing the full-load capacity of the compressor between its discharge port and the first isolation valve. Since helical-lobe compressors are less tolerant to over-pressure operation, safety valves are usually set within 10% of absolute discharge pressure, or 5 psi, whichever is lower.

Liquid-Seal Ring

The liquid-ring, or liquid-piston, compressor is shown in Figure 8.8. It has a rotor with multiple forward-turned blades that rotate about a central cone that contains inlet and discharge ports. Liquid is trapped between adjacent

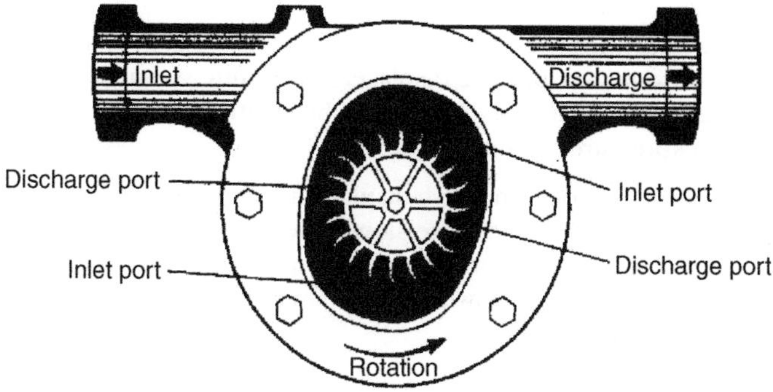

Figure 8.8 *Liquid-seal ring rotary air compressor*

blades, which drive the liquid around the inside of an elliptical casing. As the rotor turns, the liquid face moves in and out of this space due to the casing shape, creating a liquid piston. Porting in the central cone is built-in and fixed, and there are no valves.

Compression occurs within the pockets or chambers between the blades before the discharge port is uncovered. Since the port location must be designed and built for a specific compression ratio, it tends to operate above or below the design pressure (refer back to Figure 8.6).

Liquid-ring compressors are cooled directly rather than by jacketed casing walls. The cooling liquid is fed into the casing where it comes into direct contact with the gas being compressed. The excess liquid is discharged with the gas. The discharged mixture is passed through a conventional baffle or centrifugal-type separator to remove the free liquid. Because of the intimate contact of gas and liquid, the final discharge temperature can be held close to the inlet cooling water temperature. However, the discharge gas is saturated with liquid at the discharge temperature of the liquid.

The amount of liquid passed through the compressor is not critical and can be varied to obtain the desired results. The unit will not be damaged if a large quantity of liquid inadvertently enters its suction port.

Lubrication is required only in the bearings, which are generally located external to the casing. The liquid itself acts as a lubricant, sealing medium, and coolant for the stuffing boxes.

Performance

Performance of a rotary positive-displacement compressor can be evaluated using the same criteria as a positive-displacement pump. As in constant-volume machines, performance is determined by rotation speed, internal slip, and total backpressure on the compressor.

The volumetric output of rotary positive-displacement compressors can be controlled by speed changes. The more slowly the compressor turns, the lower its output volume. This feature permits the use of these compressors in load-following applications. However, care must be taken to prevent sudden, radical changes in speed.

Internal slip is simply the amount of gas that can flow through internal clearances from the discharge back to the inlet. Obviously, internal wear will increase internal slip.

Discharge pressure is relatively constant regardless of operating speed. With the exceptions of slight pressure variations caused by atmospheric changes and backpressure, a rotary positive-displacement compressor will provide a fixed discharge pressure. Backpressure, which is caused by restrictions in the discharge piping or demand from users of the compressed air or gas, can have a serious impact on compressor performance.

If backpressure is too low or demand too high, the compressor will be unable to provide sufficient volume or pressure to the downstream systems. In this instance, the discharge pressure will be noticeably lower than designed.

If the backpressure is too high or demand too low, the compressor will generate a discharge pressure higher than designed. It will continue to compress the air or gas until it reaches the unload setting on the system's relief valve or until the brake horsepower required exceeds the maximum horsepower rating of the driver.

Installation

Installation requirements for rotary positive-displacement compressors are similar to those for any rotating machine. Review the installation requirements for centrifugal pumps and compressors for foundation, pressure-relief, and other requirements. As with centrifugal compressors, rotary positive-displacement compressors must be fitted with pressure-relief devices to limit the discharge or interstage pressures to a safe maximum for the equipment served.

In applications where demand varies, rotary positive-displacement compressors require a downstream receiver tank or reservoir that minimizes the load-unload cycling frequency of the compressor. The receiver tank should have sufficient volume to permit acceptable unload frequencies for the compressor. Refer to the vendor's O&M manual for specific receiver-tank recommendations.

Operating Methods

All compressor types have moving parts, high noise levels, high pressures, and high-temperature cylinder and discharge-piping surfaces.

Rotary positive-displacement compressors should be operated as baseloaded units. They are especially sensitive to the repeated start-stop operation required by load-following applications. Generally, rotary positive-displacement compressors are designed to unload about every six to eight hours. This unload cycle is needed to dissipate the heat generated by the compression process. If the unload frequency is too great, these compressors have a high probability of failure.

There are several primary operating control inputs for rotary positive-displacement compressors. These control inputs are: discharge pressure, pressure fluctuations, and unloading frequency.

Discharge Pressure

This type of compressor will continue to compress the air volume in the downstream system until: (1) some component in the system fails; (2) the brake horsepower exceeds the driver's capacity; or (3) a safety valve opens. Therefore, the operator's primary control input should be the compressor's discharge pressure. If the discharge pressure is below the design point, it is a clear indicator that the total downstream demand is greater than the unit's capacity. If the discharge pressure is too high, the demand is too low, and excessive unloading will be required to prevent failure.

Pressure Fluctuations

Fluctuations in the inlet and discharge pressures indicate potential system problems that may adversely affect performance and reliability. Pressure fluctuations are generally caused by changes in the ambient environment, turbulent flow, or restrictions caused by partially blocked inlet filters. Any of these problems will result in performance and reliability problems if not corrected.

Unloading Frequency

The unloading function in rotary positive-displacement compressors is automatic and not under operator control. Generally, a set of limit switches, one monitoring internal temperature and one monitoring discharge pressure, are used to trigger the unload process. By design, the limit switch that monitors the compressor's internal temperature is the primary control. The secondary control, or discharge-pressure switch, is a fail-safe design to prevent overloading of the compressor.

Depending on design, rotary positive-displacement compressors have an internal mechanism designed to minimize the axial thrust caused by the instantaneous change from fully loaded to unloaded operating conditions. In some designs, a balancing piston is used to absorb the rotor's thrust during this transient. In others, oversized thrust bearings are used.

Regardless of the mechanism used, none provides complete protection from the damage imparted by the transition from load to no-load conditions. However, as long as the unload frequency is within design limits, this damage will not adversely affect the compressor's useful operating life or reliability. However, an unload frequency greater than that accommodated in the design will reduce the useful life of the compressor and may lead to premature, catastrophic failure.

Operating practices should minimize, as much as possible, the unload frequency of these compressors. Installation of a receiver tank and modification of user-demand practices are the most effective solutions to this type of problem.

Reciprocating

Reciprocating compressors are widely used by industry and are offered in a wide range of sizes and types. They vary from units requiring less than 1 hp to more than 12,000 hp. Pressure capabilities range from low vacuums at intake to special compressors capable of 60,000 psig or higher.

Reciprocating compressors are classified as constant-volume, variable-pressure machines. They are the most efficient type of compressor and can be used for partial-load, or reduced-capacity, applications.

Because of the reciprocating pistons and unbalanced rotating parts, the unit tends to shake. Therefore, it is necessary to provide a mounting that stabilizes the installation. The extent of this requirement depends on the type and size of the compressor.

Because reciprocating compressors should be supplied with clean gas, inlet filters are recommended in all applications. They cannot satisfactorily handle liquids entrained in the gas, although vapors are no problem if condensation within the cylinders does not take place. Liquids will destroy the lubrication and cause excessive wear.

Reciprocating compressors deliver a pulsating flow of gas that can damage downstream equipment or machinery. This is sometimes a disadvantage, but pulsation dampers can be used to alleviate the problem.

Configuration

Certain design fundamentals should be clearly understood before analyzing the operating condition of reciprocating compressors. These fundamentals include frame and running gear, inlet and discharge valves, cylinder cooling, and cylinder orientation.

Frame and Running Gear

Two basic factors guide frame and running gear design. The first factor is the maximum horsepower to be transmitted through the shaft and running gear to the cylinder pistons. The second factor is the load imposed on the frame parts by the pressure differential between the two sides of each piston. This is often called pin load because this full force is directly exerted on the crosshead and crankpin. These two factors determine the size of bearings, connecting rods, frame, and bolts that must be used throughout the compressor and its support structure.

Cylinder Design

Compression efficiency depends entirely upon the design of the cylinder and its valves. Unless the valve area is sufficient to allow gas to enter and leave the cylinder without undue restriction, efficiency cannot be high. Valve placement for free flow of the gas in and out of the cylinder is also important.

Both efficiency and maintenance are influenced by the degree of cooling during compression. The method of cylinder cooling must be consistent with the service intended.

The cylinders and all the parts must be designed to withstand the maximum application pressure. The most economical materials that will give the proper strength and the longest service under the design conditions are generally used.

Inlet and Discharge Valves

Compressor valves are placed in each cylinder to permit one-way flow of gas, either into or out of the cylinder. There must be one or more valve(s) for inlet and discharge in each compression chamber.

Each valve opens and closes once for each revolution of the crankshaft. The valves in a compressor operating at 700 rpm for 8 hours per day and 250 days per year will have cycled (i.e., opened and closed) 42,000 times per hour, 336,000 times per day, or 84 million times in a year. The valves have less than $\frac{1}{10}$ of a second to open, let the gas pass through, and to close.

They must cycle with a minimum of resistance for minimum power consumption. However, the valves must have minimal clearance to prevent excessive expansion and reduced volumetric efficiency. They must be tight under extreme pressure and temperature conditions. Finally, the valves must be durable under many kinds of abuse.

There are four basic valve designs used in these compressors: finger, channel, leaf, and annular ring. Within each class there may be variations in design, depending upon operating speed and size of valve required.

Finger

Figure 8.9 is an exploded view of a typical finger valve. These valves are used for smaller, air-cooled compressors. One end of the finger is fixed and the opposite end lifts when the valve opens.

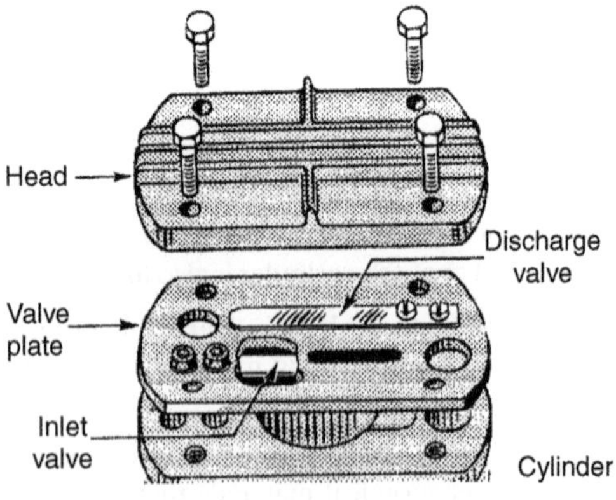

Figure 8.9 *Finger valve configuration*

Valve closed: A tight seat is formed without slamming or friction, so seat wear is at a minimum. Both channel and spring are precision made to assure a perfect fit. A gas space is formed between the bowed spring and the flat channel.

Valve opening: Channel lifts straight up in the guides without flexing. Opening is even over the full length of the port, giving uniform air velocity without turbulance. Cushioning is effected by the compression and escape of the gas between spring and channel.

Valve wide open: Gas trapped between spring and channel has been compressed and in escaping has allowed channel to float in its stop.

Figure 8.10 *Channel valve configuration*

Channel

The channel valve shown in Figure 8.10 is widely used in mid- to large-sized compressors. This valve uses a series of separate stainless steel channels. As explained in the figure, this is a cushioned valve, which adds greatly to its life.

Leaf

The leaf valve (see Figure 8.11) has a configuration somewhat like the channel valve. It is made of flat-strip steel that opens against an arched stop plate. This results in valve flexing only at its center with maximum lift. The valve operates as its own spring.

Annular Ring

Figure 8.12 shows exploded views of typical inlet and discharge annular-ring valves. The valves shown have a single ring, but larger sizes may have two or three rings. In some designs, the concentric rings are tied into a single piece by bridges.

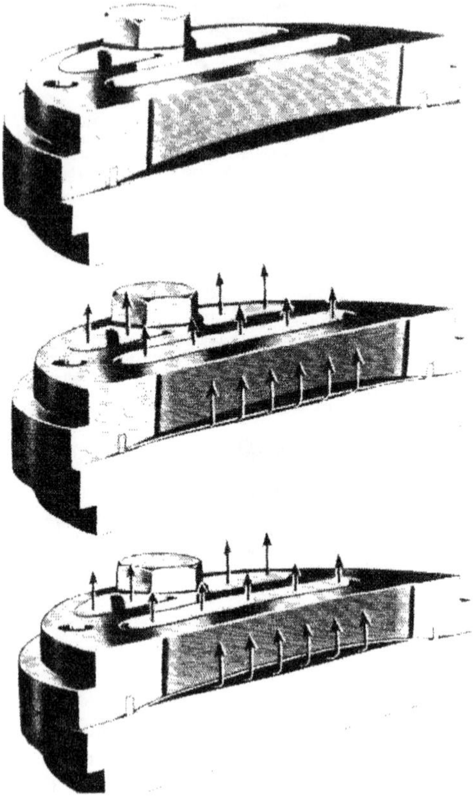

Figure 8.11 *Leaf spring configuration*

The springs and the valve move into a recess in the stop plate as the valve opens. Gas that is trapped in the recess acts as a cushion and prevents slamming. This eliminates a major source of valve and spring breakage. The valve shown was the first cushioned valve built.

Cylinder Cooling

Cylinder heat is produced by the work of compression plus friction, which is caused by the action of the piston and piston rings on the cylinder wall and packing on the rod. The amount of heat generated can be considerable,

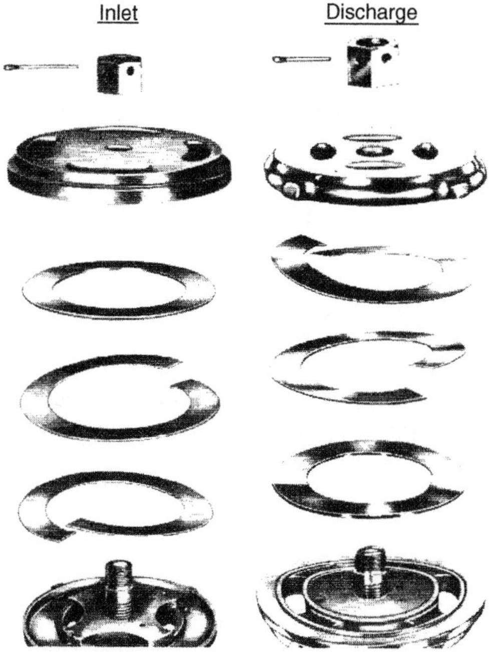

Figure 8.12 *Annular-ring valves*

particularly when moderate to high compression ratios are involved. This can result in undesirably high operating temperatures.

Most compressors use some method to dissipate a portion of this heat to reduce the cylinder wall and discharge gas temperatures. The following are advantages of cylinder cooling:

- Lowering cylinder wall and cylinder head temperatures reduces loss of capacity and horsepower per unit volume due to suction gas preheating during inlet stroke. This results in more gas in the cylinder for compression.
- Reducing cylinder wall and cylinder head temperatures removes more heat from the gas during compression, lowering its final temperature and reducing the power required.

- Reducing the gas temperature and that of the metal surrounding the valves results in longer valve service life and reduces the possibility of deposit formation.
- Reduced cylinder wall temperature promotes better lubrication, resulting in longer life and reduced maintenance.
- Cooling, particularly water-cooling, maintains a more even temperature around the cylinder bore and reduces warpage.

Cylinder Orientation

Orientation of the cylinders in a multistage or multicylinder compressor directly affects the operating dynamics and vibration level. Figure 8.13 illustrates a typical three-piston, air-cooled compressor. Since three pistons are oriented within a 120-degree arc, this type of compressor generates higher vibration levels than the opposed piston compressor illustrated in Figure 8.14.

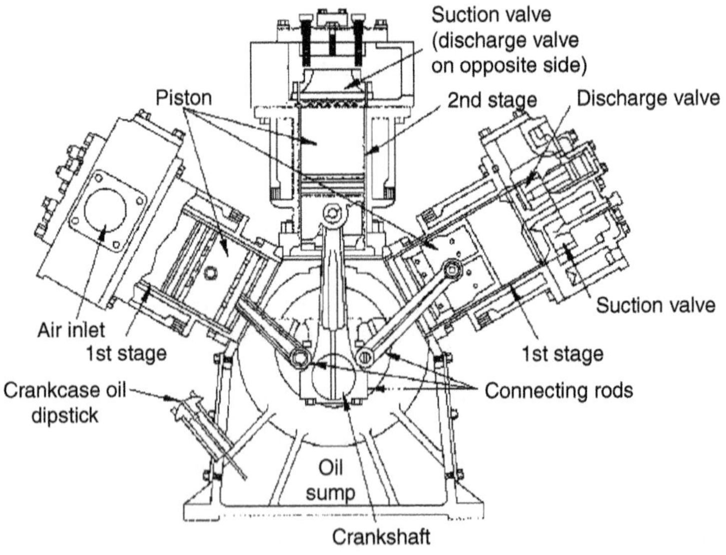

Figure 8.13 *Three-piston compressor generates higher vibration levels*

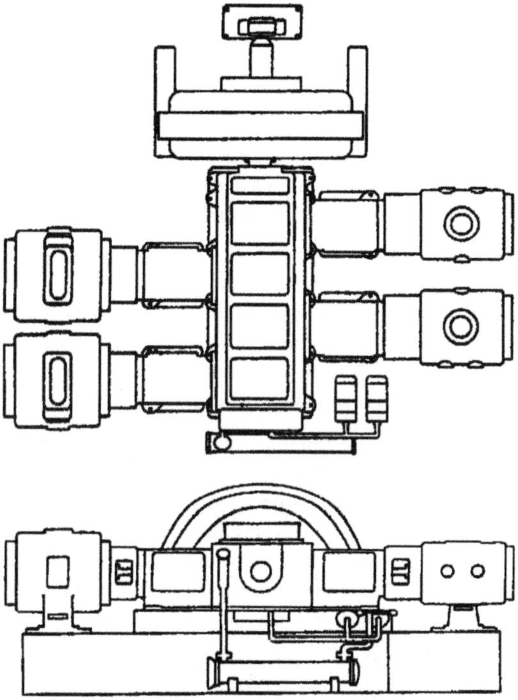

Figure 8.14 *Opposed-piston compressor balances piston forces*

Performance

Reciprocating-compressor performance is governed almost exclusively by operating speed. Each cylinder of the compressor will discharge the same volume, excluding slight variations caused by atmospheric changes, at the same discharge pressure each time it completes the discharge stroke. As the rotation speed of the compressor changes, so does the discharge volume.

The only other variables that affect performance are the inlet-discharge valves, which control flow into and out of each cylinder. Although reciprocating compressors can use a variety of valve designs, it is crucial that the valves perform reliably. If they are damaged and fail to operate at the proper time or do not seal properly, overall compressor performance will be substantially reduced.

Installation

A carefully planned and executed installation is extremely important and makes compressor operation and maintenance easier and safer. Key components of a compressor installation are location, foundation, and piping.

Location

The preferred location for any compressor is near the center of its load. However, the choice is often influenced by the cost of supervision, which can vary by location. The ongoing cost of supervision may be less expensive at a less-optimum location, which can offset the cost of longer piping.

A compressor will always give better, more reliable service when enclosed in a building that protects it from cold, dusty, damp, and corrosive conditions. In certain locations it may be economical to use a roof only, but this is not recommended unless the weather is extremely mild. Even then, it is crucial to prevent rain and wind-blown debris from entering the moving parts. Subjecting a compressor to adverse inlet conditions will dramatically reduce reliability and significantly increase maintenance requirements.

Ventilation around a compressor is vital. On a motor-driven, air-cooled unit, the heat radiated to the surrounding air is at least 65% of the power input. On a water-jacketed unit with an aftercooler and outside receiver, the heat radiated to the surrounding air may be 15 to 25% of the total energy input, which is still a substantial amount of heat. Positive outside ventilation is recommended for any compressor room where the ambient temperature may exceed 104°F.

Foundation

Because of the alternating movement of pistons and other components, reciprocating compressors often develop a shaking that alternates in direction. This force must be damped and contained by the mounting. The foundation also must support the weight load of the compressor and its driver.

There are many compressor arrangements, and the net magnitude of the moments and forces developed can vary a great deal among them. In some cases, they are partially or completely balanced within the compressors themselves. In others, the foundation must handle much of the force.

When complete balance is possible, reciprocating compressors can be mounted on a foundation just large and rigid enough to carry the weight and maintain alignment. However, most reciprocating compressors require larger, more massive foundations than other machinery.

Depending upon size and type of unit, the mounting may vary from simply bolting to the floor to attaching to a massive foundation designed specifically for the application. A proper foundation must: (1) maintain the alignment and level of the compressor and its driver at the proper elevation, and (2) minimize vibration and prevent its transmission to adjacent building structures and machinery. There are five steps to accomplish the first objective:

1 The safe weight-bearing capacity of the soil must not be exceeded at any point on the foundation base.

2 The load to the soil must be distributed over the entire area.

3 The size and proportion of the foundation block must be such that the resultant vertical load due to the compressor, block, and any unbalanced force falls within the base area.

4 The foundation must have sufficient mass and weight-bearing area to prevent its sliding on the soil due to unbalanced forces.

5 Foundation temperature must be uniform to prevent warping.

Bulk is not usually the complete solution to foundation problems. A certain weight is sometimes necessary, but soil area is usually of more value than foundation mass.

Determining if two or more compressors should have separate or single foundations depends on the compressor type. A combined foundation is recommended for reciprocating units since the forces from one unit usually will partially balance out the forces from the others. In addition, the greater mass and surface area in contact with the ground damps foundation movement and provides greater stability.

Soil quality may vary seasonally, and such conditions must be carefully considered in the foundation design. No foundation should rest partially on bedrock and partially on soil; it should rest entirely on one or the other. If placed on the ground, make sure that part of the foundation does not rest on soil that has been disturbed. In addition, pilings may be necessary to ensure stability.

Piping

Piping should easily fit the compressor connections without needing to spring or twist it to fit. It must be supported independently of the compressor and anchored, as necessary, to limit vibration and to prevent expansion strains. Improperly installed piping may distort or pull the compressor's cylinders or casing out of alignment.

Air Inlet

The intake pipe on an air compressor should be as short and direct as possible. If the total run of the inlet piping is unavoidably long, the diameter should be increased. The pipe size should be greater than the compressor's air-inlet connection.

Cool inlet air is desirable. For every 5°F of ambient air temperature reduction, the volume of compressed air generated increases by 1% with the same power consumption. This increase in performance is due to the greater density of the intake air.

It is preferable for the intake air to be taken from outdoors. This reduces heating and air conditioning costs and, if properly designed, has fewer contaminants. However, the intake piping should be a minimum of six feet above the ground and be screened or, preferably, filtered. An air inlet must be free of steam and engine exhausts. The inlet should be hooded or turned down to prevent the entry of rain or snow. It should be above the building eaves and several feet from the building.

Discharge

Discharge piping should be the full size of the compressor's discharge connection. The pipe size should not be reduced until the point along the pipeline is reached where the flow has become steady and nonpulsating. With a reciprocating compressor, this is generally beyond the aftercooler or the receiver. Pipes to handle nonpulsating flow are sized by normal methods, and long-radius bends are recommended. All discharge piping must be designed to allow adequate expansion loops or bends to prevent undue stresses at the compressor.

Drainage

Before piping is installed, the layout should be analyzed to eliminate low points where liquid could collect and to provide drains where low points cannot be eliminated. A regular part of the operating procedure must be

the periodic drainage of low points in the piping and separator
inspection of automatic drain traps.

Pressure-Relief Valves

All reciprocating compressors must be fitted with pressure relie
to limit the discharge or interstage pressures to a safe maximum
equipment served. Always install a relief valve that is capable of b
the full-load capacity of the compressor between its discharge port
first isolation valve. The safety valves should be set to open at a p
slightly higher than the normal discharge-pressure rating of the comp
For standard 100- to 115-psig two-stage air compressors, safety valv
normally set at 125 psig.

The pressure-relief safety valve is normally situated on top of the air re
voir, and there must be no restriction on its operation. The valve is usu
of the "huddling chamber" design, in which the static pressure acting on
disk area causes it to open. Figure 8.15 illustrates how such a valve fu
tions. As the valve pops, the air space within the huddling chamber betwee

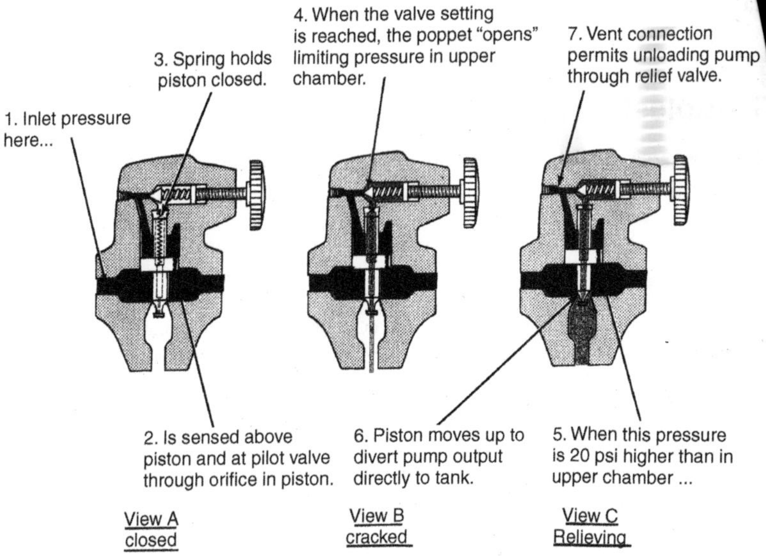
4. When the valve setting is reached, the poppet "opens" limiting pressure in upper chamber.
3. Spring holds piston closed.
7. Vent connection permits unloading pump through relief valve.
1. Inlet pressure here...
2. Is sensed above piston and at pilot valve through orifice in piston.
6. Piston moves up to divert pump output directly to tank.
5. When this pressure is 20 psi higher than in upper chamber ...
View A closed
View B cracked
View C Relieving

Figure 8.15 *Illustrates how a safety valve functions*

fills with pressurized air and builds up more disk holder. This temporary pressure increases he spring, causing the disk and its holder to fully

pressure drop (i.e., blowdown) occurs, the valve action by trapping pressurized air on top of the or lowering the blowdown ring adjusts the pressure- ng the ring increases the pressure-drop setting, while ses the setting.

hods

an be hazardous to work around because they have moving that clothing is kept away from belt drives, couplings, and fts. In addition, high-temperature surfaces around cylinders rge piping are exposed. Compressors are notoriously noisy, so tion should be worn. These machines are used to generate high- gas so, when working around them, it is important to wear safety and to avoid searching for leaks with bare hands. High-pressure leaks use severe friction burns.

roubleshooting

Compressors can be divided into three classifications: centrifugal, rotary, and reciprocating. This section identifies the common failure modes for each.

Centrifugal

The operating dynamics of centrifugal compressors are the same as for other centrifugal machine-trains. The dominant forces and vibration profiles are typically identical to pumps or fans. However, the effects of variable load and other process variables (e.g., temperatures, inlet/discharge pressure, etc.) are more pronounced than in other rotating machines. Table 8.1 identifies the common failure modes for centrifugal compressors.

Aerodynamic instability is the most common failure mode for centrifugal compressors. Variable demand and restrictions of the inlet-air flow are common sources of this instability. Even slight variations can cause dramatic changes in the operating stability of the compressor.

the periodic drainage of low points in the piping and separators, as well as inspection of automatic drain traps.

Pressure-Relief Valves

All reciprocating compressors must be fitted with pressure relief devices to limit the discharge or interstage pressures to a safe maximum for the equipment served. Always install a relief valve that is capable of bypassing the full-load capacity of the compressor between its discharge port and the first isolation valve. The safety valves should be set to open at a pressure slightly higher than the normal discharge-pressure rating of the compressor. For standard 100- to 115-psig two-stage air compressors, safety valves are normally set at 125 psig.

The pressure-relief safety valve is normally situated on top of the air reservoir, and there must be no restriction on its operation. The valve is usually of the "huddling chamber" design, in which the static pressure acting on its disk area causes it to open. Figure 8.15 illustrates how such a valve functions. As the valve pops, the air space within the huddling chamber between

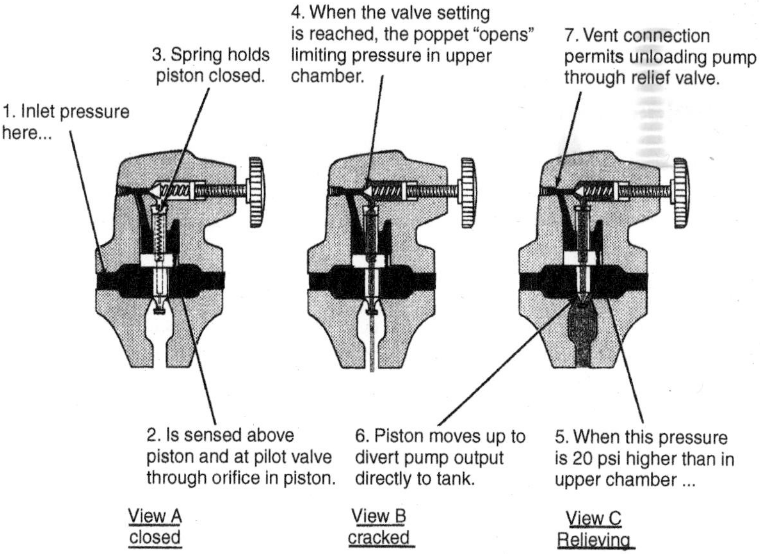
4. When the valve setting is reached, the poppet "opens" limiting pressure in upper chamber.
3. Spring holds piston closed.
7. Vent connection permits unloading pump through relief valve.
1. Inlet pressure here...
2. Is sensed above piston and at pilot valve through orifice in piston.
6. Piston moves up to divert pump output directly to tank.
5. When this pressure is 20 psi higher than in upper chamber ...
View A closed
View B cracked
View C Relieving

Figure 8.15 *Illustrates how a safety valve functions*

the seat and blowdown ring fills with pressurized air and builds up more pressure on the roof of the disk holder. This temporary pressure increases the upward thrust against the spring, causing the disk and its holder to fully pop open.

Once a predetermined pressure drop (i.e., blowdown) occurs, the valve closes with a positive action by trapping pressurized air on top of the disk holder. Raising or lowering the blowdown ring adjusts the pressure-drop setpoint. Raising the ring increases the pressure-drop setting, while lowering it decreases the setting.

Operating Methods

Compressors can be hazardous to work around because they have moving parts. Ensure that clothing is kept away from belt drives, couplings, and exposed shafts. In addition, high-temperature surfaces around cylinders and discharge piping are exposed. Compressors are notoriously noisy, so ear protection should be worn. These machines are used to generate high-pressure gas so, when working around them, it is important to wear safety glasses and to avoid searching for leaks with bare hands. High-pressure leaks can cause severe friction burns.

Troubleshooting

Compressors can be divided into three classifications: centrifugal, rotary, and reciprocating. This section identifies the common failure modes for each.

Centrifugal

The operating dynamics of centrifugal compressors are the same as for other centrifugal machine-trains. The dominant forces and vibration profiles are typically identical to pumps or fans. However, the effects of variable load and other process variables (e.g., temperatures, inlet/discharge pressure, etc.) are more pronounced than in other rotating machines. Table 8.1 identifies the common failure modes for centrifugal compressors.

Aerodynamic instability is the most common failure mode for centrifugal compressors. Variable demand and restrictions of the inlet-air flow are common sources of this instability. Even slight variations can cause dramatic changes in the operating stability of the compressor.

Table 8.1 *Common failure modes of centrifugal compressors*

	THE PROBLEM								
THE CAUSES	Excessive vibration	Compressor surges	Loss of discharge pressure	Low lube oil pressure	Excessive bearing oil drain temp.	Units do not stay in alignment	Persistent unloading	Water in lube oil	Motor trips
Bearing lube oil orifice missing or plugged				•					
Bent rotor (caused by uneven heating and cooling)	•						•		
Build-up of deposits on diffuser		•							
Build-up of deposits on rotor	•	•							
Change in system resistance		•							•
Clogged oil strainer/filter				•					
Compressor not up to speed			•						
Condensate in oil reservoir								•	
Damaged rotor	•								
Dry gear coupling	•								
Excessive bearing clearance	•								
Excessive inlet temperature			•						
Failure of both main and auxiliary oil pumps				•					
Faulty temperature gauge or switch				•	•				•

Continued

Table 8.1 *continued*

THE CAUSES	THE PROBLEM: Excessive vibration	Compressor surges	Loss of discharge pressure	Low lube oil pressure	Excessive bearing oil drain temp.	Units do not stay in alignment	Persistent unloading	Water in lube oil	Motor trips
Improperly assembled parts	•						•		•
Incorrect pressure control valve setting				•					
Insufficient flow		•							
Leak in discharge piping			•						
Leak in lube oil cooler tubes or tube sheet								•	
Leak in oil pump suction piping				•					
Liquid "slugging"	•					•			
Loose or broken bolting	•								
Loose rotor parts	•								
Oil leakage				•					
Oil pump suction plugged				•					
Oil reservoir low level				•					
Operating at low speed w/o auxiliary oil pump				•					
Operating in critical speed range	•								

Table 8.1 *continued*

THE CAUSES	THE PROBLEM: Excessive vibration	Compressor surges	Loss of discharge pressure	Low lube oil pressure	Excessive bearing oil drain temp.	Units do not stay in alignment	Persistent unloading	Water in lube oil	Motor trips
Operating in surge region	•								
Piping strain	•					•	•	•	•
Poor oil condition					•				
Relief valve improperly set or stuck open				•					
Rotor imbalance	•						•		
Rough rotor shaft journal surface					•		•		•
Shaft misalignment	•					•			
Sympathetic vibration	•						•	•	
Vibration					•				
Warped foundation or baseplate							•		•
Wiped or damaged bearings					•				•
Worn or damaged coupling	•								

Entrained liquids and solids also can affect operating life. When dirty air must be handled, open-type impellers should be used. An open design provides the ability to handle a moderate amount of dirt or other solids in the inlet-air supply. However, inlet filters are recommended for all applications, and controlled liquid injection for cleaning and cooling should be considered during the design process.

Rotary-Type, Positive Displacement

Table 8.2 lists the common failure modes of rotary-type, positive-displacement compressors. This type of compressor can be grouped into two types: sliding vane and rotary screw.

Sliding Vane Compressors

Sliding-vane compressors have the same failure modes as vane-type pumps. The dominant components in their vibration profile are running speed, vane-pass frequency, and bearing-rotation frequencies. In normal operation, the dominant energy is at the shaft's running speed. The other frequency components are at much lower energy levels. Common failures of this type of compressor occur with shaft seals, vanes, and bearings.

Shaft Seals

Leakage through the shaft's seals should be checked visually once a week or as part of every data-acquisition route. Leakage may not be apparent from the outside of the gland. If the fluid is removed through a vent, the discharge should be configured for easy inspection. Generally, more leakage than normal is the signal to replace a seal. Under good conditions, they have a normal life of 10,000 to 15,000 hours and should routinely be replaced when this service life has been reached.

Vanes

Vanes wear continuously on their outer edges and, to some degree, on the faces that slide in and out of the slots. The vane material is affected somewhat by prolonged heat, which causes gradual deterioration. Typical life expectancy of vanes in 100-psig services is about 16,000 hours of operation. For low-pressure applications, life may reach 32,000 hours.

Replacing vanes before they break is extremely important. Breakage during operation can severely damage the compressor, which requires a complete overhaul and realignment of heads and clearances.

Bearings

In normal service, bearings have a relatively long life. Replacement after about six years of operation is generally recommended. Bearing defects are usually displayed in the same manner in a vibration profile as for any rotating machine-train. Inner and outer race defects are the dominant failure modes, but roller spin also may contribute to the failure.

Table 8.2 *Common failure modes of rotary-type, positive-displacement compressors*

	THE PROBLEM									
THE CAUSES	No air/gas delivery	Insufficient discharge pressure	Insufficient capacity	Excessive wear	Excessive heat	Excessive vibration and noise	Excessive power demand	Motor trips	Elevated motor temperature	Elevated air/gas temperature
Air leakage into suction piping or shaft seal		•	•			•				
Coupling misaligned				•	•	•	•		•	
Excessive discharge pressure			•	•		•	•	•		•
Excessive inlet temperature/moisture			•							
Insufficient suction air/gas supply		•	•	•		•		•		
Internal component wear	•	•	•							
Motor or driver failure	•									
Pipe strain on compressor casing				•	•	•	•		•	
Relief valve stuck open or set wrong		•	•							
Rotating element binding				•	•	•	•	•	•	
Solids or dirt in inlet air/gas supply				•						
Speed too low		•	•					•		
Suction filter or strainer clogged	•	•	•			•			•	
Wrong direction of rotation	•	•							•	

Rotary Screw

The most common reason for compressor failure or component damage is process instability. Rotary-screw compressors are designed to deliver a constant volume and pressure of air or gas. These units are extremely susceptible to any change in either inlet or discharge conditions. A slight variation in pressure, temperature, or volume can result in instantaneous failure. The following are used as indices of instability and potential problems: rotor mesh, axial movement, thrust bearings, and gear mesh.

Rotor Mesh

In normal operation, the vibration energy generated by male and female rotor meshing is very low. As the process becomes unstable, the energy due to the rotor-meshing frequency increases, with both the amplitude of the meshing frequency and the width of the peak increasing. In addition, the noise floor surrounding the meshing frequency becomes more pronounced. This white noise is similar to that observed in a cavitating pump or unstable fan.

Axial Movement

The normal tendency of the rotors and helical timing gears is to generate axial shaft movement, or thrusting. However, the extremely tight clearances between the male and female rotors do not tolerate any excessive axial movement, and therefore, axial movement should be a primary monitoring parameter. Axial measurements are needed from both rotor assemblies. If there is any increase in the vibration amplitude of these measurements, it is highly probable that the compressor will fail.

Thrust Bearings

While process instability can affect both the fixed and float bearings, the thrust bearing is more likely to show early degradation as a result of process instability or abnormal compressor dynamics. Therefore, these bearings should be monitored closely, and any degradation or hint of excessive axial clearance should be corrected immediately.

Gear Mesh

The gear mesh vibration profile also provides an indication of prolonged compressor instability. Deflection of the rotor shafts changes the wear pattern on the helical gear sets. This change in pattern increases the backlash in the gear mesh, results in higher vibration levels, and increases thrusting.

Reciprocating, Positive Displacement

Reciprocating compressors have a history of chronic failures that include valves, lubrication system, pulsation, and imbalance. Table 8.3 identifies common failure modes and causes for this type of compressor.

Like all reciprocating machines, reciprocating compressors normally generate higher levels of vibration than centrifugal machines. In part, the increased level of vibration is due to the impact as each piston reaches top dead center and bottom dead center of its stroke. The energy levels also are influenced by the unbalanced forces generated by nonopposed pistons and looseness in the piston rods, wrist pins, and journals of the compressor. In most cases, the dominant vibration frequency is the second harmonic (2X) of the main crankshaft's rotating speed. Again, this results from the impact that occurs when each piston changes directions (i.e., two impacts occur during one complete crankshaft rotation).

Valves

Valve failure is the dominant failure mode for reciprocating compressors. Because of their high cyclic rate, which exceeds 80 million cycles per year, inlet and discharge valves tend to work harder and crack.

Lubrication System

Poor maintenance of lubrication-system components, such as filters and strainers, typically causes premature failure. Such maintenance is crucial to reciprocating compressors because they rely on the lubrication system to provide a uniform oil film between closely fitting parts (e.g., piston rings and the cylinder wall). Partial or complete failure of the lube system results in catastrophic failure of the compressor.

Pulsation

Reciprocating compressors generate pulses of compressed air or gas that are discharged into the piping that transports the air or gas to its point(s) of use. This pulsation often generates resonance in the piping system, and pulse impact (i.e., standing waves) can severely damage other machinery connected to the compressed-air system. While this behavior does not cause the compressor to fail, it must be prevented to protect other plant equipment. Note, however, that most compressed-air systems do not use pulsation dampers.

Table 8.3 *A-E in electronic files*

THE CAUSES	THE PROBLEM																							
	Air discharge temperature above normal	Carbonaceous deposits abnormal	Compressor fails to start	Compressor fails to unload	Compressor noisy or knocks	Compressor parts overheat	Crankcase oil pressure low	Crankcase water accumulation	Delivery less than rated capacity	Discharge pressure below normal	Excessive compressor vibration	Intercooler pressure above normal	Intercooler pressure below normal	Intercooler safety valve pops	Motor over-heating	Oil pumping excessive (single-acting compressor)	Operating cycle abnormally long	Outlet water temperature above normal	Piston ring, piston, cylinder wear excessive	Piston rod or packing wear excessive	Receiver pressure above normal	Receiver safety valve pops	Starts too often	Valve wear and breakage normal
Air discharge temperature too high		●																●						
Air filter defective		●																	●	●				●
Air flow to fan blocked	●	●				●																		
Air leak into pump suction							●																	
Ambient temperature too high	●	●				●									●									
Assembly incorrect																								●

Bearings need adjustment or renewal					●	●	●								●									
Belts slipping					●				●	●														
Belts too tight			●				●								●									
Centrifugal pilot valve leaks																●								
Check or discharge valve defective						●																		
Control air filter, strainer clogged				●																				
Control air line clogged																					●			
Control air pipe leaks																					●	●		
Crankcase oil pressure too high																●								
Crankshaft end play too great					●																			
Cylinder, head, cooler dirty	●	●																						
Cylinder, head, intercooler dirty						●												●						
Cylinder (piston) worn or scored	●	●			●	●		●	●	●		•*H*	•*L*	•*H*	•*L*	●	●		•*H*	•*H*				
Detergent oil being used (3)								●																
Demand too steady (2)																							●	
Dirt, rust entering cylinder		●																	●	●				●

Table 8.3 *continued*

THE CAUSES	THE PROBLEM: Air discharge temperature above normal	Carbonaceous deposits abnormal	Compressor fails to start	Compressor fails to unload	Compressor noisy or knocks	Compressor parts overheat	Crankcase oil pressure low	Crankcase water accumulation	Delivery less than rated capacity	Discharge pressure below normal	Excessive compressor vibration	Intercooler pressure above normal	Intercooler pressure below normal	Intercooler safety valve pops	Motor over-heating	Oil pumping excessive (single-acting compressor)	Operating cycle abnormally long	Outlet water temperature above normal	Piston ring, piston, cylinder wear excessive	Piston rod or packing wear excessive	Receiver pressure above normal	Receiver safety valve pops	Starts too often	Valve wear and breakage normal
Discharge line restricted	●														●									
Discharge pressure above rating	●	●			●	●			●		●	●		●	●		●	●	●	●	●	●		
Electrical conditions wrong			●												●									
Excessive number of starts															●									
Excitation inadequate			●												●									
Foundation bolts loose					●						●													

Foundation too small											●													
Foundation uneven-unit rocks					●						●													
Fuses blown			●																					
Gaskets leak	●	●			●	●			●	●		●*H*	●*L*	●*H*	●*L*		●		●*H*	●*H*				
Gauge defective							●			●		●	●								●			
Gear pump worn/defective							●																	
Grout, improperly placed											●													
Intake filter clogged	●				●	●			●	●			●		●	●	●							
Intake pipe restricted, too small, too long	●				●	●			●	●			●		●	●	●							
Intercooler, drain more often								●																
Intercooler leaks													●											
Intercooler passages clogged												●		●										
Intercooler pressure too high																		●						
Intercooler vibrating					●																			
Leveling wedges left under compressor											●													
Liquid carry-over					●			●											●	●				●

Table 8.3 *continued*

THE CAUSES	THE PROBLEM: Air discharge temperature above normal	Carbonaceous deposits abnormal	Compressor fails to start	Compressor fails to unload	Compressor noisy or knocks	Compressor parts overheat	Crankcase oil pressure low	Crankcase water accumulation	Delivery less than rated capacity	Discharge pressure below normal	Excessive compressor vibration	Intercooler pressure above normal	Intercooler pressure below normal	Intercooler safety valve pops	Motor over-heating	Oil pumping excessive (single-acting compressor)	Operating cycle abnormally long	Outlet water temperature above normal	Piston ring, piston, cylinder wear excessive	Piston rod or packing wear excessive	Receiver pressure above normal	Receiver safety valve pops	Starts too often	Valve wear and breakage normal
Location too humid and damp								●																
Low oil pressure relay open			●																					
Lubrication inadequate	●				●	●									●		●		●	●				●
Motor overload relay tripped			●																					
Motor rotor loose on shaft					●						●													
Motor too small			●												●									

New valve on worn seat																								•
"Off" time insufficient	•	•				•																		
Oil feed excessive		•			•															•				•
Oil filter or strainer clogged							•																	
Oil level too high	•	•			•	•										•								
Oil level too low						•	•																	
Oil relief valve defective							•																	
Oil viscosity incorrect		•			•	•	•								•	•			•	•				•
Oil wrong type																•								
Packing rings worn, stuck, broken																				•				
Piping improperly supported											•													
Piston or piston nut loose					•																			
Piston or ring drain hole clogged																•								
Piston ring gaps not staggered																•								
Piston rings worn, broken, or stuck	•	•			•	•		•	•	•		•*H*	•*L*	•*H*	•*L*	•	•		•*H*	•*H*				
Piston-to-head clearance too small					•																			

Table 8.3 *continued*

THE CAUSES	THE PROBLEM: Air discharge temperature above normal	Carbonaceous deposits abnormal	Compressor fails to start	Compressor fails to unload	Compressor noisy or knocks	Compressor parts overheat	Crankcase oil pressure low	Crankcase water accumulation	Delivery less than rated capacity	Discharge pressure below normal	Excessive compressor vibration	Intercooler pressure above normal	Intercooler pressure below normal	Intercooler safety valve pops	Motor over-heating	Oil pumping excessive (single-acting compressor)	Operating cycle abnormally long	Outlet water temperature above normal	Piston ring, piston, cylinder wear excessive	Piston rod or packing wear excessive	Receiver pressure above normal	Receiver safety valve pops	Starts too often	Valve wear and breakage normal
Pulley or flywheel loose					•						•													
Receiver, drain more often																							•	
Receiver too small																							•	
Regulation piping clogged				•																				
Resonant pulsation (inlet or discharge)												•	•	•	•									•
Rod packing leaks	•				•	•			•	•														

Rod packing too tight						●																		
Rod scored, pitted, worn																				●				
Rotation wrong	●	●	●																					
Runs too little (2)								●																
Safety valve defective													●	●								●		
Safety valve leaks	●				●				●	●			●		●									
Safety valve set too low														●								●		
Speed demands exceed rating																	●							
Speed lower than rating									●	●														
Speed too high	●	●				●					●				●			●						
Springs broken																								●
System demand exceeds rating	●				●				●	●			●		●		●							
System leakage excessive	●				●				●	●			●		●		●						●	
Tank ringing noise					●																			
Unloader running time too long (1)																●								
Unloader or control defective	●	●	●	●	●	●			●	●	●	●	●	●	●		●		●	●	●	●	●	●

Table 8.3 *continued*

	THE PROBLEM																							
THE CAUSES	Air discharge temperature above normal	Carbonaceous deposits abnormal	Compressor fails to start	Compressor fails to unload	Compressor noisy or knocks	Compressor parts overheat	Crankcase oil pressure low	Crankcase water accumulation	Delivery less than rated capacity	Discharge pressure below normal	Excessive compressor vibration	Intercooler pressure above normal	Intercooler pressure below normal	Intercooler safety valve pops	Motor over-heating	Oil pumping excessive (single-acting compressor)	Operating cycle abnormally long	Outlet water temperature above normal	Piston ring, piston, cylinder wear excessive	Piston rod or packing wear excessive	Receiver pressure above normal	Receiver safety valve pops	Starts too often	Valve wear and breakage normal
Unloader parts worn or dirty				●																				
Unloader setting incorrect	●	●	●		●	●			●	●		●	●	●	●		●		●	●	●	●	●	
V-belt or other misalignment					●	●					●													
Valves dirty	●	●				●						●	●											●
Valves incorrectly located	●	●			●	●			●	●		●*H*	●*L*	●*H*	●*L*		●		●*H*	●*H*				
Valves not seated in cylinder	●	●			●	●			●	●		●*H*	●*L*	●*H*	●*L*		●		●*H*	●*H*				

Valves worn or broken	●	●			●	●			●	●		●*H*	●*L*	●*H*	●*L*		●	●*H*	●*H*	●*H*				
Ventilation poor	●	●				●									●									
Voltage abnormally low			●												●									
Water inlet temperature too high	●	●				●			●			●						●						
Water jacket or cooler dirty	●	●																						
Water jackets or intercooler dirty						●						●						●						
Water quantity insufficient	●					●			●			●						●						
Wiring incorrect			●																					
Worn valve on good seat																								●
Wrong oil type		●																	●	●				
(1) Use automatic start/stop control																								
(2) Use constant speed control																								
(3) Change to nondetergent oil																								
H (In high pressure cylinder)																								
L (In low pressure cylinder)																								

Crank arrangements	Forces: Primary	Forces: Secondary	Couples: Primary	Couples: Secondary
Single crank 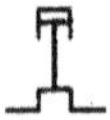	F′ without counterwts. 0.5F′ with counterwts.	F″	None	None
Two cranks at 180°				
In line cylinders 	Zero	2F″	F′D without counterwts. F′D/2 with counterwts.	None
Opposed cylinders 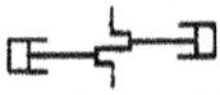	Zero	Zero	Nil	Nil
Two cranks at 90° 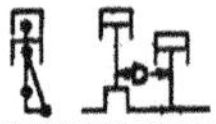	141F′ without counterwts. 0.707F′ with counterwts.	Zero	707F′D without counterwts. 0.354F′D with counterwts.	F″D
Two cylinders on one crank Cylinders at 90° 	F′ without counterwts. Zero with counterwts.	1.41F″	Nil	Nil
Two cylinders on one crank Opposed cylinders 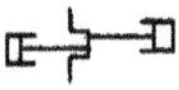	2F′ without counterwts. F′ with counterwts.	Zero	None	Nil
Three cranks at 120° 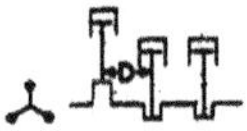	Zero	Zero	3.46F′D without counterwts. 1.73F′D with counterwts.	3.46F″D
Four cylinders				
Cranks at 180° 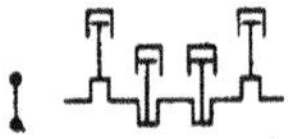	Zero	4F″	Zero	Zero
Cranks at 90° 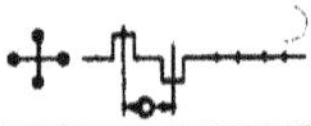	Zero	Zero	1.41F′D without counterwts. 0.707F′D with counterwts.	4.0F″D
Six cylinders 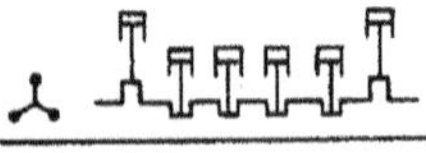	Zero	Zero	Zero	Zero

F′ = Primary inertia force in lbs.
$F' = .0000284\ RN^2W$
F″ = Secondary inertia force in lbs.
F″ = R/L F′
R = Crank radius, inches
N = R.P.M
W = Reciprocating weight of one cylinder, lbs
L = Length of connecting rod, inches
D = Cylinder center distance

Figure 8.16 *Unbalanced inertial forces and couples for various reciprocating compressors*

Each time the compressor discharges compressed air, the air tends to act like a compression spring. Because it rapidly expands to fill the discharge piping's available volume, the pulse of high-pressure air can cause serious damage. The pulsation wavelength, λ, from a compressor having a double-acting piston design can be determined by:

$$\lambda = \frac{60a}{2n} = \frac{34,050}{n}$$

Where:

λ = Wavelength, feet

a = Speed of sound = 1,135 feet/second

n = Compressor speed, revolutions/minute

For a double-acting piston design, a compressor running at 1,200 rpm will generate a standing wave of 28.4 feet. In other words, a shock load equivalent to the discharge pressure will be transmitted to any piping or machine connected to the discharge piping and located within twenty-eight feet of the compressor. Note that for a single-acting cylinder, the wavelength will be twice as long.

Imbalance

Compressor inertial forces may have two effects on the operating dynamics of a reciprocating compressor, affecting its balance characteristics. The first is a force in the direction of the piston movement, which is displayed as impacts in a vibration profile as the piston reaches top and bottom dead center of its stroke. The second effect is a couple, or moment, caused by an offset between the axes of two or more pistons on a common crankshaft. The interrelationship and magnitude of these two effects depend upon such factors as: (1) number of cranks; (2) longitudinal and angular arrangement; (3) cylinder arrangement; and (4) amount of counterbalancing possible. Two significant vibration periods result, the primary at the compressor's rotation speed (X) and the secondary at 2X.

Although the forces developed are sinusoidal, only the maximum (i.e., the amplitude) is considered in the analysis. Figure 8.16 shows relative values of the inertial forces for various compressor arrangements.

9 Control Valves

Control valves can be broken into two major classifications: process and fluid power. Process valves control the flow of gases and liquids through a process system. Fluid-power valves control pneumatic or hydraulic systems.

Process

Process-control valves are available in a variety of sizes, configurations, and materials of construction. Generally, this type of valve is classified by its internal configuration.

Configuration

The device used to control flow through a valve varies with its intended function. The more common types are: ball, gate, butterfly, and globe valves.

Ball

Ball valves (see Figure 9.1) are simple shutoff devices that use a ball to stop and start the flow of fluid downstream of the valve. As the valve stem turns to the open position, the ball rotates to a point where part or the entire hole machined through the ball is in line with the valve-body inlet and outlet. This allows fluid to pass through the valve. When the ball rotates so that the hole is perpendicular to the flow path, the flow stops.

Most ball valves are quick-acting and require a 90-degree turn of the actuator lever to fully open or close the valve. This feature, coupled with the turbulent flow generated when the ball opening is only partially open, limits the use of the ball valve. Use should be limited to strictly an "on-off" control function (i.e., fully open or fully closed) because of the turbulent-flow condition and severe friction loss when in the partially open position. They should not be used for throttling or flow-control applications.

Ball valves used in process applications may incorporate a variety of actuators to provide direct or remote control of the valve. The more common actuators are either manual or motor-operated. Manual values have a hand-wheel or lever attached directly or through a gearbox to the valve stem.

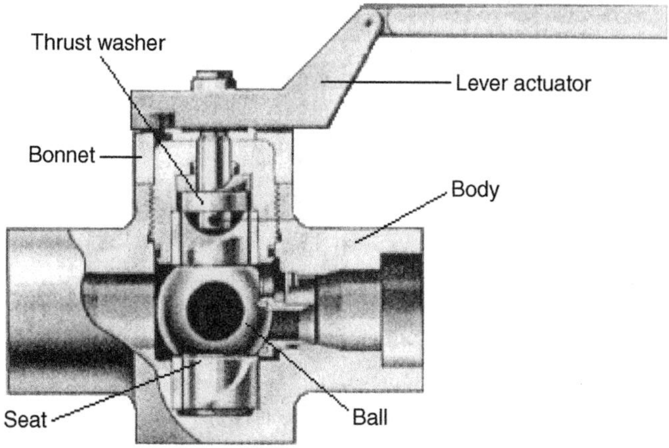

Figure 9.1 *Ball valve*

The valve is opened or closed by moving the valve stem through a 90-degree arc. Motor-controlled valves replace the handwheel with a fractional horsepower motor that can be controlled remotely. The motor-operated valve operates in exactly the same way as the manually operated valve.

Gate

Gate valves are used when straight-line, laminar fluid flow and minimum restrictions are needed. These valves use a wedge-shaped sliding plate in the valve body to stop, throttle, or permit full flow of fluids through the valve. When the valve is wide open, the gate is completely inside the valve bonnet. This leaves the flow passage through the valve fully open with no flow restrictions, allowing little or no pressure drop through the valve.

Gate valves are not suitable for throttling the flow volume unless specifically authorized for this application by the manufacturer. They generally are not suitable because the flow of fluid through a partially open gate can cause extensive damage to the valve.

Gate valves are classified as either *rising-stem* or *nonrising-stem*. In the nonrising-stem valve, which is shown in Figure 9.2, the stem is threaded into the gate. As the handwheel on the stem is rotated, the gate travels up or down the stem on the threads, while the stem remains vertically stationary.

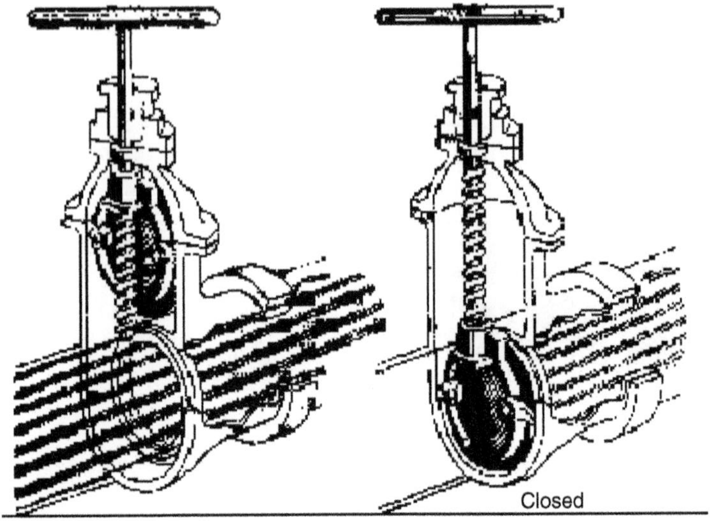

Figure 9.2 *Nonrising-stem gate valve*

This type of valve will almost always have a pointer indicator threaded onto the upper end of the stem to indicate the position of the gate.

Valves with rising stems (see Figure 9.3) are used when it is important to know by immediate inspection if the valve is open or closed, or when the threads exposed to the fluid could become damaged by fluid contamination. In this valve, the stem rises out of the valve bonnet when the valve is opened.

Butterfly

The butterfly valve has a disk-shaped element that rotates about a central shaft or stem. When the valve is closed, the disk face is across the pipe and blocks the flow. Depending upon the type of butterfly valve, the seat may consist of a bonded resilient liner, a mechanically fastened resilient liner, an insert-type reinforced resilient liner, or an integral metal seat with an O-ring inserted around the edge of the disk.

As shown in Figure 9.4, both the full open and the throttled positions permit almost unrestricted flow. Therefore, this valve does not induce turbulent flow in the partially closed position. While the design does not permit exact

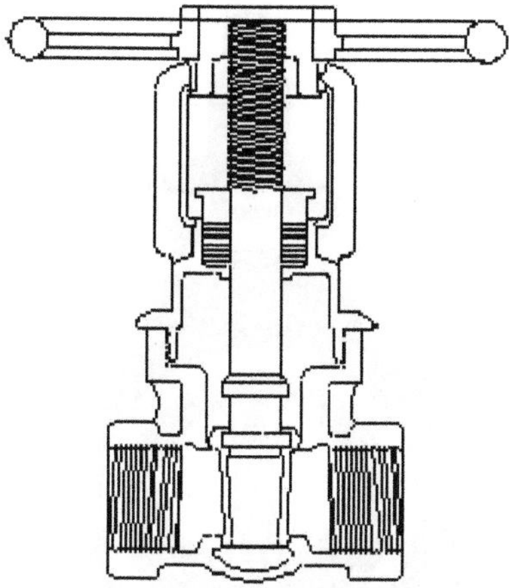

Figure 9.3 *Rising stem gate valve*

flow-control capabilities, a butterfly valve can be used for throttling flow through the valve. In addition, these valves have the lowest pressure drop of all the conventional types. For these reasons, they are commonly used in process-control applications.

Globe

The globe valve gets its name from the shape of the valve body, although other types of valves also may have globular-shaped bodies. Figure 9.5 shows three configurations of this type of valve: straight-flow, angle-flow, and cross-flow.

A disk attached to the valve stem controls flow in a globe valve. Turning the valve stem until the disk is seated, which is illustrated in View A of Figure 9.6, closes the valve. The edge of the disk and the seat are very accurately machined to form a tight seal. It is important for globe valves to be installed with the pressure against the disk face to protect the stem packing from system pressure when the valve is shut.

While this type of valve is commonly used in the fully open or fully closed position, it also may be used for throttling. However, since the seating surface is a relatively large area, it is not suitable for throttling applications where fine adjustments are required.

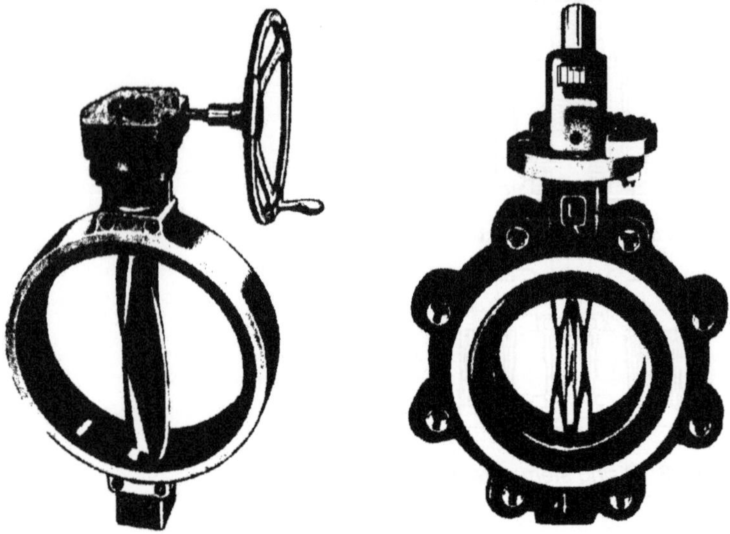

Figure 9.4 *Butterfly valves provide almost unrestricted flow*

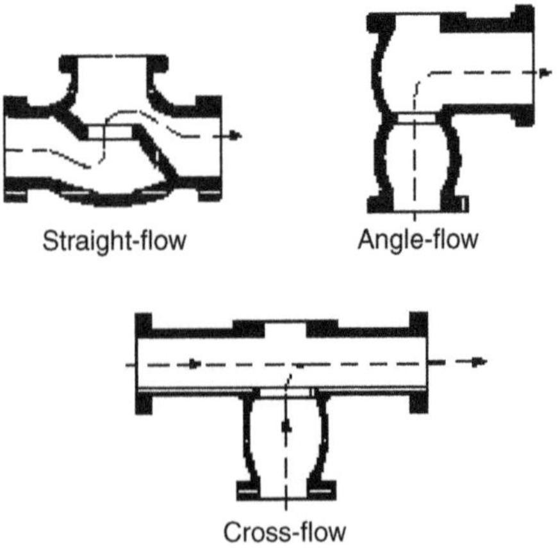

Figure 9.5 *Three globe valve configurations: straight-flow, angle-flow, and cross-flow*

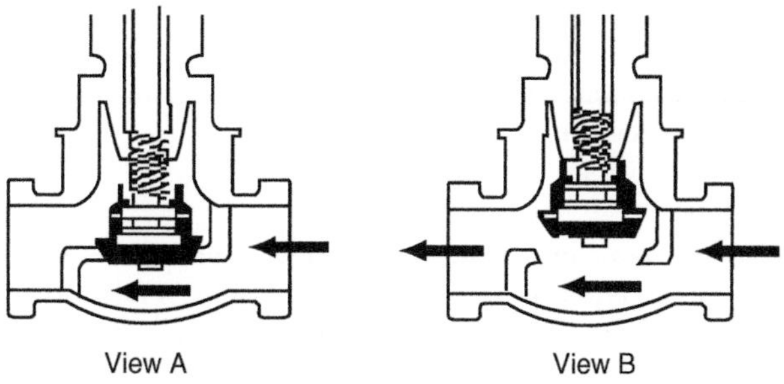

Figure 9.6 *Globe valve*

When the valve is open, as illustrated in View B of Figure 9.6, the fluid flows through the space between the edge of the disk and the seat. Since the fluid flow is equal on all sides of the center of support when the valve is open, there is no unbalanced pressure on the disk to cause uneven wear. The rate at which fluid flows through the valve is regulated by the position of the disk in relation to the valve seat.

The globe valve should never be jammed in the open position. After a valve is fully opened, the handwheel or actuating handle should be closed approximately one-half turn. If this is not done, the valve may seize in the open position making it difficult, if not impossible, to close the valve. Many valves are damaged in the manner. Another reason to partially close a globe valve is because it can be difficult to tell if the valve is open or closed. If jammed in the open position, the stem can be damaged or broken by someone who thinks the valve is closed.

Performance

Process-control valves have few measurable criteria that can be used to determine their performance. Obviously, the valve must provide a positive seal when closed. In addition, it must provide a relatively laminar flow with minimum pressure drop in the fully open position. When evaluating valves, the following criteria should be considered: capacity rating, flow characteristics, pressure drop, and response characteristics.

Capacity Rating

The primary selection criterion of a control valve is its capacity rating. Each type of valve is available in a variety of sizes to handle most typical process-flow rates. However, proper size selection is critical to the performance characteristics of the valve and the system where it is installed. A valve's capacity must accommodate variations in viscosity, temperature, flow rates, and upstream pressure.

Flow Characteristics

The internal design of process-control valves has a direct impact on the flow characteristics of the gas or liquid flowing through the valve. A fully open butterfly or gate valve provides a relatively straight, obstruction-free flow path. As a result, the product should not be affected.

Pressure Drop

Control-valve configuration impacts the resistance to flow through the valve. The amount of resistance, or pressure drop, will vary greatly, depending on type, size, and position of the valve's flow-control device (i.e. ball, gate, disk). Pressure-drop formulas can be obtained for all common valve types from several sources (e.g., *Crane, Technical Paper No. 410*).

Response Characteristics

With the exception of simple, manually controlled shutoff valves, process-control valves are generally used to control the volume and pressure of gases or liquids within a process system. In most applications, valves are controlled from a remote location through the use of pneumatic, hydraulic, or electronic actuators. Actuators are used to position the gate, ball, or disk that starts, stops, directs, or proportions the flow of gas or liquid through the valve. Therefore, the response characteristics of a valve are determined, in part, by the actuator. Three factors critical to proper valve operation are: response time, length of travel, and repeatability.

Response Time

Response time is the total time required for a valve to open or close to a specific set-point position. These positions are fully open, fully closed, and any position in between. The selection and maintenance of the actuator used to control process-control valves have a major impact on response time.

Length of Travel

The valve's flow-control device (i.e., gate, ball, or disk) must travel some distance when going from one set point to another. With a manually operated valve, this is a relatively simple operation. The operator moves the stem lever or handwheel until the desired position is reached. The only reasons a manually controlled valve will not position properly are mechanical wear or looseness between the lever or handwheel and the disk, ball, or gate. For remotely controlled valves, however, there are other variables that directly impact valve travel. These variables depend on the type of actuator that is used. There are three major types of actuators: pneumatic, hydraulic, and electronic.

Pneumatic Actuators

Pneumatic actuators, including diaphragms, air motors, and cylinders, are suitable for simple on-off valve applications. As long as there is enough air volume and pressure to activate the actuator, the valve can be repositioned over its full length of travel. However, when the air supply required to power the actuator is inadequate or the process-system pressure is too great, the actuator's ability to operate the valve properly is severely reduced.

A pneumatic (i.e., compressed air-driven) actuator is shown in Figure 9.7. This type is not suited for precision flow-control applications, because the compressibility of air prevents it from providing smooth, accurate valve positioning.

Hydraulic Actuators

Hydraulic (i.e., fluid-driven) actuators, also illustrated in Figure 9.7, can provide a positive means of controlling process valves in most applications. Properly installed and maintained, this type of actuator can provide

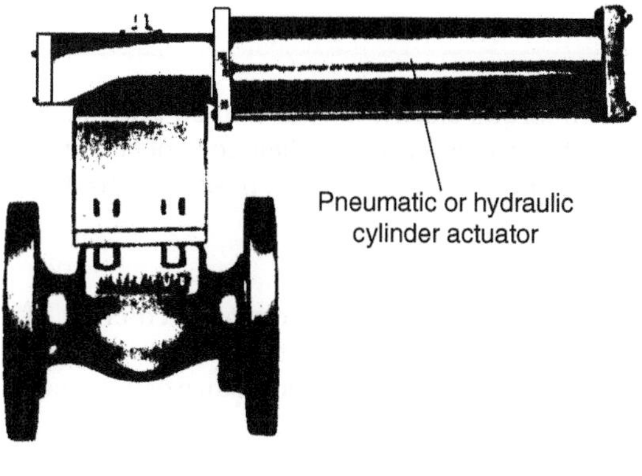

Figure 9.7 *Pneumatic or hydraulic cylinders are used as actuators*

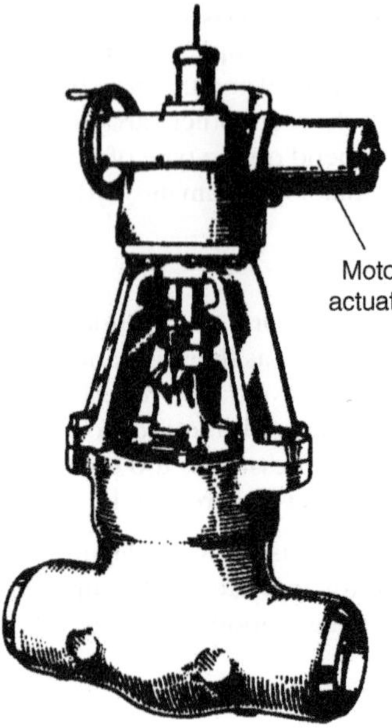

Figure 9.8 *High-torque electric motors can be used as actuators*

accurate, repeatable positioning of the control valve over its full range of travel.

Electronic Actuators

Some control valves use high-torque electric motors as their actuator (see Figure 9.8). If the motors are properly sized and their control circuits are maintained, this type of actuator can provide reliable, positive control over the full range of travel.

Repeatability

Repeatability is, perhaps, the most important performance criterion of a process-control valve. This is especially true in applications where precise flow or pressure control is needed for optimum performance of the process system.

New process-control valves generally provide the repeatability required. However, proper maintenance and periodic calibration of the valves and their actuators are required to ensure long-term performance. This is especially true for valves that use mechanical linkages as part of the actuator assembly.

Installation

Process-control valves cannot tolerate solids, especially abrasives, in the gas or liquid stream. In applications where high concentrations of particulates are present, valves tend to experience chronic leakage or seal problems because the particulate matter prevents the ball, disk, or gate from completely closing against the stationary surface.

Simply installing a valve with the same inlet and discharge size as the piping used in the process is not acceptable. In most cases, the valve must be larger than the piping to compensate for flow restrictions within the valve.

Operating Methods

Operating methods for control valves, which are designed to control or direct gas and liquid flow through process systems or fluid-power circuits, range from manual to remote, automatic operation. The key parameters that govern the operation of valves are the speed of the control movement and the impact of speed on the system. This is especially important in process systems.

Hydraulic hammer, or the shock wave generated by the rapid change in the flow rate of liquids within a pipe or vessel, has a serious and negative impact on all components of the process system. For example, instantaneously closing a large flow-control valve may generate in excess of three million foot-pounds of force on the entire system upstream of the valve. This shock wave can cause catastrophic failure of upstream valves, pumps, welds, and other system components.

Changes in flow rate, pressure, direction, and other controllable variables must be gradual enough to permit a smooth transition. Abrupt changes in valve position should be avoided. Neither the valve installation nor the control mechanism should permit complete shutoff, referred to as deadheading, of any circuit in a process system.

Restricted flow forces system components, such as pumps, to operate outside of their performance envelope. This reduces equipment reliability and sets the stage for catastrophic failure or abnormal system performance. In applications where radical changes in flow are required for normal system operation, control valves should be configured to provide an adequate bypass for surplus flow in order to protect the system.

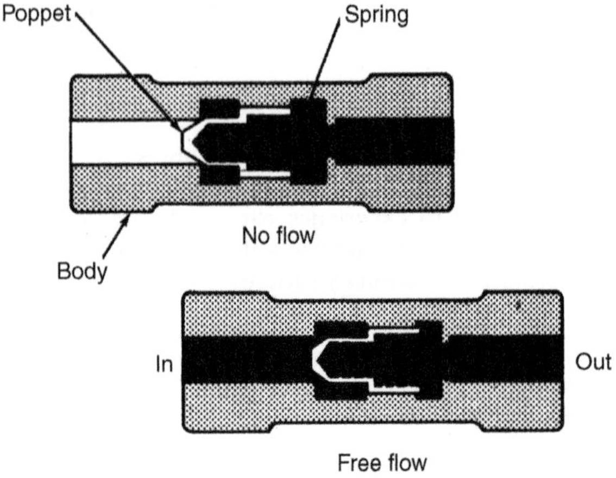

Figure 9.9 *One-way, fluid-power valve*

For example, systems that must have close control of flow should use two proportioning valves that act in tandem to maintain a balanced hydraulic or aerodynamic system. The primary, or master, valve should control flow to the downstream process. The second valve, slaved to the master, should divert excess flow to a bypass loop. This master-slave approach ensures that the pumps and other upstream system components are permitted to operate well within their operating envelope.

Fluid Power

Fluid power control valves are used on pneumatic and hydraulic systems or circuits.

Configuration

The configuration of fluid power control valves varies with their intended application. The more common configurations include: one-way, two-way, three-way, and four-way.

One-Way

One-way valves are typically used for flow and pressure control in fluid-power circuits (see Figure 9.9). Flow-control valves regulate the flow of hydraulic fluid or gases in these systems. Pressure-control valves, in the form of regulators or relief valves, control the amount of pressure transmitted downstream from the valve. In most cases, the types of valves used for flow control are smaller versions of the types of valves used in process control. These include ball, gate, globe, and butterfly valves.

Pressure-control valves have a third port to vent excess pressure and prevent it from affecting the downstream piping. The bypass, or exhaust, port has an internal flow-control device, such as a diaphragm or piston, that opens at predetermined set-points to permit the excess pressure to bypass the valve's primary discharge. In pneumatic circuits, the bypass port vents to the atmosphere. In hydraulic circuits, it must be connected to a piping system that returns to the hydraulic reservoir.

Two-Way

A two-way valve has two functional flow-control ports. A two-way, sliding spool directional control valve is shown in Figure 9.10. As the spool moves back and forth, it either allows fluid to flow through the valve or prevents it from flowing. In the open position, the fluid enters the inlet port, flows around the shaft of the spool, and through the outlet port. Because the forces in the cylinder are equal when open, the spool cannot move back and forth. In the closed position, one of the spool's pistons simply blocks the inlet port, which prevents flow through the valve.

A number of features common to most sliding-spool valves are shown in Figure 9.10. The small ports at either end of the valve housing provide a path for fluid that leaks past the spool to flow to the reservoir. This prevents pressure from building up against the ends of the pistons, which would hinder the movement of the spool. When these valves become worn, they may lose balance because of greater leakage on one side of the spool than on

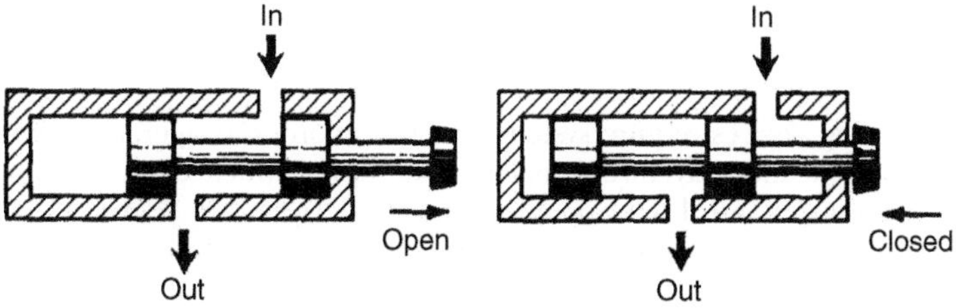

Figure 9.10 *Two-way, fluid-power valve*

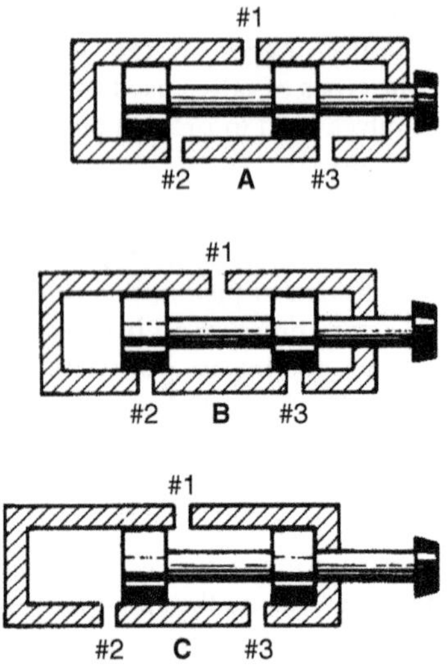

Figure 9.11 *Three-way, fluid-power valve*

the other. This can cause the spool to stick as it attempts to move back and forth. Therefore, small grooves are machined around the sliding surface of the piston. In hydraulic valves, leaking liquid encircles the piston, keeping the contacting surfaces lubricated and centered.

Three-Way

Three-way valves contain a pressure port, cylinder port, and return or exhaust port (see Figure 9.11). The three-way directional control valve is designed to operate an actuating unit in one direction. It is returned to its original position either by a spring or the load on the actuating unit.

Four-Way

Most actuating devices require system pressure in order to operate in two directions. The four-way directional control valve, which contains four ports, is used to control the operation of such devices (see Figure 9.12). The four-way valve also is used in some systems to control the operation of other valves. It is one of the most widely used directional-control valves in fluid-power systems.

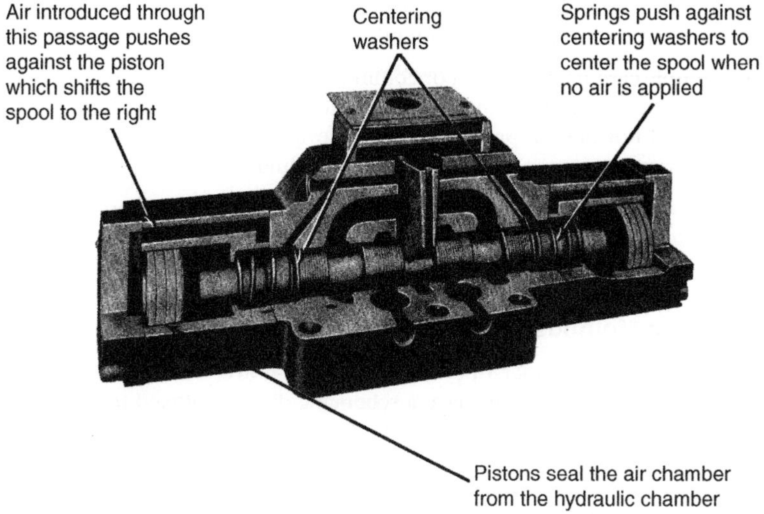
Air introduced through this passage pushes against the piston which shifts the spool to the right
Centering washers
Springs push against centering washers to center the spool when no air is applied
Pistons seal the air chamber from the hydraulic chamber

Figure 9.12 *Four-way, fluid-power valve*

The typical four-way directional control valve has four ports: pressure port, return port, and two cylinder or work (output) ports. The pressure port is connected to the main system-pressure line, and the return port is connected to the reservoir return line. The two outputs are connected to the actuating unit.

Performance

The criteria that determine performance of fluid-power valves are similar to those for process-control valves. As with process-control valves, fluid-power valves must also be selected based on their intended application and function.

Installation

When installing fluid power control valves, piping connections are made either directly to the valve body or to a manifold attached to the valve's base. Care should be taken to ensure that piping is connected to the proper valve port. The schematic diagram that is affixed to the valve body will indicate the proper piping arrangement, as well as the designed operation of the valve. In addition, the ports on most fluid power valves are generally clearly marked to indicate their intended function.

In hydraulic circuits, the return or common ports should be connected to a return line that directly connects the valve to the reservoir tank. This return line should not need a pressure-control device, but should have a check valve to prevent reverse flow of the hydraulic fluid.

Pneumatic circuits may be vented directly to atmosphere. A return line can be used to reduce noise or any adverse effect that locally vented compressed air might have on the area.

Operating Methods

The function and proper operation of a fluid-power valve are relatively simple. Most of these valves have a schematic diagram affixed to the body that clearly explains how to operate the valve.

Valves

Figure 9.13 is a schematic of a two-position, cam-operated valve. The primary actuator, or cam, is positioned on the left of the schematic and any secondary actuators are on the right. In this example, the secondary actuator consists of a spring-return and a spring-compensated limit switch. The schematic indicates that when the valve is in the neutral position (right box), flow is directed from the inlet (P) to work port A. When the cam is depressed, the flow momentarily shifts to work port B. The secondary actuator, or spring, automatically returns the valve to its neutral position when the cam returns to its extended position. In these schematics, T indicates the return connection to the reservoir.

Figure 9.14 illustrates a typical schematic of a two-position and three-position directional control valve. The boxes contain flow direction arrows that indicate the flow path in each of the positions. The schematics do not include the actuators used to activate or shift the valves between positions.

In a two-position valve, the flow path is always directed to one of the work ports (A or B). In a three-position valve, a third or neutral position is added. In this figure, a Type 2 center position is used. In the neutral position, all ports are blocked, and no flow through the valve is possible.

Figure 9.15 is the schematic for the center or neutral position of three-position directional control valves. Special attention should be given to the type of center position that is used in a hydraulic control valve. When Type 2, 3, and 6 (see Figure 9.15) are used, the upstream side of the valve must

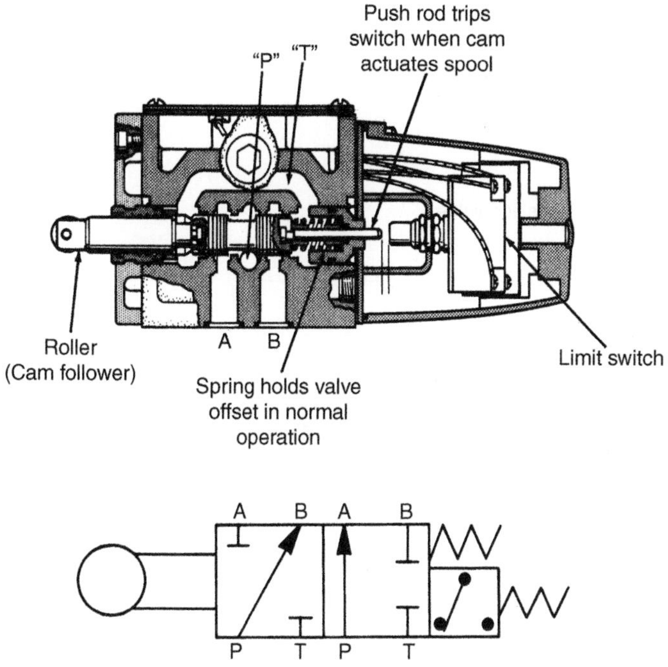

Figure 9.13 *Schematic for a cam-operated, two-position valve*

have a relief or bypass valve installed. Since the pressure port is blocked, the valve cannot relieve pressure on the upstream side of the valve. The Type 4 center position, called a motor spool, permits the full pressure and volume on the upstream side of the valve to flow directly to the return line and storage reservoir. This is the recommended center position for most hydraulic circuits.

The schematic affixed to the valve includes the primary and secondary actuators used to control the valve. Figure 9.16 provides the schematics for three actuator-controlled valves:

1 Double-solenoid, spring-centered, three-position valve

2 Solenoid-operated, spring-return, two-position valve

3 Double-solenoid, detented, two-position valve

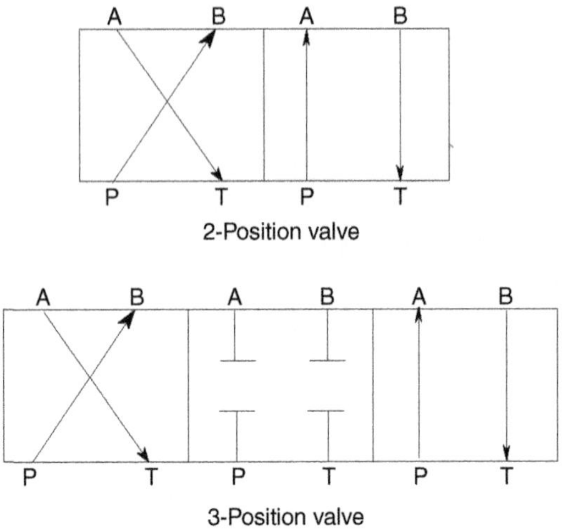

Figure 9.14 *Schematic of two-position and three-position valves*

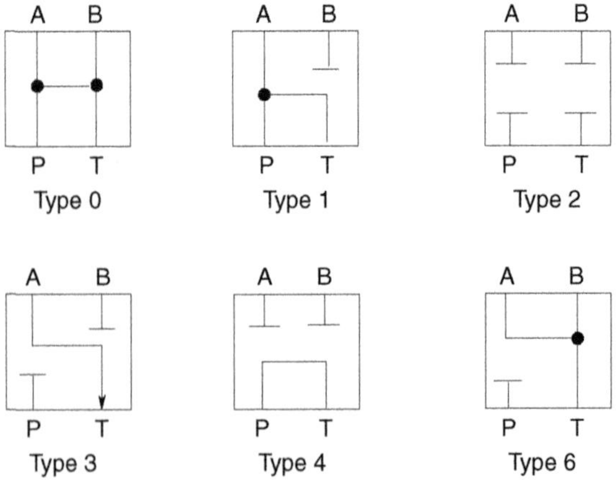

Figure 9.15 *Schematic for center or neutral configurations of three-position valves*

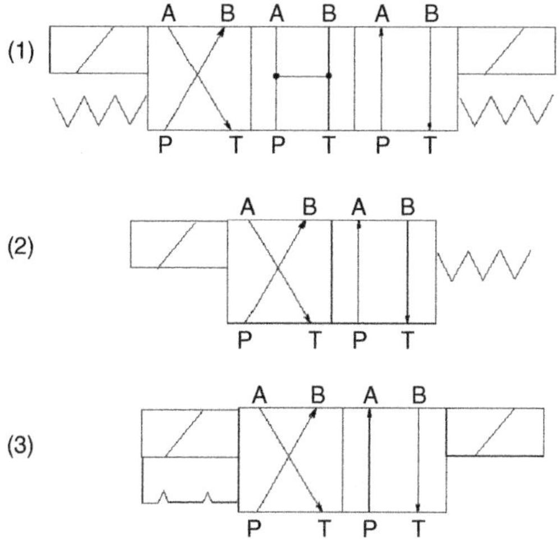

Figure 9.16 *Actuator-controlled valve schematics*

The top schematic, in Figure 9.16, represents a double-solenoid, spring-centered, three-position valve. When neither of the two solenoids is energized, the double springs ensure that the valve is in its center or neutral position. In this example, a Type 0 (see Figure 9.15) configuration is used. This neutral-position configuration equalizes the pressure through the valve. Since the pressure port is open to both work ports and the return line, pressure is equalized throughout the system. When the left or primary solenoid is energized, the valve shifts to the left-hand position and directs pressure to work port B. In this position, fluid in the A-side of the circuit returns to the reservoir. As soon as the solenoid is de-energized, the valve shifts back to the neutral or center position. When the secondary (i.e., right) solenoid is energized, the valve redirects flow to port A, and port B returns fluid to the reservoir.

The middle schematic, in Figure 9.16, represents a solenoid-operated, spring-return, two-position valve. Unless the solenoid is energized, the pressure port P is connected to work port A. While the solenoid is energized, flow is redirected to work port B. The spring return ensures that the valve is in its neutral (i.e., right) position when the solenoid is de-energized.

The bottom schematic, in Figure 9.16, represents a double-solenoid, detented, two-position valve. The solenoids are used to shift the valve between its two positions. A secondary device, called a detent, is used to hold the valve in its last position until the alternate solenoid is energized. Detent configuration varies with the valve type and manufacturer. However, all configurations prevent the valve's control device from moving until a strong force, such as that provided by the solenoid, overcomes its locking force.

Actuators

As with process-control valves, actuators used to control fluid-power valves have a fundamental influence on performance. The actuators must provide positive, real-time response to control inputs. The primary types of actuators used to control fluid-power valves are: mechanical, pilot, and solenoid.

Mechanical

The use of manually controlled mechanical valves is limited in both pneumatic and hydraulic circuits. Generally, this type of actuator is used only on isolation valves that are activated when the circuit or fluid-power system is shut down for repair or when direct operator input is required to operate one of the system components.

Manual control devices (e.g., levers, cams, or palm buttons) can be used as the primary actuator on most fluid power control valves. Normally, these actuators are used in conjunction with a secondary actuator, such as a spring return or detent, to ensure proper operation of the control valve and its circuit.

Spring returns are used in applications where the valve is designed to stay open or shut only when the operator holds the manual actuator in a particular position. When the operator releases the manual control, the spring returns the valve to the neutral position.

Valves with a detented secondary actuator are designed to remain in the last position selected by the operator until manually moved to another position. A detent actuator is simply a notched device that locks the valve in one of several preselected positions. When the operator applies force to the primary actuator, the valve shifts out of the detent and moves freely until the next detent is reached.

Pilot

Although there are a variety of pilot actuators used to control fluid-power valves, they all work on the same basic principle. A secondary source of fluid or gas pressure is applied to one side of a sealing device, such as a piston or diaphragm. As long as this secondary pressure remains within preselected limits, the sealing device prevents the control valve's flow-control mechanism (i.e., spool or poppet) from moving. However, if the pressure falls outside of the preselected window, the actuator shifts and forces the valve's primary mechanism to move to another position.

This type of actuator can be used to sequence the operation of several control valves or operations performed by the fluid-power circuit. For example, a pilot-operated valve is used to sequence the retraction of an airplane's landing gear. The doors that conceal the landing gear when retracted cannot close until the gear is fully retracted. A pilot-operated valve senses the hydraulic pressure in the gear-retraction circuit. When the hydraulic pressure reaches a pre-selected point that indicates the gear is fully retracted, the pilot-actuated valve triggers the closure circuit for the wheel-well doors.

Solenoid

Solenoid valves are widely used as actuators for fluid-power systems. This type of actuator consists of a coil that generates an electric field when energized. The magnetic forces generated by this field force a plunger that is attached to the main valve's control mechanism to move within the coil. This movement changes the position of the main valve.

In some applications, the mechanical force generated by the solenoid coil is not sufficient to move the main valve's control mechanism. When this occurs, the solenoid actuator is used in conjunction with a pilot actuator. The solenoid coil opens the pilot port, which uses system pressure to shift the main valve.

Solenoid actuators are always used with a secondary actuator to provide positive control of the main valve. Because of heat buildup, solenoid actuators must be limited to short-duration activation. A brief burst of electrical energy is transmitted to the solenoid's coil, and the actuation triggers a movement of the main valve's control mechanism. As soon as the main valve's position is changed, the energy to the solenoid coil is shut off.

This operating characteristic of solenoid actuators is important. For example, a *normally closed* valve that uses a solenoid actuation can only be open when the solenoid is energized. As soon as the electrical energy is removed from the solenoid's coil, the valve returns to the closed position. The reverse is true of a *normally open* valve. The main valve remains open, except when the solenoid is energized.

The combination of primary and secondary actuators varies with the specific application. Secondary actuators can be another solenoid or any of the other actuator types that have been previously discussed.

Troubleshooting

Although there are limited common control valve failure modes, the dominant problems are usually related to leakage, speed of operation, or complete valve failure. Table 9.1 lists the more common causes of these failures.

Table 9.1 *Common failure modes of control valves*

		THE PROBLEM						
	THE CAUSES	Valve fails to open	Valve fails to close	Leakage through valve	Leakage around stem	Excessive pressure drop	Opens/closes too fast	Opens/closes too slow
Manually actuated	Dirt/debris trapped in valve seat		•	•				
	Excessive wear		•	•				
	Galling		•	•				
	Line pressure too high	•	•	•	•	•		
	Mechanical damage	•	•					
	Not packed properly				•			
	Packed box too loose				•			
	Packing too tight	•	•					
	Threads/lever damaged	•	•					
	Valve stem bound	•	•					
	Valve undersized					•		•

Table 9.1 *continued*

	THE CAUSES	Valve fails to open	Valve fails to close	Leakage through valve	Leakage around stem	Excessive pressure drop	Opens/closes too fast	Opens/closes too slow
Pilot actuated	Dirt/debris trapped in valve seat	•	•	•				
	Galling	•	•					
	Mechanical damage (seals, seat)	•	•	•				
	Pilot port blocked/plugged	•	•	•				
	Pilot pressure too high		•				•	
	Pilot pressure too low	•		•				•
Solenoid actuated	Corrosion	•	•	•				
	Dirt/debris trapped in valve seat	•	•	•				
	Galling	•	•					
	Line pressure too high	•	•	•	•			•
	Mechanical damage	•	•	•				
	Solenoid failure	•	•					
	Solenoid wiring defective	•	•					
	Wrong type of valve (N-O, N-C)	•	•					

Special attention should be given to the valve actuator when conducting a root cause failure analysis. Many of the problems associated with both process and fluid-power control valves are really actuator problems.

In particular, remotely controlled valves that use pneumatic, hydraulic, or electrical actuators are subject to actuator failure. In many cases, these failures are the reason a valve fails to properly open, close, or seal. Even with manually controlled valves, the true root cause can be traced to an actuator problem. For example, when a manually operated process-control valve is jammed open or closed, it may cause failure of the valve mechanism. This over-torquing of the valve's sealing device may cause damage or failure of the seal, or it may freeze the valve stem. Either of these failure modes results in total valve failure.

10 Conveyors

Conveyors are used to transport materials from one location to another within a plant or facility. The variety of conveyor systems is almost infinite, but the two major classifications used in typical chemical plants are pneumatic and mechanical. Note that the power requirements of a pneumatic-conveyor system are much greater than for a mechanical conveyor of equal capacity. However, both systems offer some advantages.

Pneumatic

Pneumatic conveyors are used to transport dry, free-flowing, granular material in suspension within a pipe or duct. This is accomplished by the use of a high-velocity air stream, or by the energy of expanding compressed air within a comparatively dense column of fluidized or aerated material. Principal uses are: (1) dust collection; (2) conveying soft materials, such as flake or tow; and (3) conveying hard materials, such as fly ash, cement, and sawdust. The primary advantages of pneumatic-conveyor systems are the flexibility of piping configurations and the fact that they greatly reduce the explosion hazard. Pneumatic conveyors can be installed in almost any configuration required to meet the specific application. With the exception of the primary driver, there are no moving parts that can fail or cause injury. However, when used to transport explosive materials, there is still some potential for static charge buildup that could cause an explosion.

Configuration

A typical pneumatic-conveyor system consists of Schedule-40 pipe or ductwork, which provides the primary flow path used to transport the conveyed material. Motive power is provided by the primary driver, which can be a fan, fluidizer, or positive-displacement compressor.

Performance

Pneumatic conveyor performance is determined by the following factors: (1) primary-driver output; (2) internal surface of the piping or ductwork; and (3) condition of the transported material. Specific factors affecting performance include motive power, friction loss, and flow restrictions.

Motive Power

The motive power is provided by the primary driver, which generates the gas (typically air) velocity required to transport material within a pneumatic-conveyor system. Therefore, the efficiency of the conveying system depends on the primary driver's operating condition.

Friction Loss

Friction loss within a pneumatic-conveyor system is a primary source of efficiency loss. The piping or ductwork must be properly sized to minimize friction without lowering the velocity below the value needed to transport the material.

Flow Restrictions

An inherent weakness of pneumatic-conveyor systems is their potential for blockage. The inside surfaces must be clean and free of protrusions or other defects that can restrict or interrupt the flow of material. In addition, when a system is shut down or the velocity drops below the minimum required to keep the transported material suspended, the product will drop out or settle in the piping or ductwork. In most cases, this settled material would compress and lodge in the piping. The restriction caused by this compacted material will reduce flow and eventually result in a complete blockage of the system.

Another major contributor to flow restrictions is blockage caused by system backups. This occurs when the end point of the conveyor system (i.e., storage silo, machine, or vessel) cannot accept the entire delivered flow of material. As the transported material backs up in the conveyor piping, it compresses and forms a solid plug that must be manually removed.

Installation

All piping and ductwork should be as straight and short as possible. Bends should have a radius of at least three diameters of the pipe or ductwork. The diameter should be selected to minimize friction losses and maintain enough velocity to prevent settling of the conveyed material. Branch lines should be configured to match as closely as possible the primary flow direction and avoid 90-degree angles to the main line. The area of the main conveyor line at any point along its run should be 20 to 25% greater than the sum of all its branch lines. When vertical runs are short in proportion to the horizontal runs, the size of the riser can be restricted to provide additional velocity if needed. If the vertical runs are long, the primary or a secondary driver must provide sufficient velocity to transport the material.

Clean-outs, or drop-legs, should be installed at regular intervals throughout the system to permit foreign materials to drop out of the conveyed material. In addition, they provide the means to remove materials that drop out when the system is shut down or air velocity is lost. It is especially important to install adequate clean-out systems near flow restrictions and at the end of the conveyor system.

Operating Methods

Pneumatic-conveyor systems must be operated properly to prevent chronic problems, with the primary concern being to maintain constant flow and velocity. If either of these variables is permitted to drop below the system's design envelope, partial or complete blockage of the conveyor system will occur.

Constant velocity can be maintained only when the system is operated within its performance envelope and when regular clean-out is part of the normal operating practice. In addition, the primary driver must be in good operating condition. Any deviation in the primary driver's efficiency reduces the velocity and can result in partial or complete blockage.

The entire pneumatic-conveyor system should be completely evacuated before shutdown to prevent material from settling in the piping or ductwork. In noncontinuous applications, the conveyor system should be operated until all material within the conveyor's piping is transported to its final destination. Material that is allowed to settle will compact and partially block the piping. Over time, this will cause a total blockage of the conveyor system.

Mechanical

There are a variety of mechanical-conveyor systems used in chemical plants. These systems generally are comprised of chain- or screw-type mechanisms.

Chain

A commonly used chain-type system is a flight conveyor (e.g., Hefler conveyor), which is used to transport granular, lumpy, or pulverized materials along a horizontal or inclined path within totally enclosed ductwork.

The Hefler systems generally have lower power requirements than the pneumatic conveyor and have the added benefit of preventing product contamination. This section focuses primarily on the Hefler-type conveyor because it is one of the most commonly used systems.

Configuration

The most common chain conveyor uses a center- or double-chain configuration to provide positive transfer of material within its ductwork. Both chain configurations use hardened bars or U-shaped devices that are an integral part of the chain to drag the conveyed material through the ductwork.

Performance

Data used to determine a chain conveyor's capacity and the size of material that can be conveyed are presented in Table 10.1. Note that these data are for level conveyors. When inclined, capacity data obtained from Table 10.1 must be multiplied by the factors provided in Table 10.2.

Installation

The primary installation concerns with Hefler-type conveyor systems are the ductwork and primary-drive system.

Ductwork

The inside surfaces of the ductwork must be free of defects or protrusions that interfere with the movement of the conveyor's chain or transported product. This is especially true at the joints. The ductwork must be sized to provide adequate chain clearance but should not be large enough to have areas where the chain-drive bypasses the product.

Table 10.1 *Approximate capacities of chain conveyors*

Flight width and depth (inches)	Quantity of material (Ft^3/Ft)	Approximate capacity (short tons/ hour)	Lump size single strand (inches)	Lump size dual strand (inches)
12 × 6	0.40	60	31.5	4.0
15 × 6	0.49	73	41.5	5.0
18 × 6	0.56	84	5.0	6.0
24 × 8	1.16	174		10.0
30 × 10	1.60	240		14.0
36 × 12	2.40	360		16.0

Table 10.2 *Capacity correction factors for inclined chain conveyors*

Inclination, degrees	20	25	30	35
Factor	0.9	0.8	0.7	0.6

A long horizontal run followed by an upturn is inadvisable because of radial thrust. All bends should have a large radius to permit smooth transition and to prevent material buildup. As with pneumatic conveyors, the ductwork should include clean-out ports at regular intervals for ease of maintenance.

Primary-Drive System

Most mechanical conveyors use a primary-drive system that consists of an electric motor and a speed-increaser gearbox.

The drive-system configuration may vary, depending on the specific application or vendor. However, all configurations should include a single-point-of-failure device, such as a shear pin, to protect the conveyor. The shear pin is critical in this type of conveyor because it is prone to catastrophic failure caused by blockage or obstructions that may lock the chain. Use of the proper shear pin prevents major damage from occurring to the conveyor system.

For continuous applications, the primary-drive system must have adequate horsepower to handle a fully loaded conveyor. Horsepower requirements should be determined based on the specific product's density and the conveyor's maximum-capacity rating.

For intermittent applications, the initial startup torque is substantially greater than for a continuous operation. Therefore, selection of the drive system and the designed failure point of the shear device must be based on the maximum startup torque of a fully loaded system. If either the drive system or designed failure point is not properly sized, this type of conveyor is prone to chronic failures. The predominant failures are frequent breakage of the shear device and trips of the motor's circuit breaker caused by excessive startup amp loads.

Operating Methods

Most mechanical conveyors are designed for continuous operation and may exhibit problems in intermittent-service applications. The primary problem is the startup torque for a fully loaded conveyor. This is especially true for conveyor systems handling material that tends to compact or compress upon settling in a vessel, such as the conveyor trough.

The only positive method of preventing excessive startup torque is to ensure that the conveyor is completely empty before shutdown. In most cases, this can be accomplished by isolating the conveyor from its supply for a few minutes prior to shutdown. This time delay permits the conveyor to deliver its entire load of product before it is shut off.

In applications where it is impossible to completely evacuate the conveyor prior to shutdown, the only viable option is to jog, or step start, the conveyor. Step starting reduces the amp load on the motor and should control the torque to prevent the shear pin from failing.

If instead of step starting, the operator applies full motor load to a stationary, fully loaded conveyor, one of two things will occur: (1) the drive motor's circuit breaker will trip as a result of excessive amp load, or (2) the shear pin installed to protect the conveyor will fail. Either of these failures adversely impacts production.

Screw

The screw, or spiral, conveyor is widely used for pulverized or granular, noncorrosive, nonabrasive materials in systems requiring moderate capacities, distances not more than about 200 feet, and moderate inclines ($\leq$ 35 degrees). It usually costs substantially less than any other type of conveyor and can be made dust-tight by installing a simple cover plate.

Abrasive or corrosive materials can be handled with suitable construction of the helix and trough. Conveyors using special materials, such as hard-faced cast iron and linings or coatings, on the components that come into contact with the materials can be specified in these applications. The screw conveyor will handle lumpy material if the lumps are not large in proportion to the diameter of the screw's helix.

Screw conveyors may be inclined. A standard-pitch helix will handle material on inclines up to 35 degrees. Capacity is reduced in inclined applications and Table 10.3 provides the approximate reduction in capacity for various inclines.

Table 10.3 *Screw conveyor capacity reductions for inclined applications*

Inclination, degrees	10	15	20	25	30	35
Reduction in capacity, %	10	26	45	58	70	78

Configuration

Screw conveyors have a variety of configurations. Each is designed for specific applications and/or materials. Standard conveyors have a galvanized-steel rotor, or helix, and trough. For abrasive and corrosive materials (e.g., wet ash), both the helix and trough may be hard-faced cast iron. For abrasives, the outer edge of the helix may be faced with a renewable strip of Stellite (a cobalt alloy produced by Haynes Stellite Co.) or other similarly hard material. Aluminum, bronze, Monel, or stainless steel also may be used to construct the rotor and trough.

Short-Pitch Screw

The standard helix used for screw conveyors has a pitch approximately equal to its outside diameter. The short-pitch screw is designed for applications with inclines greater than 29 degrees.

Variable-Pitch Screw

Variable-pitch screws having the short pitch at the feed end automatically control the flow to the conveyor and correctly proportion the load down the screw's length. Screws having what is referred to as a "short section," which has either a shorter pitch or smaller diameter, are self-loading and do not require a feeder.

Cut-Flight

Cut-flight conveyors are used for conveying and mixing cereals, grains, and other light material. They are similar to normal flight or screw conveyors, and the only difference is the configuration of the paddles or screw. Notches are cut in the flights to improve the mixing and conveying efficiency when handling light, dry materials.

Ribbon Screw

Ribbon screws are used for wet and sticky materials, such as molasses, hot tar, and asphalt. This type of screw prevents the materials from building up and altering the natural frequency of the screw. A buildup can cause resonance problems and possibly catastrophic failure of the unit.

Paddle Screw

The paddle-screw conveyor is used primarily for mixing materials like mortar and paving mixtures. An example of a typical application is churning ashes and water to eliminate dust.

Performance

Process parameters, such as density, viscosity, and temperature, must be constantly maintained within the conveyor's design operating envelope.

Table 10.4 *Factor A for self-lubricating bronze bearings*

Conveyor . diameter, in	6	9	10	12	14	16	18	20	24
Factor A	54	96	114	171	255	336	414	510	690

Slight variations can affect performance and reliability. In intermittent applications, extreme care should be taken to fully evacuate the conveyor prior to shutdown. In addition, caution must be exercised when restarting a conveyor in case an improper shutdown was performed and material was allowed to settle.

Power Requirements

The horsepower requirement for the conveyor-head shaft, H, for horizontal screw conveyors can be determined from the following equation:

$$H = (ALN + CWLF) \times 10 - 6$$

Where:

A = Factor for size of conveyor (see Table 10.4)

C = Material volume, ft^3/h

F = Material factor, unitless (see Table 10.5)

L = Length of conveyor, feet

N = Conveyor rotation speed (rpm)

W = Density of material, lb/ft^3

In addition to H, the motor size depends on the drive efficiency (E) and a unitless allowance factor (G), which is a function of H. Values for G are found in Table 10.6. The value for E is usually 90%.

$$\text{Motor hp} = HG/E$$

Table 10.5 gives the information needed to estimate the power requirement: percentages of helix loading for five groups of material, maximum material density or capacity, allowable speeds for 6-inch and 20-inch diameter screws, and the factor F.

Table 10.5 *Power requirements by material group*

Material group	Max. cross section % occupied by the material	Max. density of material, lb/ft^3	Max. rpm for 6" diameter	Max. rpm for 20" diameter
1	45	50	170	110
2	38	50	120	75
3	31	75	90	60
4	25	100	70	50
5	$12\frac{1}{2}$		30	25

Group 1	F factor is 0.5 for light materials such as barley, beans, brewers grains (dry), coal (pulverized), cornmeal, cottonseed meal, flaxseed, flour, malt, oats, rice, and wheat.
Group 2	Includes fines and granular material. The values of F are: alum (pulverized), 0.6; coal (slack or fines), 0.9; coffee beans, 0.4; sawdust, 0.7; soda ash (light), 0.7; soybeans, 0.5; fly ash, 0.4.
Group 3	Includes materials with small lumps mixed with fines. Values of F are: alum, 1.4; ashes (dry), 4.0; borax, 0.7; brewers grains (wet), 0.6; cottonseed, 0.9; salt, coarse or fine, 1.2; soda ash (heavy), 0.7.
Group 4	Includes semiabrasive materials, fines, granular, and small lumps. Values of F are: acid phosphate (dry), 1.4; bauxite (dry), 1.8; cement (dry), 1.4; clay, 2.0; fuller's earth, 2.0; lead salts, 1.0; limestone screenings, 2.0; sugar (raw), 1.0; white lead, 1.0; sulfur (lumpy), 0.8; zinc oxide, 1.0.
Group 5	Includes abrasive lumpy materials, which must be kept from contact with hanger bearings. Values of F are: wet ashes, 5.0; flue dirt, 4.0; quartz (pulverized), 2.5; silica sand, 2.0; sewage sludge (wet and sandy), 6.0.

Table 10.6 *Allowance factor*

H, hp	1	1–2	2–4	4–5	5
G	2	1.5	1.25	1.1	1.0

Volumetric Efficiency

Screw-conveyor performance is also determined by the volumetric efficiency of the system. This efficiency is determined by the amount of slip or bypass generated by the conveyor. The amount of slip in a screw conveyor is primarily determined by three factors: product properties, screw efficiency, and clearance between the screw and the conveyor barrel or housing.

Product Properties

Not all materials or products have the same flow characteristics. Some have plastic characteristics and flow easily. Others do not self-adhere and tend to separate when pumped or mechanically conveyed. As a result, the volumetric efficiency is directly affected by the properties of each product. This also impacts screw performance.

Screw Efficiency

Each of the common screw configurations (i.e., short pitch, variable pitch, cut flights, ribbon, and paddle) has varying volumetric efficiencies, depending on the type of product that is conveyed. Screw designs or configurations must be carefully matched to the product to be handled by the system.

For most medium- to high-density products in a chemical plant, the variable-pitch design normally provides the highest volumetric efficiency and lowest required horsepower. Cut-flight conveyors are highly efficient for light, non-adhering products, such as cereals, but are inefficient when handling heavy, cohesive products. Ribbon conveyors are used to convey heavy liquids, such as molasses, but are not very efficient and have a high slip ratio.

Clearance

Improper clearance is the source of many volumetric-efficiency problems. It is important to maintain proper clearance between the outer ring, or diameter, of the screw and the conveyor's barrel, or housing, throughout the operating life of the conveyor. Periodic adjustments to compensate for wear, variations in product, and changes in temperature are essential. While the recommended clearance varies with specific conveyor design and the product to be conveyed, excessive clearance severely impacts conveyor performance as well.

Installation

Installation requirements vary greatly with screw-conveyor design. The vendor's Operating and Maintenance (O&M) manuals should be consulted and followed to ensure proper installation. However, as with practically all mechanical equipment, there are basic installation requirements common to all screw conveyors. Installation requirements presented here should be evaluated in conjunction with the vendor's O&M manual. If the information provided here conflicts with the vendor-supplied information, the O&M manual's recommendations should always be followed.

Foundation

The conveyor and its support structure must be installed on a rigid foundation that absorbs the torsional energy generated by the rotating screws. Because of the total overall length of most screw conveyors, a single foundation that supports the entire length and width should be used. There must be enough lateral (i.e., width) stiffness to prevent flexing during normal operation. Mounting conveyor systems on decking or suspended-concrete flooring should provide adequate support.

Support Structure

Most screw conveyors are mounted above the foundation level on a support structure that generally has a slight downward slope from the feed end to the discharge end. While this improves the operating efficiency of the conveyor, it also may cause premature wear of the conveyor and its components.

The support's structural members (i.e., I-beams and channels) must be adequately rigid to prevent conveyor flexing or distortion during normal operation. Design, sizing, and installation of the support structure must guarantee rigid support over the full operating range of the conveyor. When evaluating the structural requirements, variations in product type, density, and operating temperature must also be considered. Since these variables directly affect the torsional energy generated by the conveyor, the worst-case scenario should be used to design the conveyor's support structure.

Product-Feed System

One of the major limiting factors of screw conveyors is their ability to provide a continuous supply of incoming product. While some conveyor designs, such as those having a variable-pitch screw, provide the ability to self-feed, their installation should include a means of ensuring a constant, consistent incoming supply of product.

In addition, the product-feed system must prevent entrainment of contaminants in the incoming product. Normally, this requires an enclosure that seals the product from outside contaminants.

Operating Methods

As previously discussed, screw conveyors are sensitive to variations in incoming product properties and the operating environment. Therefore, the primary operating concern is to maintain a uniform operating envelope at all times, in particular by controlling variations in incoming product and operating environment.

Incoming-Product Variations

Any measurable change in the properties of the incoming product directly affects the performance of a screw conveyor. Therefore, the operating practices should limit variations in product density, temperature, and viscosity. If they occur, the conveyor's speed should be adjusted to compensate for them.

For property changes directly related to product temperature, preheaters or coolers can be used in the incoming-feed hopper, and heating/cooling traces can be used on the conveyor's barrel. These systems provide a means of achieving optimum conveyor performance despite variations in incoming product.

Operating-Environment Variations

Changes in the ambient conditions surrounding the conveyor system may also cause deviations in performance. A controlled environment will substantially improve the conveyor's efficiency and overall performance. Therefore, operating practices should include ways to adjust conveyor speed and output to compensate for variations. The conveyor should be protected from wind chill, radical variations in temperature and humidity, and any other environment-related variables.

11 Couplings

Couplings are designed to provide two functions: (1) to transmit torsional power between a power source and driven unit and (2) to absorb torsional variations in the drive train. They are *not* designed to correct misalignment between two shafts. While certain types of couplings provide some correction for slight misalignment, reliance on these devices to obtain alignment is not recommended.

Coupling Types

The sections to follow provide overviews of the more common coupling types: rigid and flexible. Also discussed are couplings used for special applications: floating-shaft (spacer) and fluid (hydraulic).

Rigid Couplings

A rigid coupling permits neither axial nor radial relative motion between the shafts of the driver and driven unit. When the two shafts are connected solidly and properly, they operate as a single shaft. A rigid coupling is primarily used for vertical applications, e.g., vertical pumps. Types of rigid couplings discussed in this section are flanged, split, and compression.

Flanged couplings are used where there is free access to both shafts. Split couplings are used where access is limited on one side. Both flanged and split couplings require the use of keys and keyways. Compression couplings are used when it is not possible to use keys and keyways.

Flanged Couplings

A flanged rigid coupling is comprised of two halves, one located on the end of the driver shaft and the other on the end of the driven shaft. These halves are bolted together to form a solid connection. To positively transmit torque, the coupling incorporates axially fitted keys and split circular key rings or dowels, which eliminate frictional dependency for transmission. The use of flanged couplings is restricted primarily to vertical pump shafts. A typical flanged rigid coupling is illustrated in Figure 11.1.

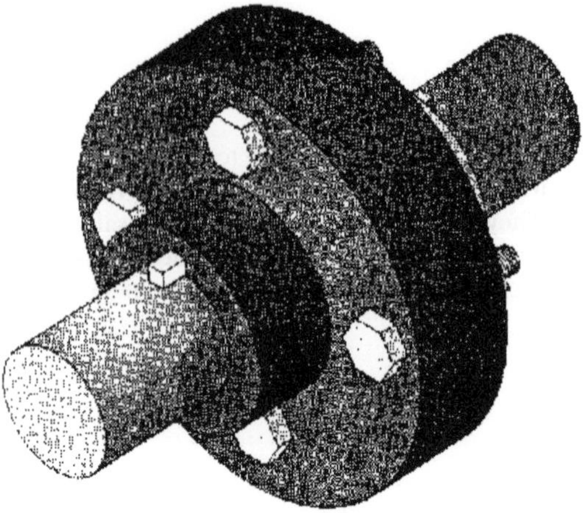

Figure 11.1 *Typical flanged rigid coupling*

Split Couplings

A split rigid coupling, also referred to as a clamp coupling, is basically a sleeve that is split horizontally along the shaft and held together with bolts. It is clamped over the adjoining ends of the driver and driven shafts, forming a solid connection. Clamp couplings are used primarily on vertical pump shafting. A typical split rigid coupling is illustrated in Figure 11.2. As with the flanged coupling, the split rigid coupling incorporates axially fitted keys and split circular key rings to eliminate frictional dependency in the transmission of torque.

Compression Coupling

A rigid compression coupling is comprised of three pieces: a compressible core and two encompassing coupling halves that apply force to the core. The core is comprised of a slotted bushing that has been machine bored to fit both ends of the shafts. It also has been machined with a taper on its external diameter from the center outward to both ends. The coupling halves are finish bored to fit this taper. When the coupling halves are bolted together, the core is compressed down on the shaft by the two halves, and the resulting frictional grip transmits the torque without the use of keys. A typical compression coupling is illustrated in Figure 11.3.

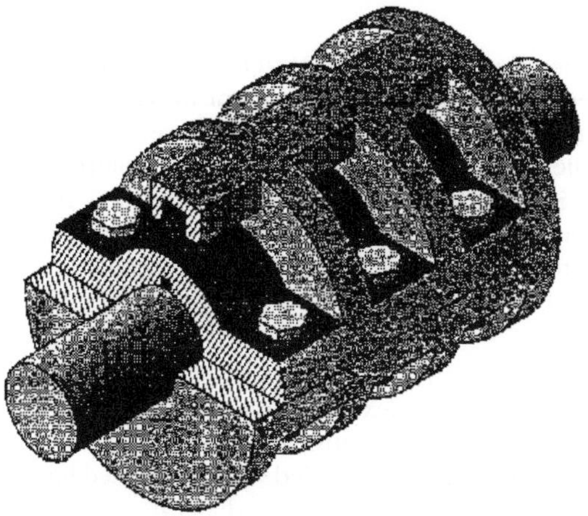

Figure 11.2 *Typical split rigid coupling*

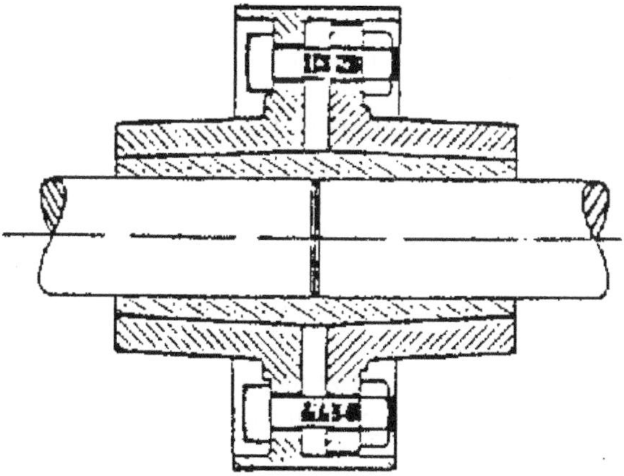

Figure 11.3 *Typical compression rigid coupling*

Flexible Couplings

Flexible couplings, which are classified as mechanical flexing, material flexing, or combination, allow the coupled shafts to slide or move relative to each other. Although clearances are provided to permit movement within specified tolerance limits, flexible couplings are not designed to compensate for major misalignments. (Shafts must be aligned to less than 0.002 inches for proper operation.) Significant misalignment creates a whipping movement of the shaft, adds thrust to the shaft and bearings, causes axial vibrations, and leads to premature wear or failure of equipment.

Mechanical Flexing

Mechanical-flexing couplings provide a flexible connection by permitting the coupling components to move or slide relative to each other. In order to permit such movement, clearance must be provided within specified limits. It is important to keep cross loading on the connected shafts at a minimum. This is accomplished by providing adequate lubrication to reduce wear on the coupling components. The most popular of the mechanical-flexing type are the chain and gear couplings.

Chain

Chain couplings provide a good means of transmitting proportionately high torque at low speeds. Minor shaft misalignment is compensated for by means of clearances between the chain and sprocket teeth and the clearance that exists within the chain itself.

The design consists of two hubs with sprocket teeth connected by a chain of the single-roller, double-roller, or silent type. A typical example of a chain coupling is illustrated in Figure 11.4.

Special-purpose components may be specified when enhanced flexibility and reduced wear is required. Hardened sprocket teeth, special tooth design, and barrel-shaped rollers are available for special needs. Light-duty drives are sometimes supplied with nonmetallic chains on which no lubrication should be used.

Gear

Gear couplings are capable of transmitting proportionately high torque at both high and low speeds. The most common type of gear coupling consists of two identical hubs with external gear teeth and a sleeve, or cover, with matching internal gear teeth. Torque is transmitted through the gear teeth,

Roller-chain coupling.

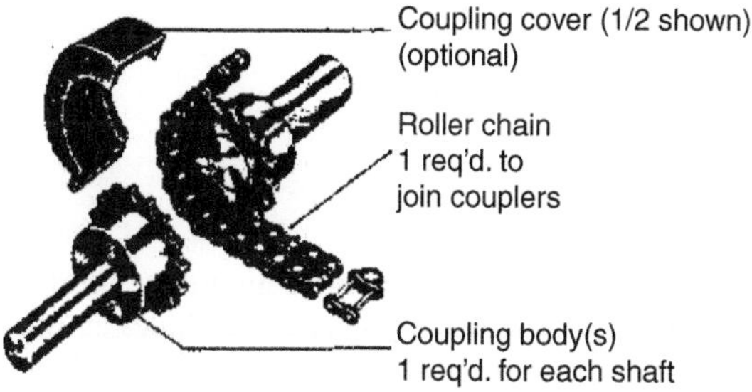

Figure 11.4 *Typical chain coupling*

whereas the necessary sliding action and ability for slight adjustments in position comes from a certain freedom of action provided between the two sets of teeth.

Slight shaft misalignment is compensated for by the clearance between the matching gear teeth. However, any degree of misalignment decreases the useful life of the coupling and may cause damage to other machine-train components such as bearings. A typical example of a gear-tooth coupling is illustrated in Figure 11.5.

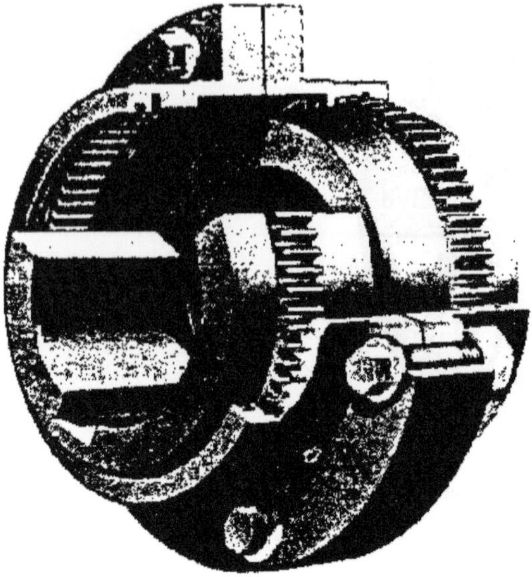

Figure 11.5 *Typical gear-tooth coupling*

Material-Flexing

Material-flexing couplings incorporate elements that accommodate a certain amount of bending or flexing. The material-flexing group includes laminated disk-ring, bellows, flexible shaft, diaphragm, and elastomeric couplings.

Various materials, such as metal, plastics, or rubber, are used to make the flexing elements in these couplings. The use of the couplings is governed by the operational fatigue limits of these materials. Practically all metals have fatigue limits that are predictable; therefore, they permit definite boundaries of operation to be established. Elastomers such as plastic or rubber, however, usually do not have a well defined fatigue limit. Their service life is determined primarily by conditions of installation and operation.

Laminated Disk-Ring

The laminated disk-ring coupling consists of shaft hubs connected to a single flexible disk, or a series of disks, that allows axial movement. The laminated disk-ring coupling also reduces heat and axial vibration that can transmit

Morflex couplings

Dropout style

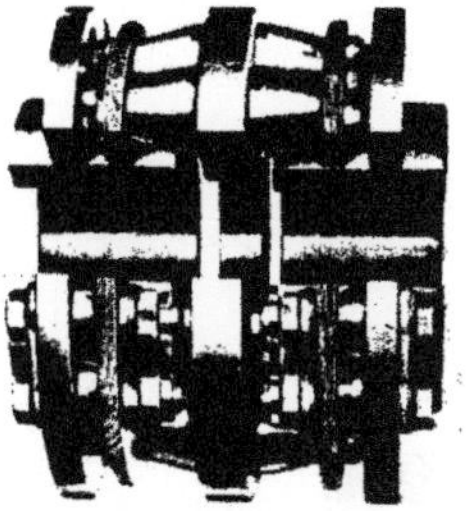

Laminated disk-ring coupling
(standard double-engagement)

Laminated disk-ring coupling
(high speed spacer type)

Figure 11.6 *Typical laminated disk-ring couplings*

between the driver and driven unit. Figure 11.6 illustrates some typical laminated disk-ring couplings.

Bellows

Bellows couplings consist of two shaft hubs connected to a flexible bellows. This design, which compensates for minor misalignment, is used at moderate rotational torque and shaft speed. This type of coupling provides flexibility to compensate for axial movement and misalignment caused by thermal expansion of the equipment components. Figure 11.7 illustrates a typical bellows coupling.

Flexible Shaft or Spring

Flexible shaft or spring couplings are generally used in small equipment applications that do not experience high torque loads. Figure 11.8 illustrates a typical flexible shaft coupling.

Figure 11.7 *Typical bellows coupling*

Figure 11.8 *Typical flexible shaft coupling*

Diaphragm

Diaphragm couplings provide torsional stiffness while allowing flexibility in axial movement. Typical construction consists of shaft hub flanges and a diaphragm spool, which provides the connection between the driver and driven unit. The diaphragm spool normally consists of a center shaft fastened to the inner diameter of a diaphragm on each end of the spool shaft. The shaft hub flanges are fastened to the outer diameter of the diaphragms to complete the mechanical connection. A typical diaphragm coupling is illustrated in Figure 11.9.

Elastomeric

Elastomeric couplings consist of two hubs connected by an elastomeric element. The couplings fall into two basic categories, one with the element

Figure 11.9 *Typical diaphragm coupling*

placed in *shear* and the other with the element placed in *compression*. The coupling compensates for minor misalignments because of the flexing capability of the elastomer. These couplings are usually applied in light- or medium-duty applications running at moderate speeds.

With the *shear-type* coupling, the elastomeric element may be clamped or bonded in place, or fitted securely to the hubs. The *compression-type* couplings may be fitted with projecting pins, bolts, or lugs to connect the components. Polyurethane, rubber, neoprene, or cloth and fiber materials are used in the manufacture of these elements.

Although elastomeric couplings are practically maintenance free, it is good practice to periodically inspect the condition of the elastomer and the alignment of the equipment. If the element shows signs of defects or wear, it should be replaced and the equipment realigned to the manufacturer's specifications. Typical elastomeric couplings are illustrated in Figure 11.10.

Combination (Metallic-Grid)

The metallic-grid coupling is an example of a combination of mechanical-flexing and material-flexing type couplings. Typical metallic-grid couplings are illustrated in Figure 11.11.

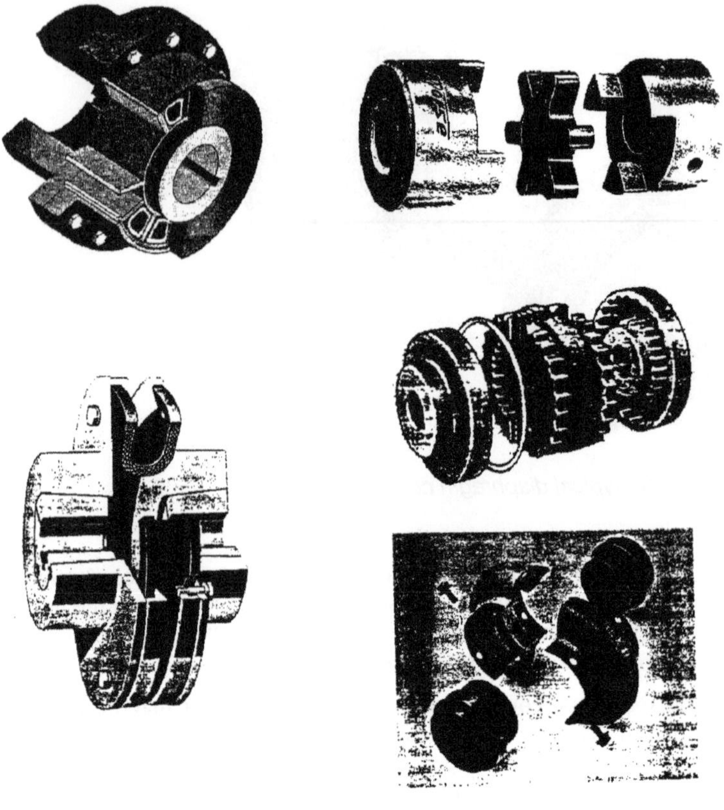

Figure 11.10 *Typical elastomeric couplings*

The metallic-grid coupling is a compact unit capable of transmitting high torque at moderate speeds. The construction of the coupling consists of two flanged hubs, each with specially grooved slots cut axially on the outer edges of the hub flanges. The flanges are connected by means of a serpentine-shaped spring grid that fits into the grooved slots. The flexibility of this grid provides torsional resilience.

Special Application Couplings

Two special application couplings are discussed in this section: (1) floating-shaft or spacer coupling and (2) hydraulic or fluid coupling.

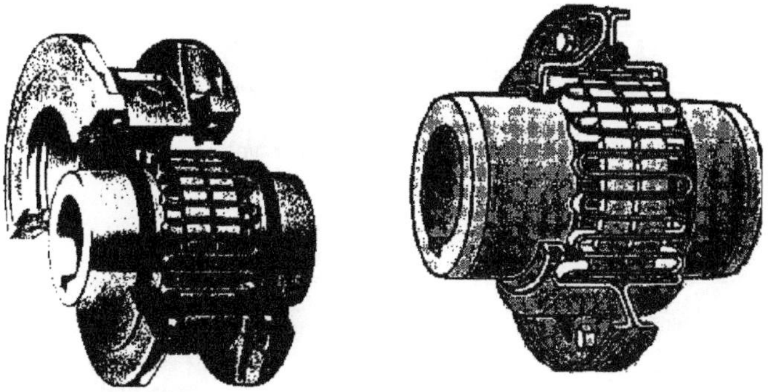

Figure 11.11 *Typical metallic-grid couplings*

Floating-Shaft or Spacer Coupling

Regular flexible couplings connect the driver and driven shafts with relatively close ends and are suitable for limited misalignment. However, allowances sometimes have to be made to accommodate greater misalignment or when the ends of the driver and driven shafts have to be separated by a considerable distance.

Such is the case, for example, with end-suction pump designs in which the power unit of the pump assembly is removed for maintenance by being axially moved toward the driver. If neither the pump nor the driver can be readily removed, they should be separated sufficiently to permit withdrawal of the pump's power unit. An easily removable flexible coupling of sufficient length (i.e., floating-shaft or spacer coupling) is required for this type of maintenance. Examples of couplings for this type of application are shown in Figure 11.12.

In addition to the maintenance application described above, this coupling (also referred to as extension or spacer sleeve coupling) is commonly used where equipment is subject to thermal expansion and possible misalignment because of high process temperatures. The purpose of this type of coupling is to prevent harmful misalignment with minimum separation of the driver and driven shaft ends. An example of a typical floating-shaft coupling for this application is shown in Figure 11.13.

The floating-shaft coupling consists of two support elements connected by a shaft. Manufacturers use various approaches in their designs for these couplings. For example, each of the two support elements may be of the

Laminated disk-ring coupling, spacer type

Gear coupling, spindle type

Gear coupling, high speed spacer type

Figure 11.12 *Typical floating-shaft or spacer couplings*

Figure 11.13 *Typical floating-shaft or spacer couplings for high-temperature applications*

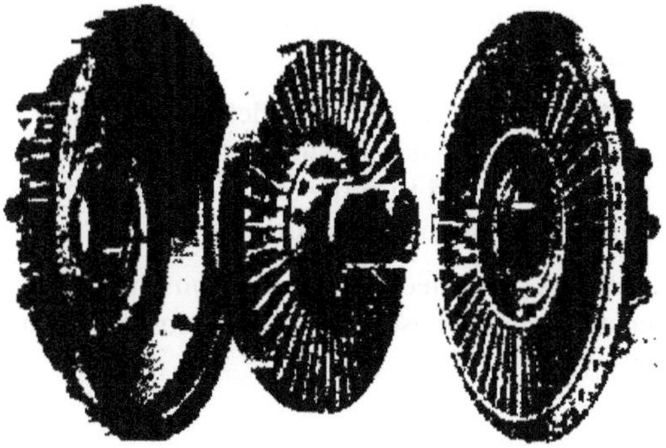

Figure 11.14 *Typical hydraulic coupling*

single-engagement type, may consist of a flexible half-coupling on one end and a rigid half-coupling on the other end, or may be completely flexible with some piloting or guiding supports.

Floating-shaft gear couplings usually consist of a standard coupling with a two-piece sleeve. The sleeve halves are bolted to rigid flanges to form two single-flex couplings. An intermediate shaft that permits the transmission of power between widely separated drive components, in turn, connects these.

Hydraulic or Fluid

Hydraulic couplings provide a soft start with gradual acceleration and limited maximum torque for fixed operating speeds. Hydraulic couplings are typically used in applications that undergo torsional shock from sudden changes in equipment loads (e.g., compressors). Figure 11.14 is an illustration of a typical hydraulic coupling.

Coupling Selection

Periodically, worn or broken couplings must be replaced. One of the most important steps in performing this maintenance procedure is to ensure that the correct replacement parts are used. After having determined the cause of failure, it is crucial to identify the correct type and size of coupling needed.

Even if *practically identical in appearance* to the original, a part still may *not* be an adequate replacement.

The manufacturer's specification number usually provides the information needed for part selection. If the part is not in stock, a cross-reference guide will provide the information needed to verify ratings and to identify a coupling that meets the same requirements as the original.

Criteria that must be considered in part selection include: equipment type, mode of operation, and cost. Each of these criteria is discussed in the sections to follow.

Equipment Type

Coupling selection should be application specific and, therefore, it is important to consider the type of equipment that it connects. For example, demanding applications such as variable, high-torque machine trains require couplings that are specifically designed to absorb radical changes in speed and torque (e.g., metallic-grid). Less demanding applications such as run-out table rolls can generally get by with elastomeric couplings. Table 11.1 lists the coupling type commonly used in a particular application.

Mode of Operation

Coupling selection is highly dependent on the mode of operation, which includes torsional characteristics, speed, and the operating envelope.

Torsional Characteristics

Torque requirements are a primary concern during the selection process. In all applications in which variable or high torque is transmitted from the driver to the driven unit, a flexible coupling rated for the maximum torque requirement must be used. Rigid couplings are not designed to absorb variations in torque and should not be used.

Speed

Two speed-related factors should be considered as part of the selection process: maximum speed and speed variation.

Maximum Speed

When selecting coupling type and size, the maximum speed rating must be considered, which can be determined from the vendor's catalog.

Table 11.1 *Coupling application overview*

Application	Coupling selection recommendation
Limited Misalignment Compensation	
Variable, high-torque machine trains operating at moderate speeds	Metallic-grid combination couplings
Run-out table rolls	Elastomeric flexible couplings
Vertical pump shafting	Flanged rigid couplings, split rigid or clamp couplings
Keys and keyways not appropriate (e.g., brass shafts)	Rigid compression couplings
Transmission of proportionately high torque at low speeds	Chain couplings (mechanical-flexing)
Transmission of proportionately high torque at both high and low speeds	Gear couplings (mechanical-flexing)
Allowance for axial movement and reduction of heat and axial vibration	Laminated disk-ring couplings (material-flexing)
Moderate rotational torque and shaft speed	Bellows couplings (material-flexing)
Small equipment that does not experience high torque loads	Flexible shaft or spring couplings (material-flexing)
Torsional stiffness while allowing flexibility in axial movement	Diaphragm material-flexing couplings
Light- or medium-duty applications running at moderate speeds	Elastomeric couplings (material-flexing)
Gradual acceleration and limited maximum torque for fixed operating speeds (e.g., compressors)	Hydraulic or fluid couplings
Variable or high torque and/or speed transmission	Flexible couplings rated for the maximum torque requirement
Greater Misalignment Compensation	
Maintenance requiring considerable distance between the driver and driven shaft ends. Misalignment results from expansion due to high process temperatures.	Floating-shaft or spacer couplings

Note: Rigid couplings are not designed to absorb variations in torque and speed and should not be used in such applications. Maximum in-service coupling speed should be at least 15% below the maximum coupling speed rating.

The maximum in-service speed of a coupling should be well below (at least 15%) the maximum speed rating. The 15% margin provides a service factor that should be sufficient to prevent coupling damage or catastrophic failure.

Speed Variation

Variation in speed equates to a corresponding variation in torque. Most variable-speed applications require some type of flexible coupling capable of absorbing these torsional variations.

Operating Envelope

The operating envelope defines the physical requirements, dimensions, and type of coupling needed in a specific application. The envelope information should include: shaft sizes, orientation of shafts, required horsepower, full range of operating torque, speed ramp rates, and any other data that would directly or indirectly affect the coupling.

Cost

Coupling cost should not be the deciding factor in the selection process, although it will certainly play a part in it. Although higher performance couplings may be more expensive, they actually may be the cost-effective solution in a particular application. Selecting the most appropriate coupling for an application not only extends coupling life, but also improves the overall performance of the machine train and its reliability.

Installation

Couplings must be installed properly if they are to operate satisfactorily. This section discusses shaft and coupling preparation, coupling installation, and alignment.

Shaft Preparation

A careful inspection of both shaft ends must be made to ensure that no burrs, nicks, or scratches are present that will damage the hubs. Potentially damaging conditions must be corrected before coupling installation. Emery cloth should be used to remove any burrs, scratches, or oxidation that may be present. A light film of oil should be applied to the shafts prior to installation.

Keys and keyways also should be checked for similar defects and to ensure that the keys fit properly. Properly sized key stock must be used with all keyways; do not use bar stock or other material.

Coupling Preparation

The coupling must be disassembled and inspected prior to installation. The location and position of each component should be noted so that it can be reinstalled in the correct order. When old couplings are removed for inspection, bolts and bolt holes should be numbered so that they can be installed in the same location when the coupling is returned to service.

Any defects, such as burrs, should be corrected before the coupling is installed. Defects on the mating parts of the coupling can cause interference between the bore and shaft, preventing proper operation of the coupling.

Coupling Installation

Once the inspection shows the coupling parts to be free of defects, the hubs can be mounted on their respective shafts. If it is necessary to heat the hubs to achieve the proper interference fit, an oil or water bath should be used. Spot heating using a flame or torch should be avoided, as it causes distortion and may adversely affect the hubs.

Care must be exercised during installation of a new coupling or the reassembly of an existing unit. Keys and keyways should be coated with a sealing compound that is resistant to the lubricant used in the coupling. Seals should be inspected to ensure that they are pliable and in good condition. They must be installed properly in the sleeve with the lip in good contact with the hub. Sleeve flange gaskets must be whole, in good condition, clean, and free of nicks or cracks. Lubrication plugs must be cleaned before being installed and must fit tightly.

The specific installation procedure is dependent on the type and mounting configuration of the coupling. However, common elements of all coupling installations include: spacing, bolting, lubrication, and the use of matching parts. The sections that follow discuss these installation elements.

Spacing

Spacing between the mating parts of the coupling must be within manufacturer's tolerances. For example, an elastomeric coupling must have a specific distance between the coupling faces. This distance determines the position of the rubber boot that provides transmission of power from the driver to the driven machine component. If this distance is not exact, the elastomer will attempt to return to its relaxed position, inducing excessive axial movement in both shafts.

Bolting

Couplings are designed to use a specific type of bolt. Coupling bolts have a hardened cylindrical body sized to match the assembled coupling width. Hardened bolts are required because standard bolts do not have the tensile strength to absorb the torsional and shearing loads in coupling applications and may fail, resulting in coupling failure and machine-train damage.

Lubrication

Most couplings require lubrication, and care must be taken to ensure that the proper type and quantity is used during the installation process. Inadequate or improper lubrication reduces coupling reliability and reduces its useful life. In addition, improper lubrication can cause serious damage to the machine train. For example, when a gear-type coupling is overfilled with grease, the coupling will lock. In most cases, its locked position will increase the vibration level and induce an abnormal loading on the bearings of both the driver and driven unit, resulting in bearing failure.

Matching Parts

Couplings are designed for a specific range of applications, and proper performance depends on the total design of the coupling system. As a result, it is generally not a good practice to mix coupling types. Note, however, that it is common practice in some steel industry applications to use coupling halves from two different types of couplings. For example, a rigid coupling half is sometimes mated to a flexible coupling half, creating a hybrid. While this approach may provide short-term power transmission, it can result in an increase in the number, frequency, and severity of machine-train problems.

Coupling Alignment

The last step in the installation process is verifying coupling and shaft alignment. With the exception of special application couplings such as spindles and jackshafts, all couplings must be aligned within relatively close tolerances (i.e., 0.001 to 0.002 inch).

Lubrication and Maintenance

Couplings require regular lubrication and maintenance to ensure optimum trouble-free service life. When proper maintenance is not conducted, premature coupling failure and/or damage to machine-train components such as bearings can be expected.

Determining Cause of Failure

When a coupling failure occurs, it is important to determine the cause of failure. Failure may result from a coupling defect, an external condition, or workmanship during installation.

Most faults are attributed to poorly machined surfaces causing out-of-specification tolerances, although defective material failures also occur. Inadequate material hardness and poor strength factors contribute to many premature failures. Other common causes are improper coupling selection, improper installation, and/or excessive misalignment.

Lubrication Requirements

Lubrication requirements vary depending on application and coupling type. Because rigid couplings do not require lubrication, this section discusses lubrication requirements for mechanical-flexing, material-flexing, and combination flexible couplings only.

Mechanical-Flexing Couplings

It is important to follow the manufacturer's instructions for lubricating mechanical-flexing couplings, which **must be lubricated internally**. Lubricant seals must be in good condition and properly fitted into place. Coupling covers contain the lubricant and prevent contaminants from entering the coupling interior. The covers are designed in two configurations, split either horizontally or vertically. Holes are provided in the covers to allow lubricant to be added without coupling disassembly.

Gear couplings are one type of mechanical-flexing coupling, and there are several ways to lubricate them: grease pack, oil fill, oil collect, and continuous oil flow. Either grease or oil can be used at speeds of 3,600 rpm to 6,000 rpm. Oil is normally used as the lubricant in couplings operating over 6,000 rpm. Grease and oil-lubricated units have end gaskets and seals, which are used to contain the lubricant and seal out the entry of contaminants. The sleeves have lubrication holes, which permit flushing and relubrication without disturbing the sleeve gasket or seals.

Material-Flexing Couplings

Material-flexing couplings are designed to be lubrication free.

Combination Couplings

Combination (metallic-grid) couplings are lubricated in the same manner as mechanical-flexing couplings.

Periodic Inspections

It is important to perform periodic inspections of all mechanical equipment and systems that incorporate rotating parts, including couplings and clutches.

Mechanical-Flexing Couplings

To maintain coupling reliability, mechanical-flexing couplings require periodic inspections on a time- or condition-based frequency established by the history of the equipment's coupling life or a schedule established by the predictive maintenance engineer. Items to be included in an inspection are listed below. If any of these items or conditions is discovered, the coupling should be evaluated to determine its remaining operational life or repaired/replaced.

- Inspect lubricant for traces of metal (indicating component wear).
- Visually inspect coupling mechanical components (roller chains and gear teeth, and grid members) for wear and/or fatigue.
- Inspect seals to ensure they are pliable and in good condition. They must be installed properly in the sleeve with the lip in good contact with the hub.
- Sleeve flange gaskets must be whole, in good condition, clean, and free of nicks or cracks.
- Lubrication plugs must be clean (to prevent the introduction of contaminants to the lubricant and machine surfaces) before being installed and must be torqued to the manufacturer's specifications.

- Setscrews and retainers must be in place and tightened to manufacturer's specifications.
- Inspect shaft hubs, keyways, and keys for cracks, breaks, and physical damage.
- Under operating conditions, perform thermographic scans to determine temperature differences on the coupling (indicates misalignment and/or uneven mechanical forces).

Material-Flexing Couplings

Although designed to be lubrication-free, material-flexing couplings also require periodic inspection and maintenance. This is necessary to ensure that the coupling components are within acceptable specification limits. Periodic inspections for the following conditions are required to maintain coupling reliability. If any of these conditions are found, the coupling should be evaluated to determine its remaining operational life or repaired/replaced.

- Inspect flexing element for signs of wear or fatigue (cracks, element dust, or particles).
- Setscrews and retainers must be in place and tightened to manufacturer's specifications.
- Inspect shaft hubs, keyways, and keys for cracks, breaks, and physical damage.
- Under operating conditions, perform thermographic scans for temperature differences on the coupling, which indicates misalignment and/or uneven mechanical forces.

Combination Couplings

Mechanical components (e.g., grid members) should be visually inspected for wear and/or fatigue. In addition to the items for mechanical-flexing couplings, the grid members on metallic-grid couplings should be replaced if any signs of wear are observed.

Rigid Couplings

The mechanical components of rigid couplings (e.g., hubs, bolts, compression sleeves and halves, keyways, and keys) should be visually inspected for cracks, breaks, physical damage, wear, and/or fatigue. Any component having any of these conditions should be replaced.

Keys, Keyways, and Keyseats

A key is a piece of material, usually metal, placed in machined slots or grooves cut into two axially oriented parts in order to mechanically lock them together. For example, keys are used in making the coupling connection between the shaft of a driver and a hub or flange on that shaft. Any rotating element whose shaft incorporates such a keyed connection is referred to as a keyed-shaft rotor. Keys provide a positive means for transmitting torque between the shaft and coupling hub when a key is properly fitted in the axial groove.

The groove into which a key is fitted is referred to as a keyseat when referring to shafts, and a keyway when referring to hubs. Keyseating is the actual machine operation of producing keyseats. Keyways are normally made on a keys eater or by a broach. Keyseats are normally made with a rotary or end mill cutter.

Figure 11.15 is an example of a keyed shaft that shows the key size versus the shaft diameter. Because of standardization and interchangeability, keys are generally proportioned with relation to shaft diameter instead of torsional load.

The effective key length, "L" is that portion of key having full bearing on hub and shaft. Note that the curved portion of the keyseat made with a

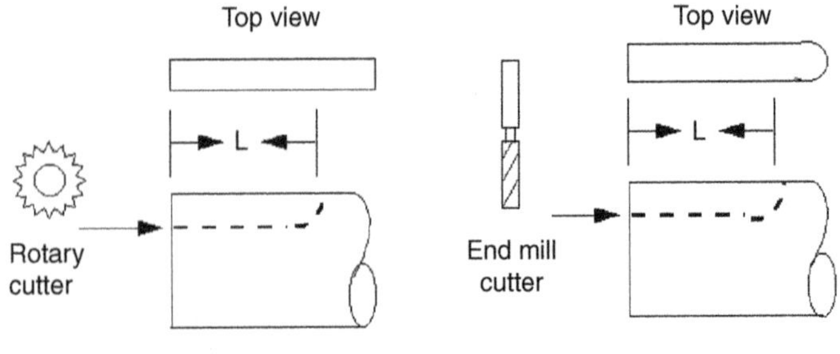

Figure 11.15 *Keyed shaft: key size versus shaft diameter*

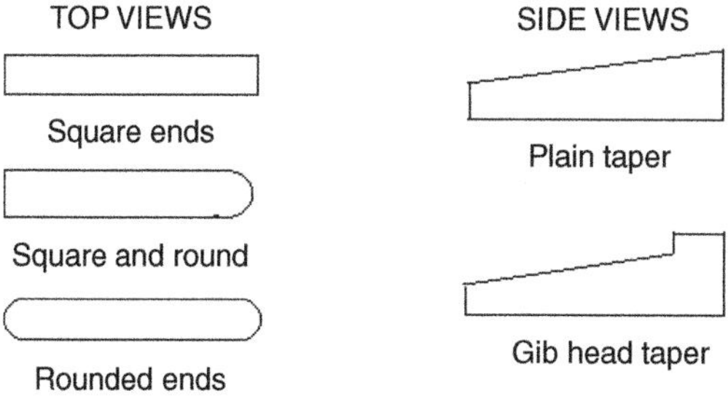

Figure 11.16 *Key shapes*

rotary cutter does not provide full key bearing, so "L" does not include this distance. The use of an end mill cutter results in a square-ended keyseat.

Figure 11.16 shows various key shapes: square ends, one square end and one round end, rounded ends, plain taper, and gib head taper. The majority of keys are square in cross section, which are preferred through $4\frac{1}{2}$" diameter shafts. For bores over $4\frac{1}{2}$" and thin wall sections of hubs, the rectangular (flat) key is used.

The ends are either square, rounded or gib-head. The gib-head is usually used with taper keys. If special considerations dictate the use of a keyway in the hub shallower than the preferred square key, it is recommended that the standard rectangular (flat) key be used.

Hub bores are usually straight, although for some special applications taper bores are sometimes specified. For smaller diameters, bores are designed for clearance fits, and a setscrew is used over the key. The major advantage of a clearance fit is that hubs can be easily assembled and disassembled. For larger diameters, the bores are designed for interference fits without setscrews. For rapid-reversing applications, interference fits are required.

The sections to follow discuss determining keyway depth and width, keyway manufacturing tolerances, key stress calculations, and shaft stress calculations.

Determining Keyway Depth and Width

The formula given below, along with Figure 11.17, Table 11.2 (square keys), and Table 11.3 (flat keys), illustrates how the depth and width of standard

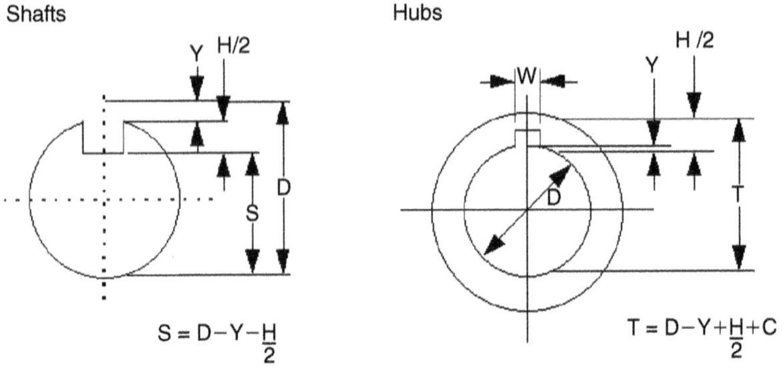

Figure 11.17 *Shaft and hub dimensions*

square and flat keys and keyways for shafts and hubs are determined.

$$Y = \frac{D - \sqrt{D^2 - W^2}}{2}$$

Where:

$C =$ Allowance or clearance for key, inches

$D =$ Nominal shaft or bore diameter, inches

$H =$ Nominal key height, inches

$W =$ Nominal key width, inches

$Y =$ Chordal height, inches

Note: Tables 11.2 and 11.3 shown below are prepared for manufacturing use. Dimensions given are for standard shafts and keyways.

Keyway Manufacturing Tolerances

Keyway manufacturing tolerances (illustrated in Figure 11.18) are referred to as offset (centrality) and lead (cross axis). Offset or centrality is referred

Table 11.2 *Standard square keys and keyways (inches)**

Diameter of holes (inclusive)	Keyways		Key stock
	Width	Depth	
5/16 to 7/16	3/32	3/64	3/32 × 3/32
1/2 to 9/16	1/8	1/16	1/8 × 1/8
5/8 to 7/8	3/16	3/32	3/16 × 3/16
1 5/16 to 11/4	1/4	1/8	1/4 × 1/4
1 5/16 to 13/8	5/16	5/32	5/16 × 5/16
1 7/16 to 13/4	3/8	3/16	3/8 × 3/8
1 13/16 to 21/4	1/2	1/4	1/2 × 1/2
2 5/16 to 23/4	5/8	5/16	5/8 × 5/8
2 13/16 to 31/4	3/4	3/8	3/4 × 3/4
3 5/16 to 33/4	7/8	7/16	7/8 × 7/8
3 13/16 to 4 1/2	1	1/2	1 × 1

Source: The Falk Corporation
*Square keys are normally used through shaft diameter $4\frac{1}{2}$"; larger shafts normally use flat keys.

to as dimension "N"; lead or cross axis is referred to as dimension "J." Both must be kept within permissible tolerances, usually 0.002 inches.

Key Stress Calculations

Calculations for shear and compressive key stresses are based on the following assumptions:

1 The force acts at the radius of the shaft.

2 The force is uniformly distributed along the key length.

3 None of the tangential load is carried by the frictional fit between shaft and bore.

The shear and compressive stresses in a key are calculated using the following equations (see Figure 11.19):

$$Ss = \frac{2T}{(d)x(w)x(L)} \qquad Sc = \frac{2T}{(d)x(h_1)x(L)}$$

Table 11.3 *Standard flat keys and keyways (inches)*

Diameter of holes (inclusive)	Keyways		Key stock
	Width	Depth	
1/2 to 9/16"	1/8	3/64	1/8 × 1/32
5/8 to 7/8"	3/16	1/16	3/16 × 1/8
15/16 to 1 1/4"	1/4	3/32	1/4 × 3/16
1 5/16 to 1 3/8"	5/16	1/8	5/16 × 1/4
1 7/16 to 1 3/4"	3/8	1/8	3/8 × 1/4
1 13/16 to 2 1/4"	1/2	3/16	1/2 × 3/8
2 5/16 to 2 3/4"	5/8	7/32	5/8 × 7/16
2 13/16 to 3 1/4"	3/4	1/4	3/4 × 1/2
3 5/16 to 3 3/4"	7/8	5/16	7/8 × 5/8
3 13/16 to 4 1/2"	1	3/8	1 × 3/4
4 9/16 to 5 1/2"	1 1/4	7/16	1 1/4 × 7/8
5 9/16 to 6 1/2"	1 1/2	1/2	1 1/2 × 1
6 9/16 to 7 1/2"	1 3/4	5/8	1 3/4 × 1 1/4
7 9/16 to 9"	2	3/4	2 × 1 1/2
9 1/16 to 11"	2 1/2	7/8	2 1/2 × 1 3/4
11 1/16 to 13"	3	1	3 × 2
13 1/16 to 15"	3 1/2	1 1/4	3 1/2 × 2 1/2
15 1/16 to 18"	4	1 1/2	4 × 3
18 1/16 to 22"	5	1 3/4	5 × 3 1/2
22 1/16 to 26"	6	4	
26 1/16 to 30"	7	5	

Source: The Falk Corporation

Where:

$d =$ Shaft diameter, inches (use average diameter for taper shafts)

$h_1 =$ Height of key in the shaft or hub that bears against the keyway, inches. Should equal h_2 for square keys. For designs where unequal portions of the key are in the hub or shaft, h_1 is the minimum portion.

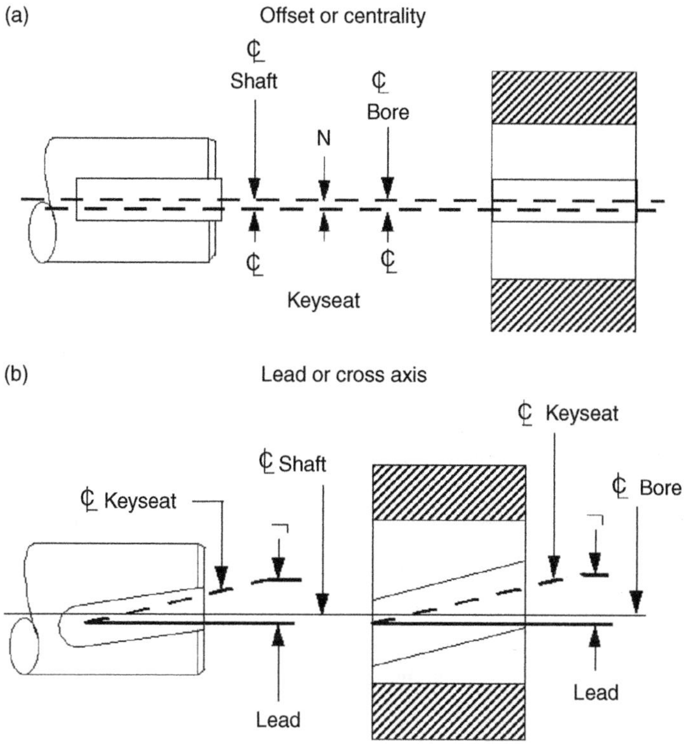

Figure 11.18 *Manufacturing tolerances: offset and lead*

hp = Power, horsepower

L = Effective length of key, inches

rpm = Revolutions per minute

Ss = Shear stress, psi

Sc = Compressive stress, psi

T = Shaft torque, lb-in or $\dfrac{\text{hp} \times 63000}{\text{rpm}}$

w = Key width, inches

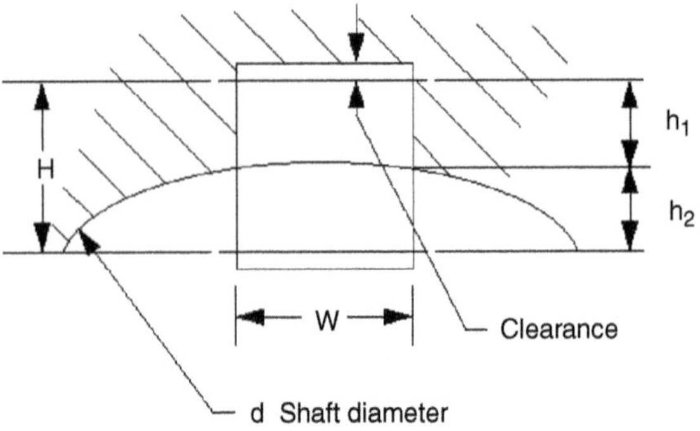

Figure 11.19 *Measurements used in calculating shear and compressive key stress*

Table 11.4 *Allowable stresses for AISI 1018 and AISI 1045*

Material	Heat treatment	Allowable stresses - psi	
		Shear	Compressive
AISI 1018	None	7,500	15,000
AISI 1045	255–300 Bhn	15,000	30,000

Source: The Falk Corporation

Key material is usually AISI 1018 or AISI 1045. Table 11.4 provides the allowable stresses for these materials.

Example: Select a key for the following conditions: 300 hp at 600 rpm; 3" diameter shaft, $\frac{3}{4}$" × $\frac{3}{4}$" key, 4" key engagement length.

$$T = Torque = \frac{hp \times 63{,}000}{rpm} = \frac{300 \times 63{,}000}{600} = 31{,}500 \text{ in-lbs}$$

$$Ss = \frac{2T}{d \times w \times L} = \frac{2 \times 31{,}500}{3 \times 3/4 \times 4} = 7{,}000 \text{ psi}$$

$$Sc = \frac{2T}{d \times h_1 \times L} = \frac{2 \times 31{,}500}{3 \times 3/8 \times 4} = 14{,}000 \text{ psi}$$

The AISI 1018 key can be used since it is within allowable stresses listed in Table 11.4 (allowable Ss = 7,500; allowable Sc = 15,000).

Note: If shaft had been 2-$\frac{3}{4}$" diameter (4" hub), the key would be $\frac{5}{8}$" × $\frac{5}{8}$", *Ss* = 9,200 psi, *Sc* = 18,400 psi, and a heat-treated key of AISI 1045 would have been required (allowable *Ss* = 15,000, allowable *Sc* = 30,000).

Shaft Stress Calculations

Torsional stresses are developed when power is transmitted through shafts. In addition, the tooth loads of gears mounted on shafts create bending stresses. Shaft design, therefore, is based on safe limits of torsion and bending.

To determine minimum shaft diameter in inches:

$$\text{Minimum shaft diameter} = \sqrt[3]{\frac{hp \times 321000}{rpm \times \text{Allowable stress}}}$$

Example:

$$hp = 300$$

$$rpm = 30$$

$$\text{Material} = 225\ \text{Brinell}$$

From Figure 11.20 at 225 Brinell, allowable torsion = 8,000 psi

$$\text{Minimum shaft diameter} = \sqrt[3]{\frac{300 \times 321000}{30 \times 8000}} = \sqrt[3]{402} = 7.38 \text{ inches}$$

From Table 11.5, note that the cube of 7$\frac{1}{4}$" is 381, which is too small (i.e., <402) for this example. The cube of 7$\frac{1}{2}$" is 422, which is large enough.

To determine shaft stress, psi:

$$\text{Shaft stress} = \frac{hp \times 321{,}000}{rpm \times d^3}$$

Example: Given 7$\frac{1}{2}$" shaft for 300 hp at 30 rpm

$$\text{Shaft stress} = \frac{300 \times 321{,}000}{30 \times (7 - 1/2)^3} = 7{,}600 \text{ psi}$$

Note: The 7$\frac{1}{4}$" diameter shaft would be stressed to 8420 psi.

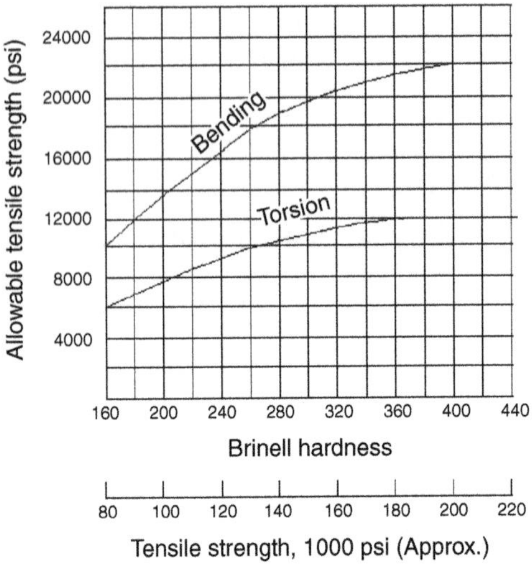

Figure 11.20 *Allowable stress as a function of Brinell hardness*

Table 11.5 *Shaft diameters (inches) and their cubes (cubic inches)*

D	D^3	D	D^3	D	D^3
1	1.00	5	125.0	9	729
1 1/4	1.95	5 1/4	145	9 1/2	857
1 1/2	3.38	5 1/2	166.4	10	1000
1 3/4	5.36	5 3/4	190.1	10 1/2	1157
2	8.00	6	216	11	1331
2 1/4	11.39	6 1/4	244	11 1/2	1520
2 1/2	15.63	6 1/2	275	12	1728
2 3/4	20.80	6 3/4	308	12 1/2	1953
3	27.00	7	343	13	2197
3 1/4	34.33	7 1/4	381	14	2744
3 1/2	42.88	7 1/2	422	15	3375
3 3/4	52.73	7 3/4	465	16	4096
4	64.00	8	512	17	4913
4 1/4	76.77	8 1/4	562	18	5832
4 1/2	91.13	8 1/2	614	19	6859
4 3/4	107.2	8 3/4	670	20	8000

Source: The Falk Corporation

12 Dust Collectors

The basic operations performed by dust-collection devices are: (1) separating particles from the gas stream by deposition on a collection surface, (2) retaining the deposited particles on the surface until removal, and (3) removing the deposit from the surface for recovery or disposal.

The separation step requires: (1) application of a force that produces a differential motion of the particles relative to the gas, and (2) sufficient gas-retention time for the particles to migrate to the collecting surface. Most dust-collections systems are comprised of a pneumatic-conveying system and some device that separates suspended particulate matter from the conveyed air stream. The more common systems use either filter media (e.g., fabric bags) or cyclonic separators to separate the particulate matter from air.

Baghouses

Fabric-filter systems, commonly called *bag-filter* or *baghouse systems*, are dust-collection systems in which dust-laden air is passed through a bag-type filter. The bag collects the dust in layers on its surface, and the dust layer itself effectively becomes the filter medium. Because the bag's pores are usually much larger than those of the dust-particle layer that forms, the initial efficiency is very low. However, it improves once an adequate dust layer forms. Therefore, the potential for dust penetration of the filter media is extremely low except during the initial period after startup, bag change, or during the fabric-cleaning, or blow-down, cycle.

The principle mechanisms of disposition in dust collectors are: (1) gravitational deposition, (2) flow-line interception, (3) inertial deposition, (4) diffusion deposition, and (5) electrostatic deposition. During the initial operating period, particle deposition takes place mainly by inertial and flow-line interception, diffusion, and gravity. Once the dust layer has been fully established, sieving is probably the dominant deposition mechanism.

Configuration

A baghouse system consists of the following: a pneumatic-conveyor system, filter media, a back-flush cleaning system, and a fan or blower to provide airflow.

Pneumatic Conveyor

The primary mechanism for conveying dust-laden air to a central collection point is a system of pipes or ductwork that functions as a pneumatic conveyor. This system gathers dust-laden air from various sources within the plant and conveys it to the dust-collection system.

Dust-Collection System

Design and configuration of the dust-collection system varies with the vendor and the specific application. Generally, a system consists of either a single large hopper-like vessel or a series of hoppers with a fan or blower affixed to the discharge manifold. Inside the vessel is an inlet manifold that directs the incoming air or gas to the dirty side of the filter media or bag. A plenum, or divider plate, separates the dirty and clean sides of the vessel.

Filter media, usually long cylindrical tubes or bags, are attached to the plenum. Depending on the design, the dust-laden air or gas may flow into the cylindrical filter bag and exit to the clean side, or it may flow through the bag from its outside and exit through the tube's opening. Figure 12.1 illustrates a typical baghouse configuration.

Fabric-filter designs fall into three types, depending on the method of cleaning used: (1) shaker-cleaned, (2) reverse-flow-cleaned, and (3) reverse-pulse-cleaned.

Shaker-Cleaned Filter

The open lower ends of shaker-cleaned filter bags are fastened over openings in the tube sheet that separate the lower, dirty-gas inlet chamber from the upper clean-gas chamber. The bags are suspended from supports, which are connected to a shaking device.

The dirty gas flows upward into the filter bag, and the dust collects on the inside surface. When the pressure drop rises to a predetermined upper limit due to dust accumulation, the gas flow is stopped and the shaker is operated. This process dislodges the dust, which falls into a hopper located below the tube sheet.

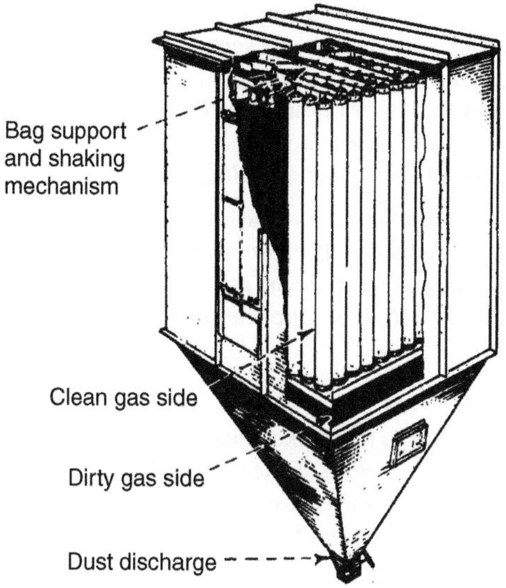

Figure 12.1 *A typical baghouse*

For continuous operation, the filter must be constructed with multiple compartments. This is necessary so that individual compartments can be sequentially taken offline for cleaning while the other compartments continue to operate.

Ordinary shaker-cleaned filters may be cleaned every fifteen minutes to eight hours, depending on the service conditions. A manometer connected across the filter is used to determine the pressure drop, which indicates when the filter should be shaken. Fully automatic filters may be shaken every two minutes, but bag maintenance is greatly reduced if the time between shakings can be increased to 15 to 20 minutes.

The determining factor in the frequency of cleaning is the pressure drop. A differential-pressure switch can serve as the actuator in automatic cleaning applications. Cyclone precleaners are sometimes used to reduce the dust load on the filter or to remove large particles before they enter the bag.

It is essential to stop the gas flow through the filter during shaking in order for the dust to fall off. With very fine dust, it may be necessary to equalize the pressure across the cloth. In practice, this can be accomplished without interrupting continuous operation by removing one section from service at a time. With automatic filters, this operation involves closing the dirty-gas inlet dampers, shaking the filter units either pneumatically or mechanically, and reopening the dampers. In some cases, a reverse flow of clean gas through the filter is used to augment the shaker-cleaning process.

The gas entering the filter must be kept above its dew point to avoid water-vapor condensation on the bags, which will cause plugging. However, fabric filters have been used successfully in steam atmospheres, such as those encountered in vacuum dryers. In these applications, the housing is generally steam-cased.

Reverse-Flow-Cleaned Filter

Reverse-flow-cleaned filters are similar to the shaker-cleaned design, except the shaker mechanism is eliminated. As with shaker-cleaned filters, compartments are taken offline sequentially for cleaning. The primary use of reverse-flow cleaning is in units using fiberglass-fabric bags at temperatures above 150°C (300°F).

After the dirty-gas flow is stopped, a fan forces clean gas through the bags from the clean-gas side. The superficial velocity of the gas through the bag is generally 1.5 to 2.0 feet per minute, or about the same velocity as the dirty-gas inlet flow. This flow of clean gas partially collapses the bag and dislodges the collected dust, which falls to the hopper. Rings are usually sewn into the bags at intervals along their length to prevent complete collapse, which would obstruct the fall of the dislodged dust.

Reverse-Pulse-Cleaned Filter

In the reverse-pulse-cleaned filter, the bag forms a sleeve drawn over a cylindrical wire cage, which supports the fabric on the clean-gas side (i.e., inside) of the bag. The dust collects on the outside of the bag.

A venturi nozzle is located in the clean-gas outlet from each bag, which is used for cleaning. A jet of high-velocity air is directed through the venturi nozzle and into the bag, which induces clean gas to pass through the fabric to the dirty side. The high-velocity jet is released in a short pulse, usually about 100 milliseconds, from a compressed air line by a solenoid-controlled valve. The pulse of air and clean gas expand the bag and dislodge the collected dust. Rows of bags are cleaned in a timed sequence by programmed operation of the solenoid valves. The pressure of the pulse must be sufficient to dislodge the dust without cessation of gas flow through the baghouse.

It is common practice to clean the bags online without stopping the flow of dirty gas into the filter. Therefore, reverse-pulse bag filters are often built without multiple compartments. However, investigations have shown that a large fraction of the dislodged dust redeposits on neighboring bags rather than falling to the dust hopper.

As a result, there is a growing trend to offline clean reverse-pulse filters by using bags with multiple compartments. These sections allow the outlet-gas plenum serving a particular section to be closed off from the clean-gas exhaust, thereby stopping the flow of inlet gas. On the dirty side of the tube sheet, the isolated section is separated by partitions from the neighboring sections where filtration continues. Sections of the filter are cleaned in rotation as with shaker and reverse-flow filters.

Some manufacturers design bags for use with relatively low-pressure air (i.e., 15 psi) instead of the normal 100 psi air. This allows them to eliminate the venturi tubes for clean-gas induction. Others have eliminated the separate jet nozzles located at the individual bags in favor of a single jet to pulse air into the outlet-gas plenum.

Reverse-pulse filters are typically operated at higher filtration velocities (i.e., air-to-cloth ratios) than shaker or reverse-flow designs. Filtration velocities may range from three to fifteen feet per minute in reverse-pulse applications, depending on the dust being collected. However, the most the commonly used range is four to five feet per minute.

The frequency of cleaning depends on the nature and concentration of the dust. Typical cleaning intervals vary from about two to fifteen minutes. However, the cleaning action of the pulse is so effective that the dust layer may be completely removed from the surface of the fabric. Consequently, the fabric itself must serve as the principal filter medium for a substantial part of the filtration cycle, which decreases cleaning efficiency. Because of this, woven fabrics are unsuitable for use in these devices, and felt-type fabrics are used instead. With felt filters, although the bulk of the dust is still removed, the fabric provides an adequate level of dust collection until the dust layer reforms.

Cleaning System

As discussed in the preceding section, filter bags must be periodically cleaned to prevent excessive buildup of dust and to maintain an acceptable pressure drop across the filters. Two of the three designs discussed, reverse-flow and reverse-pulse, depend on an adequate supply of clean air or gas to provide this periodic cleaning. Two factors are critical in these systems: the clean-gas supply and the proper cleaning frequency.

Clean-Gas Supply

Most applications that use the *reverse-flow* cleaning system use ambient air as the primary supply of clean gas. A large fan or blower draws ambient air into the clean side of the filter bags. However, unless inlet filters properly condition the air, it may contain excessive dirt loads that can affect the bag life and efficiency of the dust-collection system.

In *reverse-pulse* applications, most plants rely on plant-air systems as the source for the high-velocity pulses required for cleaning. In many cases, however, the plant-air system is not sufficient for this purpose. Although the pulses required are short (i.e., 100 milliseconds or less), the number and frequency can deplete the supply. Therefore, care must be taken to ensure that both sufficient volume and pressure are available to achieve proper cleaning.

Cleaning Frequency

Proper operation of a baghouse, regardless of design, depends on frequent cleaning of the filter medium. The system is designed to operate within a specific range of pressure drops that defines clean and fully loaded filter media. The cleaning frequency must assure that the maximum recommended pressure drop is not exceeded.

This can be a real problem for baghouses that rely on automatic timers to control cleaning frequency. The use of a timing function to control cleaning frequency is not recommended unless the dust load is known to be consistent. A better approach is to use differential-pressure gauges to physically measure the pressure drop across the filter medium to trigger the cleaning process based on preset limits.

Fan or Blower

All baghouse designs use some form of fan, blower, or centrifugal compressor to provide the dirty-air flow required for proper operation. In most cases, these units are installed on the clean side of the baghouse to draw the dirty air through the filter media.

Since these units provide the motive power required to transport and collect the dust-laden air, their operating condition is critical to the baghouse system. The type and size of air-moving unit varies with the baghouse type and design. Refer to the O&M manuals, for specific design criteria for these critical units.

Performance

The primary measure of baghouse-system performance is its ability to consistently remove dust and other particulate matter from the dirty-air stream. Pressure drop and collection efficiency determine the effectiveness of these systems.

Pressure Drop

The filtration, or superficial face, velocities used in fabric filters are generally in the range of one to ten feet per minute, depending on the type of fabric, fabric supports, and cleaning methods used. In this range, pressure drops conform to Darcy's law for streamline flow in porous media, which states that the pressure drop is directly proportional to the flow rate. The pressure drop across the fabric media and the dust layer may be expressed by:

$$\Delta p = K_1 V_f + K_2 \omega V_f$$

Where:

Δp = pressure drop (inches of water)
V_f = superficial velocity through filter (feet/minute)
ω = dust loading on filter (lbm/ft^2)
K_1 = resistance coefficient for *conditioned* fabric (inches of water/foot/minute)
K_2 = resistance coefficient for dust layer (inches of water/lbm/foot/minute)

Conditioned fabric maintains a relatively consistent dust-load deposit following a number of filtration and cleaning cycles. K_1 may be more than 10 times the value of the resistance coefficient for the original clean fabric. If the depth of the dust layer on the fabric is greater than about $\frac{1}{16}$" (which corresponds to a fabric dust loading on the order of 0.1 lbm/ft^2), the pressure drop across the fabric, including the dust in the pores, is usually negligible relative to that across the dust layer alone.

In practice, K_1 and K_2 are measured directly in filtration experiments. These values can be corrected for temperature by multiplying by the ratio of the gas viscosity at the desired condition to the gas viscosity at the original experimental condition.

Collection Efficiency

Under controlled conditions (e.g., in the laboratory), the inherent collection efficiency of fabric filters approaches 100%. In actual operation, it is determined by several variables, in particular the properties of the dust to be removed, choice of filter fabric, gas velocity, method of cleaning, and cleaning cycle. Inefficiency usually results from bags that are poorly installed, torn, or stretched from excessive dust loading and excessive pressure drop.

Installation

Most baghouse systems are provided as complete assemblies by the vendor. While the unit may require some field assembly, the vendor generally provides the structural supports, which in most cases are adequate. The only controllable installation factors that may affect performance are the foundation and connections to pneumatic conveyors and other supply systems.

Foundation

The foundation must support the weight of the baghouse. In addition, it must absorb the vibrations generated by the cleaning system. This is especially true when using the shaker-cleaning method, which can generate vibrations that can adversely affect the structural supports, foundation, and adjacent plant systems.

Connections

Efficiency and effectiveness depend on leak-free connections throughout the system. Leaks reduce the system's ability to convey dust-laden air to the baghouse. One potential source for leaks is improperly installed filter bags. Because installation varies with the type of bag and baghouse design, consult the vendor's O&M manual for specific instructions.

Operating Methods

The guidelines provided in the vendor's O&M manual should be the primary reference for proper baghouse operation. Vendor-provided information should be used because there are not many common operating guidelines among the various configurations. The only general guidelines that are applicable to most designs are cleaning frequency and inspection and replacement of filter media.

Cleaning

As previously indicated, most bag-type filters require a precoat of particulates before they can effectively remove airborne contaminants. However, particles can completely block air flow if the filter material becomes overloaded. Therefore, the primary operating criterion is to maintain the efficiency of the filter media by controlling the cleaning frequency.

Most systems use a time-sequence to control the cleaning frequency. If the particulate load entering the baghouse is constant, this approach would be valid. However, the incoming load generally changes constantly. As a result, the straight time-sequence methodology does not provide the most efficient mode of operation.

Operators should monitor the differential-pressure gauges that measure the total pressure drop across the filter media. When the differential pressure reaches the maximum recommended level (data provided by the vendor), the operator should override any automatic timer controls and initiate the cleaning sequence.

Inspecting and Replacing Filter Media

Filter media used in dust-collection systems are prone to damage and abrasive wear. Therefore, regular inspection and replacement is needed to ensure continuous, long-term performance. Any damaged, torn, or improperly sealed bags should be removed and replaced. One of the more common problems associated with baghouses is improper installation of filter media. Therefore, it is important to follow the instructions provided by the vendor. If the filter bags are not properly installed and sealed, overall efficiency and effectiveness are significantly reduced.

Cyclone Separators

A widely used type of dust-collection equipment is the cyclone separator. A cyclone is essentially a settling chamber in which gravitational acceleration is replaced by centrifugal acceleration. Dust-laden air or gas enters a cylindrical or conical chamber tangentially at one or more points and leaves through a central opening. The dust particles, by virtue of their inertia, tend to move toward the outside separator wall from where they are led into a receiver. Under common operating conditions, the centrifugal separating force or acceleration may range from five times gravity in very large diameter, low-resistance cyclones to 2,500 times gravity in very small, high-resistance units.

Within the range of their performance capabilities, cyclones are one of the least expensive dust-collection systems. Their major limitation is that, unless very small units are used, efficiency is low for particles smaller than five microns. Although cyclones may be used to collect particles larger than 200 microns, gravity-settling chambers or simple inertial separators are usually satisfactory and less subject to abrasion.

Configuration

The internal configuration of a cyclone separator is relatively simple. Figure 12.2 illustrates a typical cross-section of a cyclone separator, which

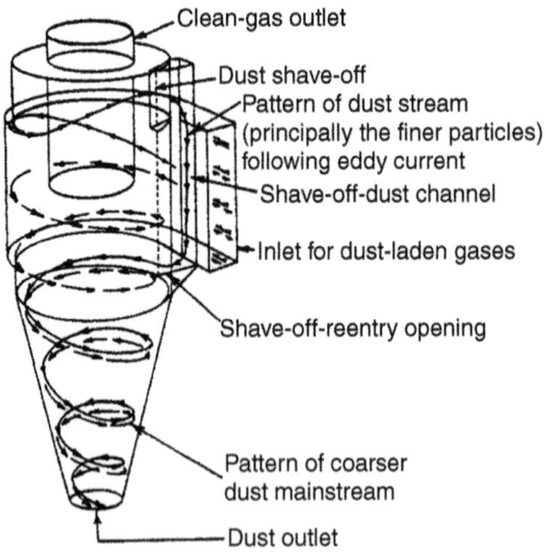

Figure 12.2 *Flow pattern through a typical cyclone separator*

consists of the following segments:

- Inlet area that causes the gas to flow tangentially;
- Cylindrical transition area;
- Decreasing taper that increases the air velocity as the diameter decreases;
- Central return tube to direct the dust-free air out the discharge port.

Particulate material is forced to the outside of the tapered segment and collected in a drop-leg located at the dust outlet. Most cyclones have a rotor-lock valve affixed to the bottom of the drop-leg. This is a motor-driven valve that collects the particulate material and discharges it into a disposal container.

Performance

Performance of a cyclone separator is determined by flow pattern, pressure drop, and collection efficiency.

Flow Pattern

The path the gas takes in a cyclone is through a double vortex that spirals the gas downward at the outside and upward at the inside. When the gas enters the cyclone, the tangential component of its velocity, V_{ct}, increases with the decreasing radius as expressed by:

$$V_{ct} \approx r^{-n}$$

In this equation, r is the cyclone radius and n is dependent on the coefficient of friction. Theoretically, in the absence of wall friction, n should equal 1.0. Actual measurements, however, indicate that n ranges from 0.5 to 0.7 over a large portion of the cyclone radius. The spiral velocity in a cyclone may reach a value several times the average inlet-gas velocity.

Pressure Drop

The pressure drop and the friction loss through a cyclone are most conveniently expressed in terms of the velocity head based on the immediate inlet area. The inlet velocity head, h_{vt}, which is expressed in inches of water, is related to the average inlet-gas velocity and density by:

$$h_{vt} = 0.0030\rho V_c^2$$

Where:

h_{vt} = Inlet-velocity head (inches of water)
ρ = Gas density (lb/ft^3)
V_c = Average inlet-gas velocity (ft/sec)

The cyclone friction loss, F_{cv}, is a direct measure of the static pressure and power that a fan must develop. It is related to the pressure drop by:

$$F_{cv} = \Delta p_{cv} + 1 - \left(\frac{4A_c}{\pi D_e^2}\right)^2$$

Where:

F_{cv} = Friction loss (inlet-velocity heads)
Δp_{cv} = Pressure drop through the cyclone (inlet-velocity heads)
A_c = Area of the cyclone (square feet)
D_e = Diameter of the gas exit (feet)

The friction loss through cyclones may range from 1 to 20 inlet-velocity heads, depending on its geometric proportions. For a cyclone of specific geometric proportions, F_{cv} and Δp_{cv} are essentially constant and independent of the actual cyclone size.

Collection Efficiency

Since cyclones rely on centrifugal force to separate particulates from the air or gas stream, particle mass is the dominant factor that controls efficiency. For particulates with high densities (e.g., ferrous oxides), cyclones can achieve 99% or better removal efficiencies, regardless of particle size. Lighter particles (e.g., tow or flake) dramatically reduce cyclone efficiency.

These devices are generally designed to meet specific pressure-drop limitations. For ordinary installations operating at approximately atmospheric pressure, fan limitations dictate a maximum allowable pressure drop corresponding to a cyclone inlet velocity in the range of 20 to 70 feet per second. Consequently, cyclones are usually designed for an inlet velocity of 50 feet per second.

Varying operating conditions change dust-collection efficiency only by a small amount. The primary design factor that controls collection efficiency is cyclone diameter. A small-diameter unit operating at a fixed pressure drop

Table 12.1 *Common failure modes of baghouses*

THE CAUSES	THE PROBLEM: Continuous release of dust-laden air	Intermittent release of dust-laden air	Loss of plant air pressure	Blow-down ineffective	Insufficient capacity	Excessive differential pressure	Fan/blower motor trips	Fan has high vibration	Premature bag failures	Differential pressure too low	Chronic plugging of bags
Bag material incompatible for application									●		●
Bag plugged						●	●	●			
Bag torn or improperly installed	●							●	●	●	
Baghouse undersized					●		●				●
Blow-down cycle interval too long						●	●				
Blow-down cycle time failed or damaged						●	●				
Blow-down nozzles plugged						●					
Blow-down pilot valve failed to open (solenoid failure)			●			●					
Dust load exceeds capacity											●
Excessive demand			●								
Fan/blower not operating properly					●						
Improper or inadequate lubrication								●			
Leaks in ductwork or baghouse	●				●						
Misalignment of fan and motor								●			
Moisture content too high											●

Table 12.1 *continued*

THE CAUSES	THE PROBLEM: Continuous release of dust-laden air	Intermittent release of dust-laden air	Loss of plant air pressure	Blow-down ineffective	Insufficient capacity	Excessive differential pressure	Fan/blower motor trips	Fan has high vibration	Premature bag failures	Differential pressure too low	Chronic plugging of bags
Not enough blow-down air (pressure and volume)			•	•		•					
Not enough dust layer on filter bags	•	•						•		•	
Piping/valve leaks			•								
Plate-out (dust build-up on fan's rotor)								•			
Plenum cracked or seal defective	•		•							•	
Rotor imbalanced								•			
Ruptured blow-down diaphrams			•	•		•					
Suction ductwork blocked or plugged					•						

has a higher efficiency than a large-diameter unit. Reducing the gas-outlet duct diameter also increases the collection efficiency.

Installation

As in any other pneumatic-conveyor system, special attention must be given to the piping or ductwork used to convey the dust-laden air or gas. The inside surfaces must be smooth and free of protrusions that affect the flow pattern. All bends should be gradual and provide a laminar-flow path for the gas.

Table 12.2 *Common failure modes of cyclonic separators*

	THE PROBLEM								
THE CAUSES	Continuous release of dust-laden air	Intermittent release of dust-laden air	Cyclone plugs in inlet chamber	Cyclone plugs in dust removal section	Rotor-lock valve fails to turn	Excessive differential pressure	Differential pressure too low	Rotor-lock valve leaks	Fan has high vibration
Clearance set wrong								•	
Density and size distribution of dust too high				•	•	•			•
Density and size distribution of dust too low	•	•							
Dust load exceeds capacity	•	•			•				•
Excessive moisture in incoming air			•						
Foreign object lodged in valve					•				
Improper drive-train adjustments					•				
Improper lubrication					•				
Incoming air velocity too high						•			
Incoming air velocity too low	•	•	•				•		
Internal wear or damage								•	
Large contaminates in incoming air stream			•		•				
Prime mover (fan, blower) malfunctioning	•	•				•	•		•
Rotor-lock valve turning too slow	•	•		•					
Seals damaged									

Troubleshooting

This section identifies common problems and their causes for baghouse and cyclonic separator dust-collection systems.

Baghouses

Table 12.1 lists the common failure modes for baghouses. This guide may be used for all such units that use fabric filter bags as the primary dust-collection media.

Cyclonic Separators

Table 12.2 identifies the failure modes and their causes for cyclonic separators. Since there are no moving parts within a cyclone, most of the problems associated with this type of system can be attributed to variations in process parameters, such as flow rate, dust load, dust composition (i.e., density, size, etc.), and ambient conditions (i.e., temperature, humidity, etc.).

13 Fans, Blowers, and Fluidizers

Technically, fans and blowers are two separate types of devices that have a similar function. However, the terms are often used interchangeably to mean any device that delivers a quantity of air or gas at a desired pressure. Differences between these two devices are their rotating elements and their discharge-pressure capabilities. Fluidizers are identical to single-stage, screw-type compressors or blowers.

Centrifugal Fans

Centrifugal fans are one of the most common machines used in industry. They utilize a rotating element with blades, vanes, or propellers to extract or deliver a specific volume of air or gas. The rotating element is mounted on a rotating shaft that must provide the energy required to overcome inertia, friction, and other factors that restrict or resist air or gas flow in the application. They are generally low-pressure machines designed to overcome friction and either suction or discharge-system pressure.

Configuration

The type of rotating element or wheel that is used to move the air or gas can classify centrifugal fans. The major classifications are propeller and axial. Axial fans also can be further differentiated by the blade configurations.

Propeller

This type of fan consists of a propeller, or paddle wheel, mounted on a rotating shaft within a ring, panel, or cage. The most widely used propeller fans are found in light- or medium-duty functions, such as ventilation units where air can be moved in any direction. These fans are commonly used in wall mountings to inject air into, or exhaust air from, a space. Figure 13.1 illustrates a belt-driven propeller fan appropriate for medium-duty applications.

This type of fan has a limited ability to boost pressure. Its use should be limited to applications where the total resistance to flow is less than

Figure 13.1 *Belt-driven propeller fan for medium duty applications*

one inch of water. In addition, it should not be used in corrosive environments or where explosive gases are present.

Axial

Axial fans are essentially propeller fans that are enclosed within a cylindrical housing or shroud. They can be mounted inside ductwork or a vessel housing to inject or exhaust air or gas.

These fans have an internal motor mounted on spokes or struts to centralize the unit within the housing. Electrical connections and grease fittings are mounted externally on the housing. Arrow indicators on the housing show the direction of airflow and rotation of the shaft, which enables the unit to be correctly installed in the ductwork. Figure 13.2 illustrates an inlet end of a direct-connected, tube-axial fan.

This type of fan should not be used in corrosive or explosive environments, since the motor and bearings cannot be protected. Applications where concentrations of airborne abrasives are present should also be avoided.

Axial fans use three primary types of blades or vanes: backward-curved, forward-curved, and radial. Each type has specific advantages and disadvantages.

Figure 13.2 *Inlet end of a direct-connected tube-axial fan*

Backward-Curved Blades

The backward-curved blade provides the highest efficiency and lowest sound level of all axial-type, centrifugal fan blades. Advantages include:

- Moderate to high volumes
- Static pressure range up to approximately 30 inches of water (gauge)
- Highest efficiency of any type of fan
- Lowest noise level of any fan for the same pressure and volumetric requirements
- Self-limiting brake horsepower (BHP) characteristics (Motors can be selected to prevent overload at any volume, and the BHP curve rises to a peak and then declines as volume increases.)

The limitations of backward-curved blades are:

- Weighs more and occupies considerably more space than other designs of equal volume and pressure
- Large wheel width
- Not to be used in dusty environments or where sticky or stringy materials are used because residues adhering to the blade surface cause imbalance and eventual bearing failure

Forward-Curved Blades

This design is commonly referred to as a squirrel-cage fan. The unit has a wheel with a large number of wide, shallow blades; a very large intake area relative to the wheel diameter; and a relatively slow operational speed. The advantages of forward-curved blades include:

- Excellent for any volume at low to moderate static pressure using clean air
- Occupies approximately same space as backward-curved blade fan
- More efficient and much quieter during operation than propeller fans for static pressures above approximately one inch of water (gauge)

The limitations of forward-curved blades include:

- Not as efficient as backward-curved blade fans
- Should not be used in dusty environments or handle sticky or stringy materials that could adhere to the blade surface
- BHP increases as this fan approaches maximum volume, as opposed to backward-curved blade centrifugal fans, which experience a decrease in BHP as they approach maximum volume

Radial Blades

Industrial exhaust fans fall into this category. The design is rugged and may be belt-driven or directly driven by a motor. The blade shape varies considerably from flat surfaces to various bent configurations to increase efficiency slightly or to suit particular applications. The advantages of radial-blade fans include:

- Best suited for severe duty, especially when fitted with flat radial blades
- Simple construction that lends itself to easy field maintenance
- Highly versatile industrial fan that can be used in extremely dusty environments as well as clean air
- Appropriate for high-temperature service
- Handles corrosive or abrasive materials.

The limitations of radial-blade fans include:

- Lowest efficiency in centrifugal-fan group
- Highest sound level in centrifugal-fan group
- BHP increases as fan approaches maximum volume

Performance

A fan is inherently a constant-volume machine. It operates at the same volumetric flow rate (i.e., cubic feet per minute) when operating in a fixed system at a constant speed, regardless of changes in air density. However, the pressure developed and the horsepower required varies directly with the air density.

The following factors affect centrifugal-fan performance: brake horsepower, fan capacity, fan rating, outlet velocity, static efficiency, static pressure, tip speed, mechanical efficiency, total pressure, velocity pressure, natural frequency, and suction conditions. Some of these factors are used in the mathematical relationships that are referred to as Fan Laws.

Brake Horsepower

Brake horsepower (BHP) is the power input required by the fan shaft to produce the required volumetric flow rate (cfm) and pressure.

Fan Capacity

The fan capacity (FC) is the volume of air moved per minute by the fan (cfm). Note: the density of air is 0.075 pounds per cubic foot at atmospheric pressure and 68°F.

Fan Rating

The fan rating predicts the fan's performance at one operating condition, which includes the fan size, speed, capacity, pressure, and horsepower.

Outlet Velocity

The outlet velocity (OV, feet per minute) is the number of cubic feet of gas moved by the fan per minute divided by the inside area of the fan outlet, or discharge area, in square feet.

Static Efficiency

Static efficiency (SE) is *not* the true mechanical efficiency, but it is convenient to use in comparing fans. This is calculated by the following equation:

$$\text{Static Efficiency (SE)} = \frac{0.000157 \times \text{FC} \times \text{SP}}{\text{BHP}}$$

Static Pressure

Static pressure (SP) generated by the fan can exist whether the air is in motion or is trapped in a confined space. SP is always expressed in inches of water (gauge).

Tip Speed

The tip speed (TS) is the peripheral speed of the fan wheel in feet per minute (fpm).

$$\text{Tip Speed} = \text{Rotor Diameter} \times \pi \times \text{rpm}$$

Mechanical Efficiency

True mechanical efficiency (ME) is equal to the total input power divided by the total output power.

Total Pressure

Total pressure (TP), inches of water (gauge) is the sum of the velocity pressure and static pressure.

Velocity Pressure

Velocity pressure (VP) is produced by the fan when the air is moving. Air having a velocity of 4,000 fpm exerts a pressure of one inch of water (gauge) on a stationary object in its flow path.

Natural Frequency

General-purpose fans are designed to operate below their first natural frequency. In most cases, the fan vendor will design the rotor-support system so that the rotating element's first critical speed is between 10 and 15% above the rated running speed. While this practice is questionable, it is acceptable if the design speed and rotating-element mass are maintained. However, if either of these two factors changes, there is a high probability that serious damage or premature failure will result.

Inlet-Air Conditions

As with centrifugal pumps, fans require stable inlet conditions. Ductwork should be configured to ensure an adequate volume of clean air or gas, stable inlet pressure, and laminar flow. If the supply air is extracted from the environment, it is subject to variations in moisture, dirt content, barometric pressure, and density. However, these variables should be controlled as much as possible. As a minimum, inlet filters should be installed to minimize the amount of dirt and moisture that enters the fan.

Excessive moisture and particulates have an extremely negative impact on fan performance and cause two major problems: abrasion or tip wear and plate-out. High concentrations of particulate matter in the inlet air act as abrasives that accelerate fan-rotor wear. In most cases, however, this wear is restricted to the high-velocity areas of the rotor, such as the vane or blade tips, but can affect the entire assembly.

Plate-out is a much more serious problem. The combination of particulates and moisture can form "glue" that binds to the rotor assembly. As this contamination builds up on the rotor, the assembly's mass increases, which reduces its natural frequency. If enough plate-out occurs, the fan's rotational speed may coincide with the rotor's reduced natural frequency. With a strong energy source like the running speed, excitation of the rotor's natural frequency can result in catastrophic fan failure. Even if catastrophic failure does not occur, the vibration energy generated by the fan may cause bearing damage.

Fan Laws

The mathematical relationships referred to as *fan laws* can be useful when applied to fans operating in a fixed system or to geometrically similar fans.

However, caution should be exercised when using these relationships. They only apply to *identical* fans and applications. The basic laws are:

- Volume in cubic feet per minute (cfm) varies directly with the rotating speed (rpm)
- Static pressure varies with the rotating speed squared (rpm^2)
- Brake horsepower (BHP) varies with the speed cubed (rpm^3)

The fan-performance curves shown in Figures 13.3 and 13.4 show the performance of the same fan type designed for different volumetric-flow rates, operating in the same duct system handling air at the same density.

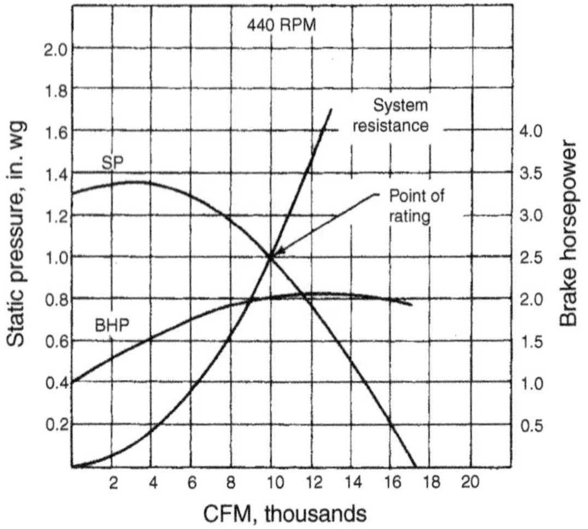

Figure 13.3 *Fan-performance curve 1*

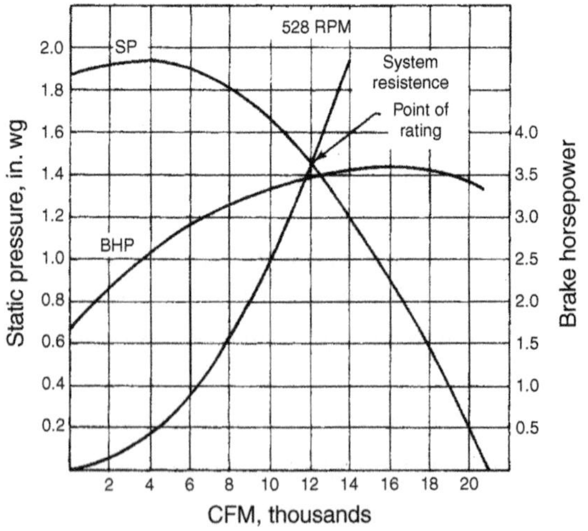

Figure 13.4 *Fan-performance curve 2*

Curve 1 is for a fan designed to handle 10,000 cfm in a duct system whose calculated system resistance is determined to be 1" water (gauge). This fan will operate at the point where the fan pressure (SP) curve intersects the system resistance curve (TSH). This intersection point is called the *Point of Rating*. The fan will operate at this point provided the fan's speed remains constant and the system's resistance does not change. The system-resistance curve illustrates that the resistance varies as the square of the volumetric flow rate (cfm). The BHP of the fan required for this application is 2 hp.

Curve 2 illustrates the situation if the fan's design capacity is increased by 20%, increasing output from 10,000 to 12,000 cfm. Applying the fan laws, the calculations are:

$$\begin{aligned}
\text{New rpm} &= 1.2 \times 440 \\
&= 528 \text{ rpm (20\% increase)} \\
\text{New SP} &= 1.2 \times 1.2 \times 1'' \text{ water (gauge)} \\
&= 1.44'' \text{ (44\% increase)} \\
\text{New TSH} &= \text{New SP} = 1.44'' \\
\text{New BHP} &= 1.2 \times 1.2 \times 1.2 \times 2 \\
&= 1.73 \times 2 \\
&= 3.46 \text{ hp (73\% increase)}
\end{aligned}$$

The curve representing the system resistance is the same in both cases, since the system has not changed. The fan will operate at the same relative point of rating and will move the increased volume through the system. The mechanical and static efficiencies are unchanged.

The increased brake horsepower (BHP) required to drive the fan is a very important point to note. If a 2-hp motor drove the Curve 1 fan, the Curve 2 fan needs a 3.5-hp motor to meet its volumetric requirement.

Centrifugal-fan selection is based on rating values such as air flow, rpm, air density, and cost. Table 13.1 is a typical rating table for a centrifugal fan. Table 13.2 provides air-density ratios.

Installation

Proper fan installation is critical to reliable operation. Suitable foundations, adequate bearing-support structures, properly sized ductwork, and flow-control devices are the primary considerations.

Foundations

As with any other rotating machine, fans require a rigid, stable foundation. With the exception of in-line fans, they must have a concrete footing or pad that is properly sized to provide a stable footprint and prevent flexing of the rotor-support system.

Bearing-Support Structures

In most cases, with the exception of in-line configurations, fans are supplied with a vendor-fabricated base. Bases normally consist of fabricated metal stands that support the motor and fan housing. The problem with

Table 13.1 *Typical rating table for a centrifugal fan*

CFM	OV	1/4" SP		3/8" SP		1/2" SP		5/8" SP		3/4" SP	
		RPM	BHP	RPM	BHP	RPM	BHP	RPM	BHP	RPM	BHP
7458	800	262	0.45	289	0.60	314	0.75	337	0.92	360	1.09
8388	900	281	0.55	308	0.72	330	0.89	351	1.06	372	1.25
9320	1000	199	0.66	325	0.85	347	1.04	368	1.23	387	1.43
10252	1100	319	0.79	343	1.00	365	1.21	385	1.42	403	1.63
11184	1200	338	0.93	362	1.17	383	1.40	402	1.63	420	1.85
12118	1300	358	1.10	381	1.35	402	1.81	421	1.85	438	2.10
13048	1400	379	1.29	401	1.56	421	1.83	439	2.10	456	2.37
13980	1500	401	1.50	420	1.78	440	2.08	458	2.37	475	2.66
14912	1600	422	1.74	441	2.03	459	2.35	477	2.67	494	2.98
15844	1700	444	2.01	462	2.32	479	2.65	496	2.98	513	3.32
18776	1800	467	2.31	483	2.63	499	2.97	516	3.33	532	3.68
17708	1900	489	2.65	504	2.98	520	3.33	536	3.70	551	4.07
18640	2000	512	3.02	526	3.36	541	3.72	556	4.10	571	4.49
19572	2100	535	3.43	548	3.77	562	4.45	576	4.53	590	4.95
20504	2200	558	3.87	570	4.23	584	4.61	597	5.02	610	5.43
21436	2300	582	4.36	593	4.72	605	5.12	618	5.54	631	5.95
22368	2400	605	4.89	616	5.26	627	5.67	640	6.10	652	6.54
23300	2500	628	5.46	639	5.85	650	6.26	661	6.70	673	7.16
24232	2600	652	6.09	662	6.48	672	6.90	683	7.34	694	7.81
25164	2700	576	6.75	685	7.15	695	7.58	705	8.04	715	8.52
26096	2800	700	7.47	708	7.88	718	8.32	727	8.78	738	9.27
27028	2900	723	8.24	732	8.66	741	9.11	750	9.58	760	10.08

Continued

Table 13.1 *continued*

7/8" SP		1" SP		11/4" SP		11/2" SP		13/4" SP	
RPM	BHP	RPM	BHP	RPM	BHP	RPM	BHP	RPM	BHP
382	1.27	403	1.46	444	1.85	483	2.28	520	2.73
393	1.44	413	1.63	451	2.05	488	2.49	523	2.96
406	1.63	425	1.83	461	2.27	493	2.72	529	3.21
421	1.84	439	2.06	473	2.51	305	2.99	537	3.50
438	2.08	454	2.31	486	2.79	517	3.29	547	3.81
455	2.34	471	2.55	501	3.09	531	3.62	359	4.16
473	2.63	488	2.90	517	3.43	545	3.98	572	4.54
491	2.94	506	3.23	534	3.75	581	4.37	587	4.96
509	3.28	524	3.56	552	4.19	578	4.79	603	5.41
528	3.64	542	3.97	570	4.61	395	5.25	619	5.80
547	4.03	561	4.38	588	5.06	613	5.74	637	6.42
556	4.45	580	4.81	606	5.54	631	6.25	654	6.98
585	4.89	599	5.28	625	6.05	649	6.81	672	7.57
604	5.36	618	5.78	644	6.59	668	7.40	690	8.19
624	5.87	637	6.30	663	7.16	686	8.01	708	8.85
644	6.41	657	6.86	682	7.77	705	8.65	727	9.54
664	6.99	677	7.46	701	6.41	724	9.35	746	10.27
685	7.63	687	8.10	721	9.09	743	10.07	765	11.04
706	8.30	717	8.77	740	9.80	762	10.83	784	11.84
727	9.01	738	9.53	760	10.58	782	11.63	803	12.69
748	9.78	759	10.30	780	11.35	801	12.46	822	13.57
770	10.60	780	11.13	800	12.20	821	13.35	841	14.49

most of the fabricated bases is that they lack the rigidity and stiffness to prevent flexing or distortion of the fan's rotating element. The typical support structure is comprised of relatively light-gauge material ($\frac{3}{16}$") and does not have the cross-bracing or structural stiffeners needed to prevent distortion of the rotor assembly. Because of this limitation, many plants fill the support structure with concrete or other solid material.

However, this approach does little to correct the problem. When the concrete solidifies, it pulls away from the sides of the support structure. Without direct bonding and full contact with the walls of the support structure, stiffness is not significantly improved.

The best solution to this problem is to add cross-braces and structural stiffeners. If they are properly sized and affixed to the support structure, the stiffness can be improved and rotor distortion reduced.

Ductwork

Ductwork should be sized to provide minimum friction loss throughout the system. Bends, junctions with other ductwork, and any change of direction should provide a clean, direct flow path. All ductwork should be airtight and leak-free to ensure proper operation.

Flow-Control Devices

Fans should always have inlet and outlet dampers or other flow-control devices, such as variable-inlet vanes. Without them, it is extremely difficult to match fan performance to actual application demand. The reason for this difficulty is that there are a number of variables (e.g., usage, humidity,

Table 13.2 *Air-density ratios*

	Altitude, ft above sea level					
	0	1,000	2,000	3,000	4,000	5,000
Air temp.,	Barometric pressure, in. mercury					
°F	29.92	28.86	27.82	26.82	25.84	24.90
70	1.000	0.964	0.930	0.896	0.864	0.832
100	0.946	0.912	0.880	0.848	0.818	0.787
150	0.869	0.838	0.808	0.770	0.751	0.723
200	0.803	0.774	0.747	0.720	0.694	0.668
250	0.747	0.720	0.694	0.669	0.645	0.622
300	0.697	0.672	0.648	0.624	0.604	0.580
350	0.654	0.631	0.608	0.586	0.565	0.544
400	0.616	0.594	0.573	0.552	0.532	0.513
450	0.582	0.561	0.542	0.522	0.503	0.484
500	0.552	0.532	0.513	0.495	0.477	0.459
550	0.525	0.506	0.488	0.470	0.454	0.437
600	0.500	0.482	0.465	0.448	0.432	0.416
650	0.477	0.460	0.444	0.427	0.412	0.397
700	0.457	0.441	0.425	0.410	0.395	0.380

Continued

Table 13.2 *continued*

	Altitude, ft above sea level						
	6,000	7,000	8,000	9,000	10,000	15,000	20,000
Air temp.,	Barometric pressure, in. mercury						
°F	23.98	23.09	22.22	21.39	20.58	16.89	13.75
70	0.801	0.772	0.743	0.714	0.688	0.564	0.460
100	0.758	0.730	0.703	0.676	0.651	0.534	0.435
150	0.696	0.671	0.646	0.620	0.598	0.490	0.400
200	0.643	0.620	0.596	0.573	0.552	0.453	0.369
250	0.598	0.576	0.555	0.533	0.514	0.421	0.344
300	0.558	0.538	0.518	0.498	0.480	0.393	0.321
350	0.524	0.505	0.486	0.467	0.450	0.369	0.301
400	0.493	0.476	0.458	0.440	0.424	0.347	0.283
450	0.466	0.449	0.433	0.416	0.401	0.328	0.268
500	0.442	0.426	0.410	0.394	0.380	0.311	0.254
550	0.421	0.405	0.390	0.375	0.361	0.296	0.242
600	0.400	0.386	0.372	0.352	0.344	0.282	0.230
650	0.382	0.368	0.354	0.341	0.328	0.269	0.219
700	0.366	0.353	0.340	0.326	0.315	0.258	0.210

and temperature) directly affecting the input-output demands for each fan application. Flow-control devices provide the means to adjust fan operation for actual conditions. Figure 13.5 shows an outlet damper with streamlined blades and linkage arranged to move adjacent blades in opposite directions for even throttling.

Airflow controllers must be inspected frequently to ensure that they are fully operable and operate in unison with each other. They also must close tightly. Ensure that the control indicators show the precise position of the vanes in all operational conditions. The "open" and "closed" positions should be permanently marked and visible at all times. Periodic lubrication of linkages is required.

Turn-buckle screws on the linkages for adjusting flow rates should never be moved without first measuring the distance between the set-point markers on each screw. This is important if the adjustments do not produce the desired effect and you wish to return to the original settings.

Figure 13.5 *Outlet damper with streamlined blades and linkage arranged to move adjacent blades in opposite directions for even throttling*

Operating Methods

Because fans are designed for stable, steady state operation, variations in speed or load may have an adverse effect on their operating dynamics. The primary operating method that should be understood is output control. Two methods can be used to control fan output: dampers and fan speed.

Dampers

Dampers can be used to control the output of centrifugal fans within the effective control limits. Centrifugal fans have a finite acceptable control range, typically about 15% below and 15% above their design point. Control variations outside this range severely affect the reliability and useful life of the fan.

The recommended practice is to use an inlet damper rather than a discharge damper for this control function whenever possible. Restricting the inlet with suction dampers can effectively control the fan's output. When using dampers to control fan performance, however, caution should be exercised to ensure that any changes remain within the fan's effective control range.

Fan Speed

Varying fan speed is an effective means of controlling a fan's performance. As defined by the Fan Laws, changing the rotating speed of the fan can directly control both volume and pressure. *However, caution must be used when changing fan speed.* All rotating elements, including fan rotors, have one or more critical speeds. When the fan's speed coincides with one of the critical speeds, the rotor assembly becomes extremely unstable and could fail catastrophically.

In most general-purpose applications, fans are designed to operate between 10 and 15% below their first critical speed. If speed is increased on these fans, there is a good potential for a critical-speed problem. Other applications have fans that are designed to operate between their first and second critical speeds. In this instance, any change up or down may cause the speed to coincide with one of the critical speeds.

Blowers

A blower uses mating helical lobes or screws and is used for the same purpose as a fan. They are normally moderate- to high-pressure devices. Blowers are almost identical both physically and functionally to positive-displacement compressors.

Fluidizers

Fluidizers are identical to single-stage, screw-type compressors or blowers. They are designed to provide moderate- to high-pressure transfer of non-abrasive, dry materials.

Troubleshooting Fans, Blowers, and Fluidizers

Tables 13.3 and 13.4 (below) list the common failure modes for fans, blowers, and fluidizers. Typical problems with these devices include: (1) output below rating, (2) vibration and noise, and (3) overloaded driver bearings.

Centrifugal Fans

Centrifugal fans are extremely sensitive to variations in either suction or discharge conditions. In addition to variations in ambient conditions (i.e., temperature, humidity, etc.), control variables can have a direct effect on fan performance and reliability.

Most of the problems that limit fan performance and reliability are either directly or indirectly caused by improper application, installation, operation, or maintenance. However, the majority are caused by misapplication or poor operating practices. Table 13.4 lists failure modes of centrifugal fans and their causes. Some of the more common failures are aerodynamic instability, plate-out, speed changes, and lateral flexibility.

Aerodynamics Instability

Generally, the control range of centrifugal fans is about 15% above and 15% below its best efficiency point (BEP). When fans are operated outside this range, they tend to become progressively unstable, which causes the fan's rotor assembly and shaft to deflect from their true centerline. This deflection increases the vibration energy of the fan and accelerates the wear rate of bearings and other drive-train components.

Plate-Out

Dirt, moisture, and other contaminants tend to adhere to the fan's rotating element. This buildup, called plate-out, increases the mass of the rotor assembly and decreases its critical speed, the point where the phenomenon referred to as resonance occurs. This occurs because the additional mass affects the rotor's natural frequency. Even if the fan's speed does not change, the change in natural frequency may cause its critical speed (note that

Table 13.3 *Common failure modes of centrifugal fans*

	THE PROBLEM									
THE CAUSES	Insufficient discharge pressure	Intermittent operation	Insufficient capacity	Overheated bearings	Short bearing life	Overload on driver	High vibration	High noise levels	Power demand excessive	Motor trips
Abnormal end thrust				•			•			
Aerodynamic instability		•	•	•	•		•	•		
Air leaks in system	•	•	•							
Bearings improperly lubricated						•	•	•		•
Bent shaft				•	•	•	•		•	
Broken or loose bolts or setscrews				•			•			
Damaged motor							•			
Damaged wheel	•		•	•						
Dampers or variable-inlet not properly adjusted	•		•							
Dirt in bearings				•			•			
Excessive belt tension				•			•			•
External radiated heat				•						
Fan delivering more than rated capacity						•	•			
Fan wheel or driver imbalanced				•			•			

Continued

Table 13.3 *continued*

	THE PROBLEM									
THE CAUSES	Insufficient discharge pressure	Intermittent operation	Insufficient capacity	Overheated bearings	Short bearing life	Overload on driver	High vibration	High noise levels	Power demand excessive	Motor trips
Foreign material in fan causing imbalance (plate-out)				•			•	•		
Incorrect direction of rotation	•		•			•	•			
Insufficient belt rension							•	•		
Loose dampers or variable-inlet vanes							•			
Misaligment of bearings, coupling, wheel, or belts				•		•	•	•	•	
Motor improperly wired						•	•	•		•
Packing too tight or defective stuffing box						•	•		•	•
Poor fan inlet or outlet conditions	•		•							
Specific gravity or density above design						•	•		•	
Speed too high		•		•	•	•	•			•
Speed too low	•	•	•					•		•
Too much grease in ball bearings				•						
Total system head greater than design	•		•	•		•			•	

THE CAUSES	THE PROBLEM									
	Insufficient discharge pressure	Intermittent operation	Insufficient capacity	Overheated bearings	Short bearing life	Overload on driver	High vibration	High noise levels	Power demand excessive	Motor trips
Total system head less than design		•					•			•
Unstable foundation		•		•			•	•		
Vibration transmitted to fan from outside sources				•			•	•		
Wheel binding on fan housing				•		•	•	•		•
Wheel mounted backward on shaft	•		•							
Worn bearings							•	•		
Worn coupling							•			
120-Cycle magnetic hum							•	•		

machines may have more than one) to coincide with the actual rotor speed. If this occurs, the fan will resonate, or experience severe vibration, and may catastrophically fail. The symptoms of plate-out are often confused with those of mechanical imbalance because both dramatically increase the vibration associated with the fan's running speed.

The problem of plate-out can be resolved by regularly cleaning the fan's rotating element and internal components. Removal of buildup lowers the rotor's mass and returns its natural frequency to the initial, or design point. In extremely dirty or dusty environments, it may be advisable to install an automatic cleaning system that uses high-pressure air or water to periodically remove any build-up that occurs.

Speed Changes

In applications where a measurable fan-speed change can occur (i.e., V-belt or variable-speed drives), care must be taken to ensure that the selected speed does not coincide with any of the fan's critical speeds. For general-purpose fans, the actual running speed is designed to be between 10 and 15% below the first critical speed of the rotating element. If the sheave ratio of a V-belt drive or the actual running speed is increased above the design value, it may coincide with a critical speed.

Some fans are designed to operate between critical speeds. In these applications, the fan must transition through the first critical speed to reach its operating speed. These transitions must be made as quickly as possible to prevent damage. If the fan's speed remains at or near the critical speed for any extended period of time, serious damage can occur.

Lateral Flexibility

By design, the structural support of most general-purpose fans lacks the mass and rigidity needed to prevent flexing of the fan's housing and rotating assembly. This problem is more pronounced in the horizontal plane but is also present in the vertical direction. If support-structure flexing is found to be the root cause or a major contributing factor to the problem, it can be corrected by increasing the stiffness and/or mass of the structure. However, do not fill the structure with concrete. As it dries, concrete pulls away from the structure and does little to improve its rigidity.

Blowers or Positive-Displacement Fans

Blowers, or positive-displacement fans, have the same common failure modes as rotary pumps and compressors. Table 13.4 lists the failure modes that most often affect blowers and fluidizers. In particular, blower failures occur due to process instability, caused by start/stop operation and demand variations, and mechanical failures due to close tolerances.

Process Instability

Blowers are very sensitive to variations in their operating envelope. As little as a one-psig change in downstream pressure can cause the blower to become extremely unstable. The probability of catastrophic failure or severe

Table 13.4 *Common failure modes of blowers and fluidizers*

	THE PROBLEM									
THE CAUSES	No air/gas delivery	Insufficient discharge pressure	Insufficient capacity	Excessive wear	Excessive hear	Excessive vibration and noise	Excessive power demand	Motor trips	Elevated motor temperature	Elevated air/gas temperature
Air leakage into suction piping or shaft seal		•	•			•				
Coupling misaligned				•	•	•	•		•	
Excessive discharge pressure			•	•		•	•	•		•
Excessive inlet temperature/moisture			•							
Insufficient suction air/gas supply		•	•	•		•		•		
Internal component wear	•	•	•							
Motor or driver failure	•									
Pipe strain on blower casing				•	•	•	•		•	
Relief valve stuck open or set wrong		•	•							
Rotating element binding				•	•	•	•	•	•	
Solids or dirt in inlet air/gas supply				•						
Speed too low		•	•					•		
Suction filter or strainer clogged	•	•	•			•			•	
Wrong direction of rotation	•	•							•	

damage to blower components increases in direct proportion to the amount and speed of the variation in demand or downstream pressure.

Start/stop Operation

The transients caused by frequent start/stop operation also have a negative effect on blower reliability. Conversely, blowers that operate constantly in a stable environment rarely exhibit problems. The major reason is the severe axial thrusting caused by the frequent variations in suction or discharge pressure caused by the start/stop operation.

Demand Variations

Variations in pressure and volume demands have a serious impact on blower reliability. Since blowers are positive-displacement devices, they generate a constant volume and a variable pressure that is dependent on the downstream system's back-pressure. If demand decreases, the blower's discharge pressure continues to increase until: (1) a downstream component fails and reduces the back-pressure, or (2) the brake horsepower required to drive the blower is greater than the motor's locked rotor rating. Either of these results in failure of the blower system. The former may result in a reportable release, while the latter will cause the motor to trip or burn out.

Frequent variations in demand greatly accelerate the wear rate of the thrust bearings in the blower. This can be directly attributed to the constant, instantaneous axial thrusting caused by variations in the discharge pressure required by the downstream system.

Mechanical Failures

Because of the extremely close clearances that must exist within the blower, the potential for serious mechanical damage or catastrophic failure is higher than with other rotating machinery. The primary failure points include: thrust bearing, timing gears, and rotor assemblies.

In many cases, these mechanical failures are caused by the instability discussed in the preceding sections, but poor maintenance practices are another major cause.

14 Gears and Gearboxes

A gear is a form of disc, or wheel, that has teeth around its periphery for the purpose of providing a positive drive by meshing the teeth with similar teeth on another gear or rack.

Spur Gears

The *spur gear* might be called the basic gear, since all other types have been developed from it. Its teeth are straight and parallel to the center bore line, as shown in Figure 14.1. Spur gears may run together with other spur gears or parallel shafts, with internal gears on parallel shafts, and with a *rack*. A rack such as the one illustrated in Figure 14.2 is in effect a straight-line gear. The smallest of a pair of gears (Figure 14.3) is often called a pinion.

The involute profile or form is the one most commonly used for gear teeth. It is a curve that is traced by a point on the end of a taut line unwinding from a circle. The larger the circle, the straighter the curvature; and for a rack, which is essentially a section of an infinitely large gear, the form is straight or flat. The generation of an involute curve is illustrated in Figure 14.4.

The involute system of spur gearing is based on a rack having straight or flat sides. All gears made to run correctly with this rack will run with each other.

Figure 14.1 *Example of a spur gear*

Figure 14.2 *Rack or straight-line gear*

Figure 14.3 *Typical spur gears*

The sides of each tooth incline toward the center top at an angle called the *pressure angle*, shown in Figure 14.5.

The 14.5-degree pressure angle was standard for many years. In recent years, however, the use of the 20-degree pressure angle has been growing, and today 14.5-degree gearing is generally limited to replacement work. The principal reasons are that a 20-degree pressure angle results in a gear tooth with greater strength and wear resistance and permits the use of pinions with fewer teeth. The effect of the pressure angle on the tooth of a rack is shown in Figure 14.6.

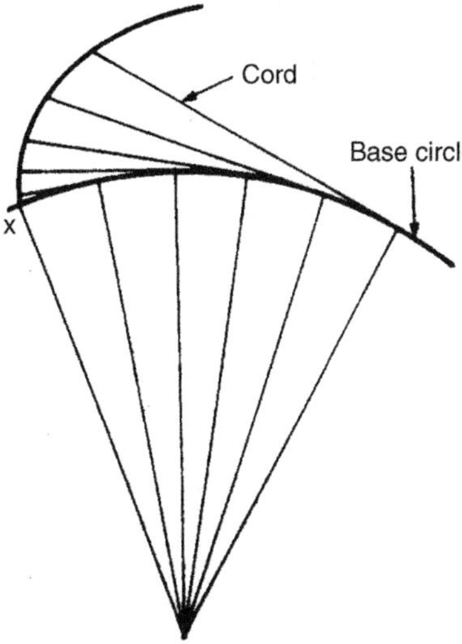

Figure 14.4 *Involute curve*

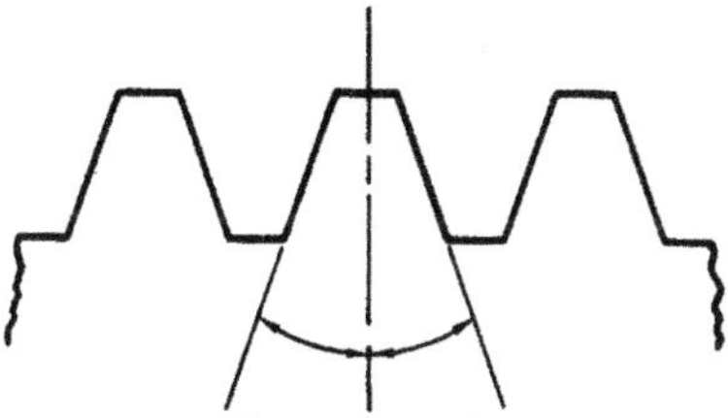

Figure 14.5 *Pressure angle*

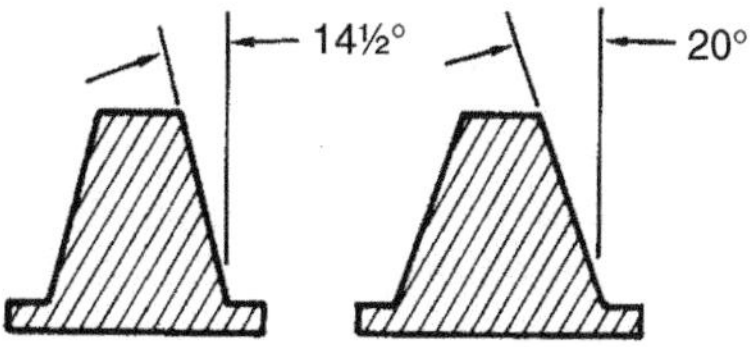

Figure 14.6 *Different pressure angles on gear teeth*

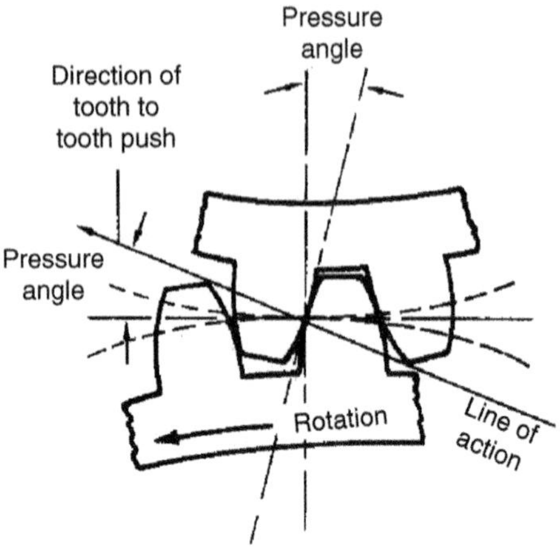

Figure 14.7 *Relationship of the pressure angle to the line of action*

It is extremely important that the pressure angle be known when gears are mated, as all gears that run together must have the same pressure angle. The pressure angle of a gear is the angle between the line of action and the line tangent to the pitch circles of mating gears. Figure 14.7 illustrates the relationship of the pressure angle to the line of action and the line tangent to the pitch circles.

Pitch Diameter and Center Distance

Pitch circles have been defined as the imaginary circles that are in contact when two standard gears are in correct mesh. The diameters of these circles are the pitch diameters of the gears. The center distance of the two gears, therefore, when correctly meshed, is equal to one half of the sum of the two pitch diameters, as shown in Figure 14.8.

This relationship may also be stated in an equation, and may be simplified by using letters to indicate the various values, as follows:

C = center distance

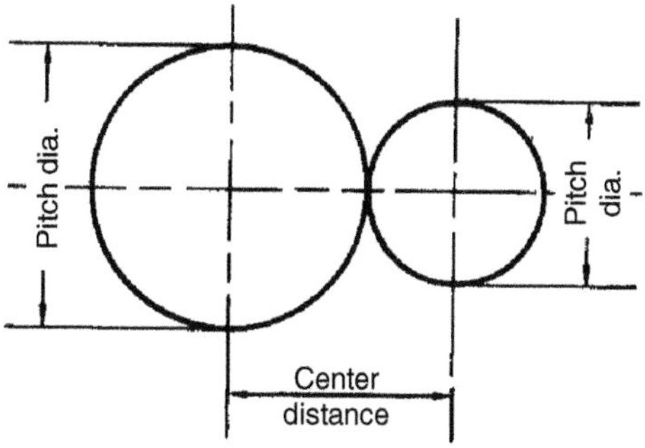

Figure 14.8 *Pitch diameter and center distance*

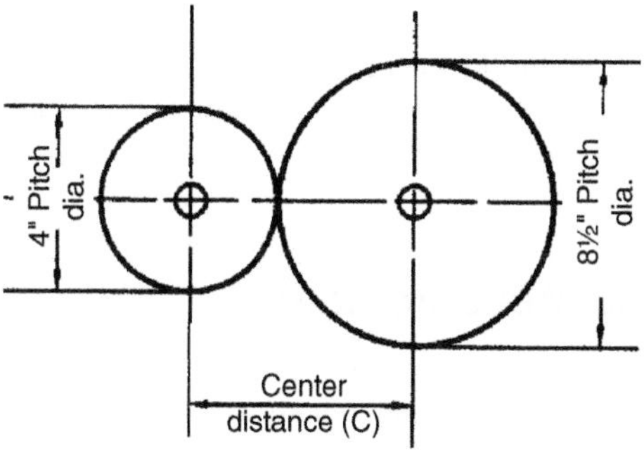

Figure 14.9 *Determining center distance*

D_1 = first pitch diameter

D_2 = second pitch diameter

$$C = \frac{D_1 + D_2}{2} \qquad D_1 = 2C - D_2 \qquad D_2 = 2C - D_1$$

Example: The center distance can be found if the pitch diameters are known (illustration in Figure 14.9).

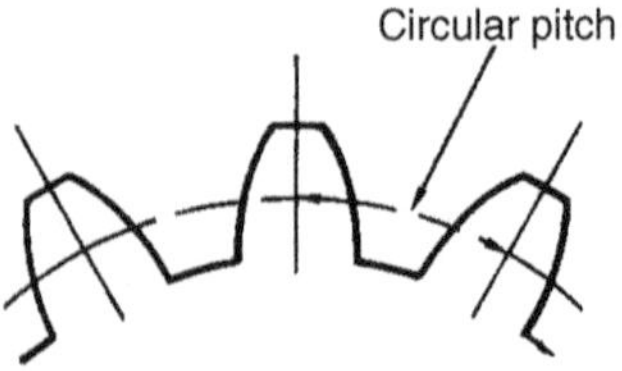

Figure 14.10 *Circular pitch*

Circular Pitch

A specific type of pitch designates the size and proportion of gear teeth. In gearing terms, there are two specific types of pitch: *circular pitch* and *diametrical pitch*. Circular pitch is simply the distance from a point on one tooth to a corresponding point on the next tooth, measured along the pitch line or circle, as illustrated in Figure 14.10. Large-diameter gears are frequently made to circular pitch dimensions.

Diametrical Pitch and Measurement

The diametrical pitch system is the most widely used, as practically all common-size gears are made to diametrical pitch dimensions. It designates the size and proportions of gear teeth by specifying the number of teeth in the gear for each inch of the gear's pitch diameter. For each inch of pitch diameter, there are pi (π) inches, or 3.1416 inches, of pitch-circle circumference. The diametric pitch number also designates the number of teeth for each 3.1416 inches of pitch-circle circumference. Stated in another way, the *diametrical pitch* number specifies the number of teeth in 3.1416 inches along the pitch line of a gear.

For simplicity of illustration, a whole-number pitch-diameter gear (4 inches), is shown in Figure 14.11.

Figure 14.11 illustrates that the *diametrical pitch* number specifying the number of teeth per inch of pitch diameter must also specify the number of teeth per 3.1416 inches of pitch-line distance. This may be more easily visualized and specifically dimensioned when applied to the rack in Figure 14.12.

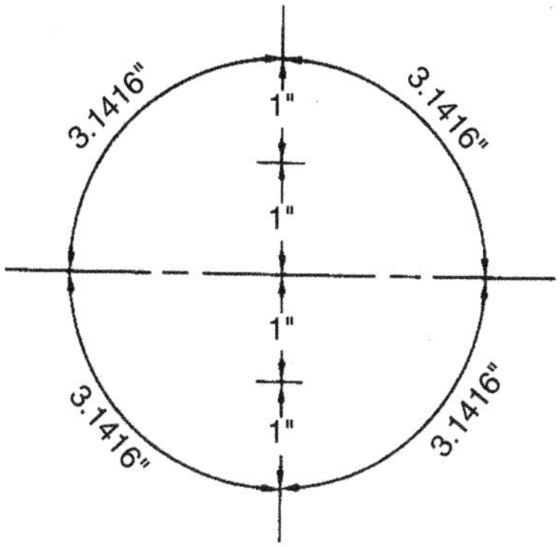

Figure 14.11 *Pitch diameter and diametrical pitch*

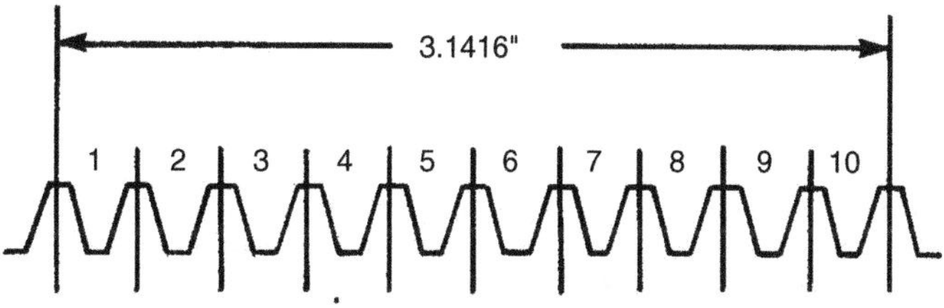

Figure 14.12 *Number of teeth in 3.1416 inches*

Because the pitch line of a rack is a straight line, a measurement can be easily made along it. In Figure 14.12, it is clearly shown that there are 10 teeth in 3.1416 inches; therefore the rack illustrated is a 10 diametrical pitch rack.

A similar measurement is illustrated in Figure 14.13, along the pitch line of a gear. The diametrical pitch being the number of teeth in 3.1416 inches of pitch line, the gear in this illustration is also a 10 diametrical pitch gear.

In many cases, particularly on machine repair work, it may be desirable for the mechanic to determine the diametrical pitch of a gear. This may be done very easily without the use of precision measuring tools, templates, or gauges. Measurements need not be exact because diametrical pitch numbers

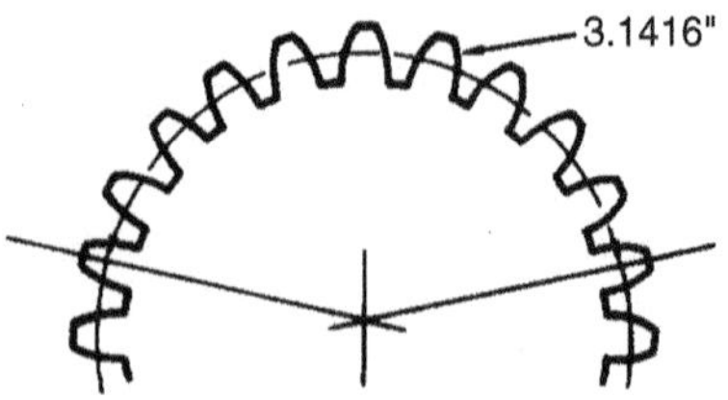

Figure 14.13 *Number of teeth in 3.1416 inches on the pitch circle*

are usually whole numbers. Therefore, if an approximate calculation results in a value close to a whole number, that whole number is the diametrical pitch number of the gear.

The following two methods may be used to determine the approximate diametrical pitch of a gear. A common steel rule, preferably flexible, is adequate to make the required measurements.

Method 1

Count the number of teeth in the gear, add 2 to this number, and divide by the outside diameter of the gear. Scale measurement of the gear to the closest fractional size is adequate accuracy.

Figure 14.14 illustrates a gear with 56 teeth and an outside measurement of $5\frac{13}{16}$ inches. Adding 2 to 56 gives 58; dividing 58 by $5\frac{13}{16}$ gives an answer of $9\frac{31}{32}$. Since this is approximately 10, it can be safely stated that the gear is a 10 diametrical pitch gear.

Method 2

Count the number of teeth in the gear and divide this number by the measured pitch diameter. The pitch diameter of the gear is measured from the root or bottom of a tooth space to the top of a tooth on the opposite side of the gear.

Figure 14.15 illustrates a gear with 56 teeth. The pitch diameter measured from the bottom of the tooth space to the top of the opposite tooth is $5\frac{5}{8}$ inches. Dividing 56 by $5\frac{5}{8}$ gives an answer of $9\frac{15}{16}$ inches, or approximately 10. This method also indicates that the gear is a 10 diametrical pitch gear.

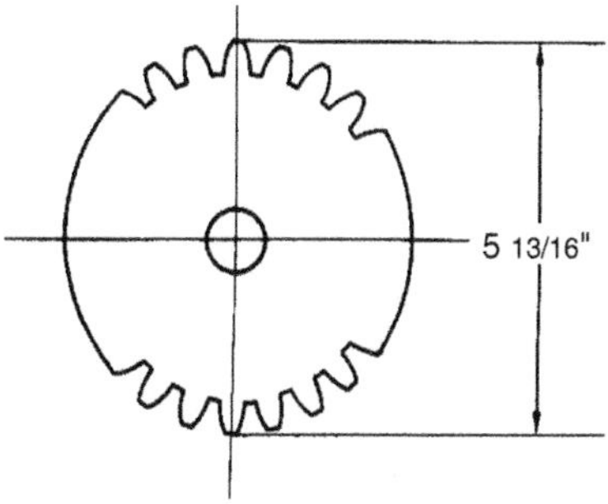

Figure 14.14 *Using method 1 to approximate the diametrical pitch. In this method the outside diameter of the gear is measured.*

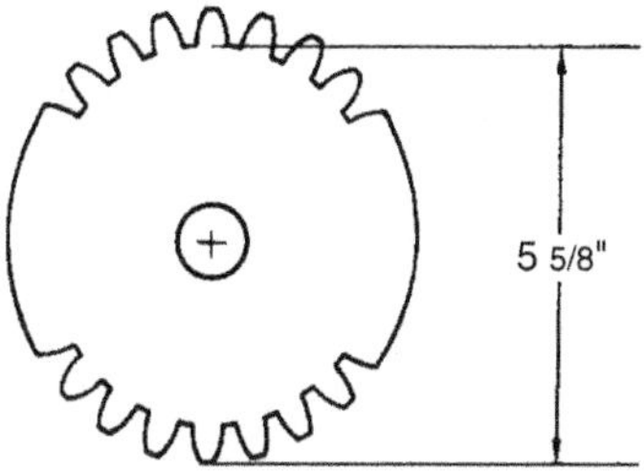

Figure 14.15 *Using method 2 to approximate the diametrical pitch. This method uses the pitch diameter of the gear.*

Pitch Calculations

Diametrical pitch, usually a whole number, denotes the ratio of the number of teeth to a gear's pitch diameter. Stated another way, it specifies the number of teeth in a gear for each inch of pitch diameter. The relationship of pitch diameter, diametrical pitch, and number of teeth can be stated mathematically as follows.

$$P = \frac{N}{D} \qquad D = \frac{N}{P} \qquad N = D \times P$$

Where:

D = pitch diameter

P = diametrical pitch

N = number of teeth

If any two values are known, the third may be found by substituting the known values in the appropriate equation.

Example 1: What is the *diametrical* pitch of a 40-tooth gear with a 5-inch pitch diameter?

$$P = \frac{N}{P} \quad \text{or} \quad P = \frac{40}{5} \quad \text{or} \quad P = 8 \text{ diametrical pitch}$$

Example 2: What is the *pitch diameter* of a 12 diametrical pitch gear with 36 teeth?

$$D = \frac{N}{P} \quad \text{or} \quad D = \frac{36}{12} \quad \text{or} \quad D = 3'' \text{ pitch diameter}$$

Example 3: How many teeth are there in a 16 *diametrical pitch* gear with a pitch diameter of $3\frac{3}{4}$ inches?

$$N = D \times P \quad \text{or} \quad N = 3\tfrac{3}{4} \times 16 \quad \text{or} \quad N = 60 \text{ teeth}$$

Circular pitch is the distance from a point on a gear tooth to the corresponding point on the next gear tooth measured along the pitch line. Its value is equal to the circumference of the pitch circle divided by the number of teeth in the gear. The relationship of the circular pitch to the *pitch-circle circumference, number of teeth,* and the *pitch diameter* may also be stated mathematically as follows:

$$\text{Circumference of pitch circle} = \pi D$$

$$p = \frac{\pi D}{N} \qquad D = \frac{pN}{\pi} \qquad N = \frac{\pi D}{p}$$

Where:

D = pitch diameter

N = number of teeth

p = circular pitch

π = pi, or 3.1416

If any two values are known, the third may be found by substituting the known values in the appropriate equation.

Example 1: What is the *circular pitch* of a gear with 48 teeth and a *pitch diameter* of 6"?

$$p = \frac{\pi D}{N} \quad \text{or} \quad \frac{3.1416 \times 6}{48} \quad \text{or} \quad \frac{3.1416}{8} \quad \text{or } p = .3927 \text{ inches}$$

Example 2: What is the *pitch diameter* of a .500" *circular-pitch* gear with 128 teeth?

$$D = \frac{pN}{\pi} \quad \text{or} \quad \frac{.5 \times 128}{3.1416} \quad D = 20.371 \text{ inches}$$

The list that follows offers just a few names of the various parts given to gears. These parts are shown in Figures 14.16 and 14.17.

- Addendum: Distance the tooth projects above, or outside, the pitch line or circle.

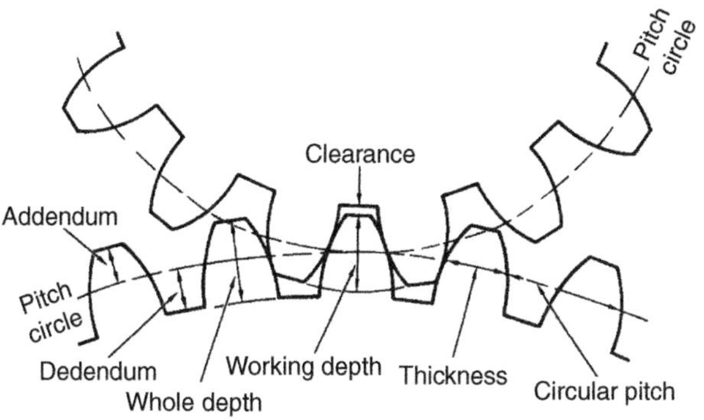

Figure 14.16 *Names of gear parts*

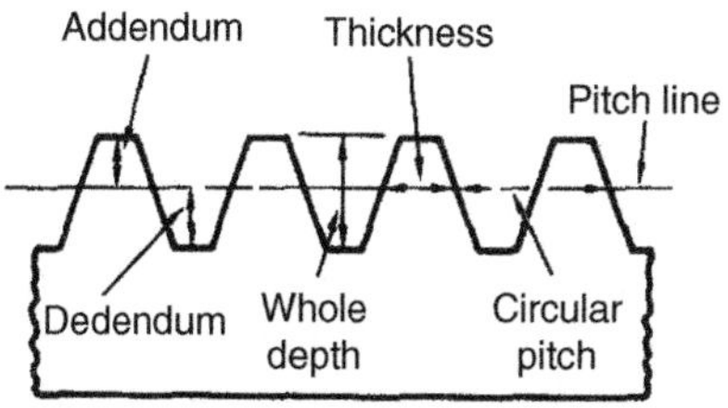

Figure 14.17 *Names of rack parts*

- Dedendum: Depth of a tooth space below, or inside, the pitch line or circle.
- Clearance: Amount by which the dedendum of a gear tooth exceeds the addendum of a matching gear tooth.
- Whole Depth: The total height of a tooth or the total depth of a tooth space.
- Working Depth: The depth of tooth engagement of two matching gears. It is the sum of their addendums.
- Tooth Thickness: The distance along the pitch line or circle from one side of a gear tooth to the other.

Tooth Proportions

The full *depth involute system* is the gear system in most common use. The formulas (with symbols) shown below are used for calculating tooth proportions of full-depth involute gears. Diametrical pitch is given the symbol P as before.

$$\text{Addendum, a} = \frac{1}{P}$$

$$\text{Whole depth, Wd} = \frac{20 + 0.002}{P} \text{(20P or smaller)}$$

$$\text{Dedendum, Wd} = \frac{2.157}{P} \text{(larger than 20P)}$$

$$\text{Whole depth, b} = \text{Wd} - \text{a}$$

$$\text{Clearance, c} = \text{b} - \text{a}$$

$$\text{Tooth thickness, t} = \frac{1.5708}{P}$$

Backlash

Backlash in gears is the play between teeth that prevents binding. In terms of tooth dimensions, it is the amount by which the width of tooth spaces exceeds the thickness of the mating gear teeth. Backlash may also be described as the distance, measured along the pitch line, that a gear

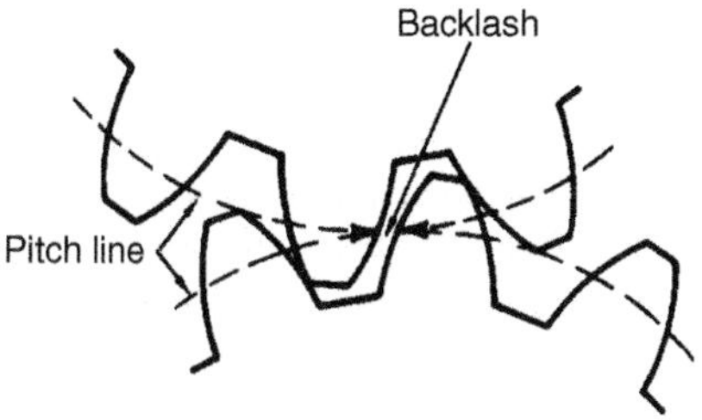

Figure 14.18 *Backlash*

will move when engaged with another gear that is fixed or immovable, as illustrated in Figure 14.18.

Normally there must be some backlash present in gear drives to provide running clearance. This is necessary, as binding of mating gears can result in heat generation, noise, abnormal wear, possible overload, and/or failure of the drive. A small amount of backlash is also desirable because of the dimensional variations involved in practical manufacturing tolerances. Backlash is built into standard gears during manufacture by cutting the gear teeth thinner than normal by an amount equal to one-half the required figure. When two gears made in this manner are run together, at standard center distance, their allowances combine, provided the full amount of backlash is required.

On nonreversing drives or drives with continuous load in one direction, the increase in backlash that results from tooth wear does not adversely affect operation. However, on reversing drive and drives where timing is critical, excessive backlash usually cannot be tolerated.

Other Gear Types

Many styles and designs of gears have been developed from the spur gear. While they are all commonly used in industry, many are complex in design and manufacture. Only a general description and explanation of principles will be given, as the field of specialized gearing is beyond the scope of this book. Commonly used styles will be discussed sufficiently to provide the millwright or mechanic with the basic information necessary to perform installation and maintenance work.

Bevel and Miter

Two major differences between bevel gears and spur gears are their shape and the relation of the shafts on which they are mounted. The shape of a spur gear is essentially a cylinder, while the shape of a bevel gear is a cone. Spur gears are used to transmit motion between parallel shafts, while bevel gears transmit motion between angular or intersecting shafts. The diagram in Figure 14.19 illustrates the bevel gear's basic cone shape. Figure 14.20 shows a typical pair of bevel gears.

Special bevel gears can be manufactured to operate at any desired shaft angle, as shown in Figure 14.21. Miter gears are bevel gears with the same number of teeth in both gears operating on shafts at right angles or at 90 degrees, as shown in Figure 14.22.

A typical pair of straight miter gears is shown in Figure 14.23. Another style of miter gears having spiral rather than straight teeth is shown in Figure 14.24. The spiral-tooth style will be discussed later.

The diametrical pitch number, as with spur gears, establishes the tooth size of bevel gears. Because the tooth size varies along its length, it must be measured at a given point. This point is the outside part of the gear where the tooth is the largest. Because each gear in a set of bevel gears must have the same angles and tooth lengths, as well as the same diametrical pitch, they are manufactured and distributed only in mating pairs. Bevel gears,

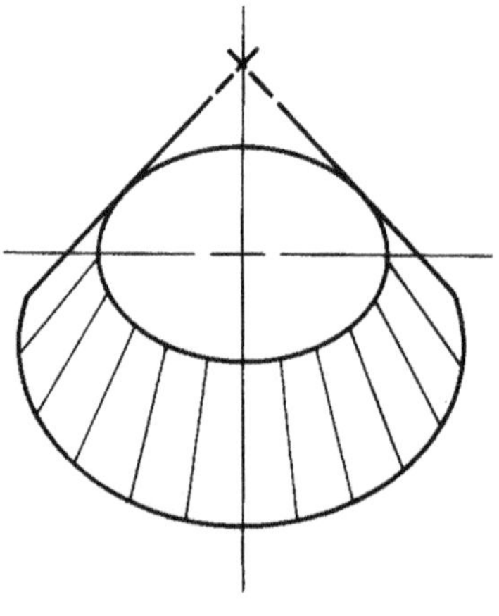

Figure 14.19 *Basic shape of bevel gears*

Figure 14.20 *Typical set of bevel gears*

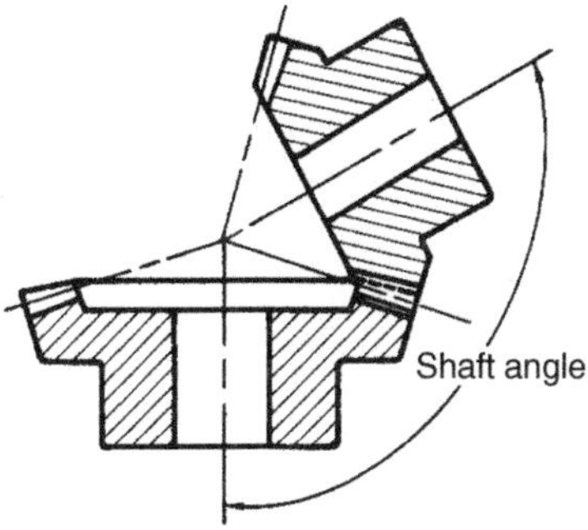

Figure 14.21 *Shaft angle, which can be at any degree*

like spur gears, are manufactured in both the 14.5-degree and 20-degree pressure-angle designs.

Helical

Helical gears are designed for parallel-shaft operation like the pair in Figure 14.25. They are similar to spur gears except that the teeth are cut at an angle to the centerline. The principal advantage of this design is the quiet, smooth action that results from the sliding contact of the meshing teeth. A disadvantage, however, is the higher friction and wear that accompanies

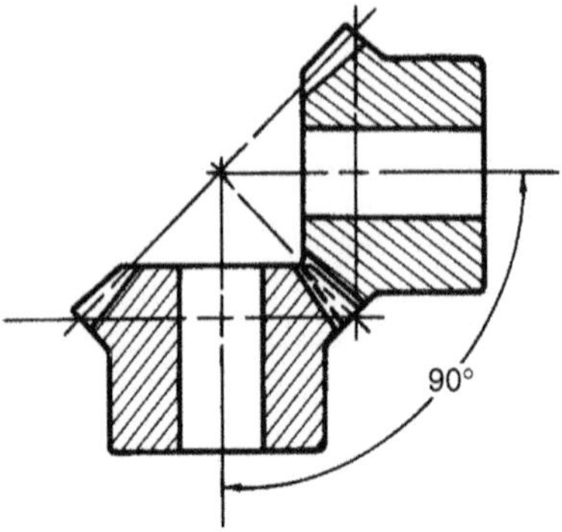

Figure 14.22 *Miter gears, which are shown at 90 degrees*

Figure 14.23 *Typical set of miter gears*

this sliding action. The angle at which the gear teeth are cut is called the *helix angle* and is illustrated in Figure 14.25.

It is very important to note that the helix angle may be on either side of the gear's centerline. Or if compared to the helix angle of a thread, it may be either a "right-hand" or a "left-hand" helix. The hand of the helix is the same regardless of how it is viewed. Figure 14.26 illustrates a helical gear as

Figure 14.24 *Miter gears with spiral teeth*

viewed from opposite sides; changing the position of the gear cannot change the hand of the tooth's helix angle. A pair of helical gears, as illustrated in Figure 14.27, must have the same pitch and helix angle but must be of opposite hands (one right hand and one left hand).

Helical gears may also be used to connect nonparallel shafts. When used for this purpose, they are often called "spiral" gears or crossed-axis helical gears. This style of helical gearing is shown in Figure 14.28.

Worm

The worm and worm gear, illustrated in Figure 14.29, are used to transmit motion and power when a high-ratio speed reduction is required. They provide a steady quiet transmission of power between shafts at right angles. The worm is always the driver and the worm gear the driven member. Like helical gears, worms and worm gears have "hand." The hand is determined by the direction of the angle of the teeth. Thus, in order for a worm and worm gear to mesh correctly, they must be the same hand.

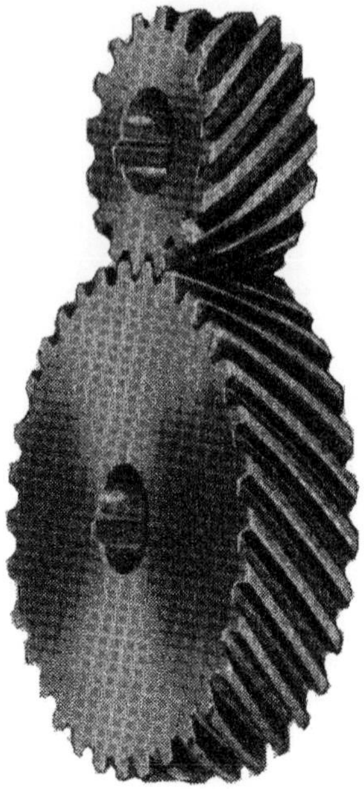

Figure 14.25 *Typical set of helical gears*

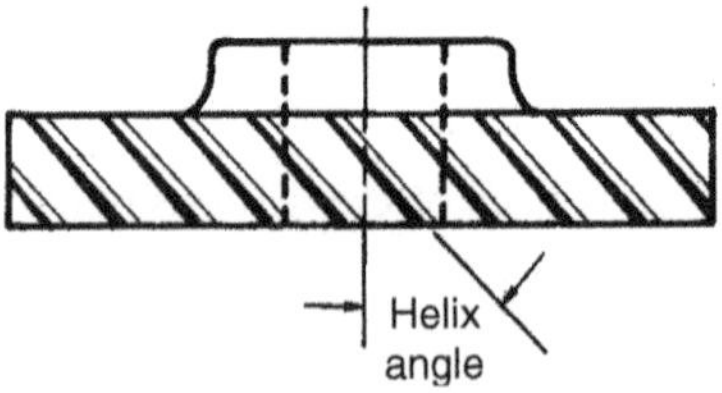

Figure 14.26 *Illustrating the angle at which the teeth are cut*

The most commonly used worms have either one, two, three, or four separate threads and are called single, double, triple, and quadruple thread worms. The number of threads in a worm is determined by counting the number of starts or entrances at the end of the worm. The thread of the

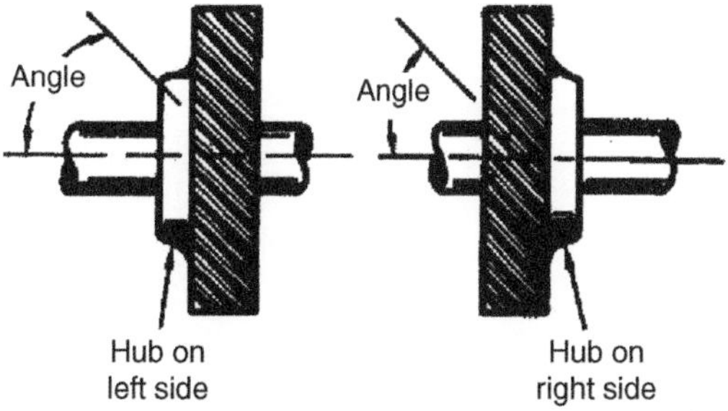

Figure 14.27 *Helix angle of the teeth the same regardless of side from which the gear is viewed*

Figure 14.28 *Typical set of spiral gears*

Figure 14.29 *Typical set of worm gears*

worm is an important feature in worm design, as it is a major factor in worm ratios. The ratio of a mating worm and worm gear is found by dividing the number of teeth in the worm gear by the number of threads in the worm.

Herringbone

To overcome the disadvantage of the high end thrust present in helical gears, the herringbone gear, illustrated in Figure 14.30, was developed. It consists simply of two sets of gear teeth, one right-hand and one left-hand, on the same gear. The gear teeth of both hands cause the thrust of one set to cancel out the thrust of the other. Thus, the advantage of helical gears is obtained, and quiet, smooth operation at higher speeds is possible. Obviously they can only be used for transmitting power between parallel shafts.

Gear Dynamics and Failure Modes

Many machine trains utilize gear drive assemblies to connect the driver to the primary machine. Gears and gearboxes typically have several vibration spectra associated with normal operation. Characterization of a gearbox's

Figure 14.30 *Herringbone gear*

vibration signature box is difficult to acquire but is an invaluable tool for diagnosing machine-train problems. The difficulty is that: (1) it is often difficult to mount the transducer close to the individual gears and (2) the number of vibration sources in a multigear drive results in a complex assortment of gear mesh, modulation, and running frequencies. Severe drive-train vibrations (gearbox) are usually due to resonance between a system's natural frequency and the speed of some shaft. The resonant excitation arises from, and is proportional to, gear inaccuracies that cause small periodic fluctuations in pitch-line velocity. Complex machines usually have many resonance zones within their operating speed range because each shaft can excite a system resonance. At resonance these cyclic excitations may cause large vibration amplitudes and stresses.

Basically, forcing torque arising from gear inaccuracies is small. However, under resonant conditions torsional amplitude growth is restrained only by damping in that mode of vibration. In typical gearboxes this damping is often small and permits the gear-excited torque to generate large vibration amplitudes under resonant conditions.

One other important fact about gear sets is that all gear-sets have a designed preload and create an induced load (thrust) in normal operation. The direction, radial or axial, of the thrust load of typical gear-sets will provide some insight into the normal preload and induced loads associated with each type of gear.

To implement a predictive maintenance program, a great deal of time should be spent understanding the dynamics of gear/gearbox operation and the frequencies typically associated with the gearbox. As a minimum, the following should be identified.

Gears generate a unique dynamic profile that can be used to evaluate gear condition. In addition, this profile can be used as a tool to evaluate the operating dynamics of the gearbox and its related process system.

Gear Damage

All gear sets create a frequency component, called *gear mesh*. The fundamental gear mesh frequency is equal to the number of gear teeth times the running speed of the shaft. In addition, all gear sets will create a series of sidebands or modulations that will be visible on both sides of the primary gear mesh frequency. In a normal gear set, each of the sidebands will be spaced at exactly the 1X or running speed of the shaft and the profile of the entire gear mesh will be symmetrical.

Normal Profile

In addition, the sidebands will always occur in pairs, one below and one above the gear mesh frequency. The amplitude of each of these pairs will be identical. For example, the sideband pair indicated, as −1 and +1 in Figure 14.31, will be spaced at exactly input speed and have the same amplitude.

If the gear mesh profile were split by drawing a vertical line through the actual mesh, i.e., the number of teeth times the input shaft speed, the two halves would be exactly identical. Any deviation from a symmetrical gear

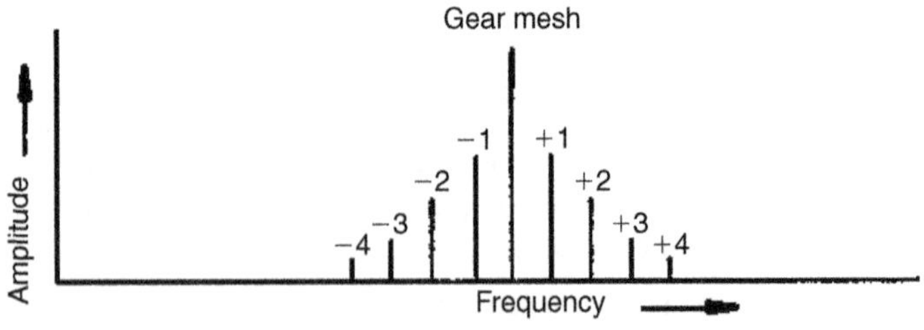

Figure 14.31 *Normal profile is symmetrical*

mesh profile is indicative of a gear problem. However, care must be exercised to ensure that the problem is internal to the gears and induced by outside influences. External misalignment, abnormal induced loads and a variety of other outside influences will destroy the symmetry of the gear mesh profile. For example, the single reduction gearbox used to transmit power to the mold oscillator system on a continuous caster drives two eccentrics. The eccentric rotation of these two cams is transmitted directly into the gearbox and will create the appearance of eccentric meshing of the gears. The spacing and amplitude of the gear mesh profile will be destroyed by this abnormal induced load.

Excessive Wear

Figure 14.32 illustrates a typical gear profile with worn gears. Note that the spacing between the sidebands becomes erratic and is no longer spaced at the input shaft speed. The sidebands will tend to vary between the input and output speeds but will not be evenly spaced.

In addition to gear tooth wear, center-to-center distance between shafts will create an erratic spacing and amplitude. If the shafts are too close together, the spacing will tend to be at input shaft speed, but the amplitude will drop drastically. Because the gears are deeply meshed, i.e., below the normal pitch line, the teeth will maintain contact through the entire mesh. This loss of clearance will result in lower amplitudes but will exaggerate any tooth profile defect that may be present.

If the shafts are too far apart, the teeth will mesh above the pitch line. This type of meshing will increase the clearance between teeth and amplify the

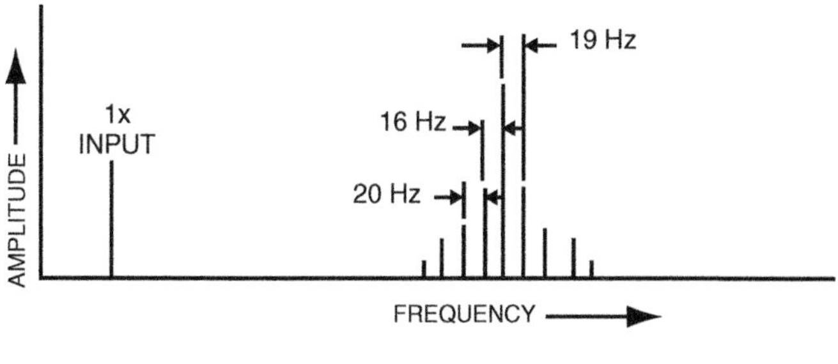

Figure 14.32 *Wear or excessive clearance changes sideband spacing*

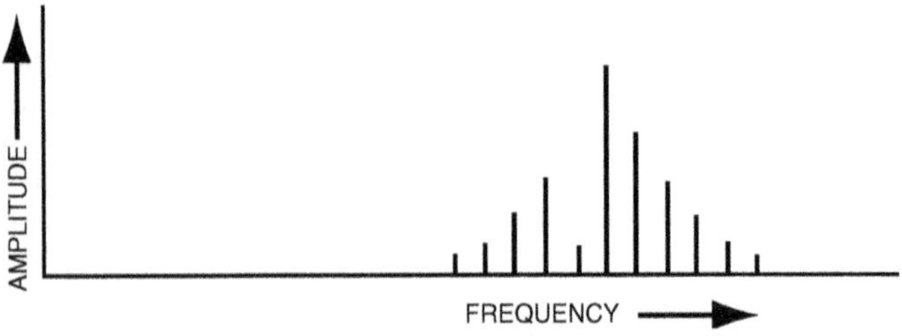

Figure 14.33 *A broken tooth will produce an asymmetrical sideband profile*

energy of the actual gear mesh frequency and all of its sidebands. In addition, the load bearing characteristics of the gear teeth will be greatly reduced. Since the pressure is focused on the tip of each tooth, there is less cross-section and strength in the teeth. The potential for tooth failure is increased in direct proportion to the amount of excess clearance between shafts.

Cracked or Broken Tooth

Figure 14.33 illustrates the profile of a gear set with a broken tooth. As the gear rotates, the space left by the chipped or broken tooth will increase the mechanical clearance between the pinion and bullgear. The result will be a low amplitude sideband that will occur to the left of the actual gear mesh frequency. When the next, undamaged teeth mesh, the added clearance will result in a higher energy impact.

The resultant sideband, to the right of the mesh frequency, will have much higher amplitude. The paired sidebands will have nonsymmetrical amplitude that represents this disproportional clearance and impact energy.

If the gear set develops problems, the amplitude of the gear mesh frequency will increase, and the symmetry of the sidebands will change. The pattern illustrated in Figure 14.34 is typical of a defective gear set. Note the asymmetrical relationship of the sidebands.

Common Characteristics

You should have a clear understanding of the types of gears generally utilized in today's machinery, how they interact, and the forces they generate on a rotating shaft. There are two basic classifications of gear drives: (1) shaft centers parallel, and (2) shaft centers not parallel. Within these two classifications are several typical gear types.

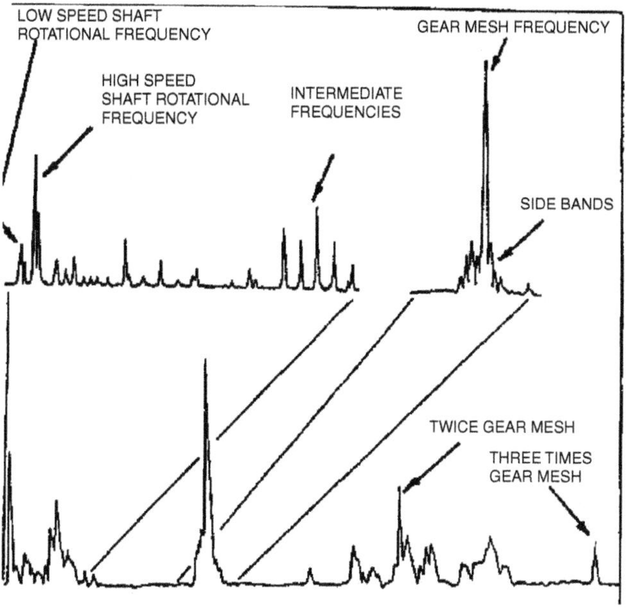

Figure 14.34 *Typical defective gear mesh signature*

Shaft Centers Parallel

There are four basic gear types that are typically used in this classification. All are mounted on parallel shafts and, unless an idler gear in also used, will have opposite rotation between the drive and driven gear (if the drive gear has a clockwise rotation, then the driven gear will have a counterclockwise rotation). The gear sets commonly used in machinery include the following:

Spur Gears

The shafts are in the same plane and parallel. The teeth are cut straight and parallel to the axis of the shaft rotation. No more than two sets of teeth are in mesh at one time, so the load is transferred from one tooth to the next tooth rapidly. Usually spur gears are used for moderate- to low-speed applications. Rotation of spur gear sets is opposite unless one or more idler gears are included in the gearbox. Typically, spur gear sets will generate a radial load (preload) opposite the mesh on their shaft support bearings and little or no axial load.

Backlash is an important factor in proper spur gear installation. A certain amount of backlash must be built into the gear drive allowing for tolerances in concentricity and tooth form. Insufficient backlash will cause early failure due to overloading.

As indicated in Figure 14.11, spur gears by design have a preload opposite the mesh and generate an induced load, or *tangential force*, TF, in the direction of rotation. This force can be calculated as:

$$\text{TF} = \frac{126{,}000 * \text{hp}}{D_\text{P} * \text{rpm}}$$

In addition, a spur gear will generate a Separating Force, S_{TF}, that can be calculated as:

$$S_{\text{TF}} = \text{TF} * \tan\phi$$

Where:

TF = Tangential force

hp = Input horsepower to pinion or gear

D_P = Pitch diameter of pinion or gear

rpm = Speed of pinion or gear

ϕ = Pinion or gear tooth pressure angle

Helical Gears

The shafts are in the same plane and parallel but the teeth are cut at an angle to the centerline of the shafts. Helical teeth have an increased length of contact, run more quietly and have a greater strength and capacity than spur gears. Normally the angle created by a line through the center of the tooth and a line parallel to the shaft axis is 45 degrees. However, other angles may be found in machinery. Helical gears also have a preload by design; the critical force to be considered, however, is the thrust load (axial) generated in normal operation; see Figure 14.12.

$$\text{TF} = \frac{126{,}000 * \text{hp}}{D_\text{P} * \text{RPM}}$$

$$S_{\text{TF}} = \frac{\text{TF} * \tan\phi}{\cos\lambda}$$

$$T_{\text{TF}} = \text{TF} * \tan\lambda$$

Where:

TF = tangential force

S_{TF} = separating force

T_{TF} = thrust force

hp = input horsepower to pinion or gear

D_P = pitch diameter of pinion or gear

rpm = speed of pinion or gear

ϕ = pinion or gear tooth pressure angle

λ = pinion or gear helix angle

Herringbone Gears

Commonly called "double helical" because they have teeth cut with right and left helix angles, they are used for heavy loads at medium to high speeds. They do not have the inherent thrust forces that are present in helical gear sets. Herringbone gears, by design, cancel the axial loads associated with a single helical gear. The typical loads associated with herringbone gear sets are the radial side-load created by gear mesh pressure and a tangential force in the direction of rotation.

Internal Gears

Internal gears can only be run with an external gear of the same type, pitch, and pressure angle. The preload and induced load will depend on the type of gears used. Refer to spur or helical for axial and radial forces.

Troubleshooting

One of the primary causes of gear failure is the fact that, with few exceptions, gear sets are designed for operation in one direction only. Failure is often caused by inappropriate bidirectional operation of the gearbox or backward installation of the gear set. Unless specifically manufactured for bidirectional operation, the "nonpower" side of the gear's teeth is not finished. Therefore, this side is rougher and does not provide the same tolerance as the finished "power" side.

Note that it has become standard practice in some plants to reverse the pinion or bullgear in an effort to extend the gear set's useful life. While this

Table 14.1 *Common failure modes of gearboxes and gear sets*

	THE PROBLEM								
THE CAUSES	Gear failures	Variations in torsional power	Insufficient power output	Overheated bearings	Short bearing life	Overload on driver	High vibration	High noise levels	Motor trips
Bent shaft				•	•	•	•		
Broken or loose bolts or setscrews				•			•		
Damaged motor						•	•		•
Eliptical gears		•	•			•	•		
Exceeds motor's brake horsepower rating			•			•			
Excessive or too little backlash	•	•							
Excessive torsional loading	•	•	•	•	•	•			•
Foreign object in gearbox	•						•	•	•
Gear set not suitable for application	•		•			•	•		
Gears mounted backward on shafts			•				•	•	
Incorrect center-to-center distance between shafts							•	•	
Incorrect direction of rotation			•			•	•		
Lack of or improper lubrication	•	•		•	•		•	•	•
Misalignment of gears or gearbox	•	•		•	•		•	•	
Overload	•		•	•	•	•			
Process induced misalignment	•	•		•	•				
Unstable foundation		•		•			•	•	
Water or chemicals in gearbox	•								
Worn bearing						•	•		
Worn couplings						•			

Source: Integrated Systems Inc.

practice permits longer operation times, the torsional power generated by a reversed gear set is not as uniform and consistent as when the gears are properly installed.

Gear overload is another leading cause of failure. In some instances, the overload is constant, which is an indication that the gearbox is not suitable for the application. In other cases, the overload is intermittent and only occurs when the speed changes or when specific production demands cause a momentary spike in the torsional load requirement of the gearbox.

Misalignment, both real and induced, is also a primary root cause of gear failure. The only way to assure that gears are properly aligned is to "hard blue" the gears immediately following installation. After the gears have run for a short time, their wear pattern should be visually inspected. If the pattern does not conform to vendor's specifications, alignment should be adjusted.

Poor maintenance practices are the primary source of real misalignment problems. Proper alignment of gear sets, especially large ones, is not an easy task. Gearbox manufacturers do not provide an easy, positive means to assure that shafts are parallel and that the proper center-to-center distance is maintained.

Induced misalignment is also a common problem with gear drives. Most gearboxes are used to drive other system components, such as bridle or process rolls. If misalignment is present in the driven members (either real or process induced), it also will directly affect the gears. The change in load zone caused by the misaligned driven component will induce misalignment in the gear set. The effect is identical to real misalignment within the gearbox or between the gearbox and mated (i.e., driver and driven) components.

Visual inspection of gears provides a positive means to isolate the potential root cause of gear damage or failures. The wear pattern or deformation of gear teeth provides clues as to the most likely forcing function or cause. The following sections discuss the clues that can be obtained from visual inspection.

Normal Wear

Figure 14.35 illustrates a gear that has a normal wear pattern. Note that the entire surface of each tooth is uniformly smooth above and below the pitch line.

Figure 14.35 *Normal wear pattern*

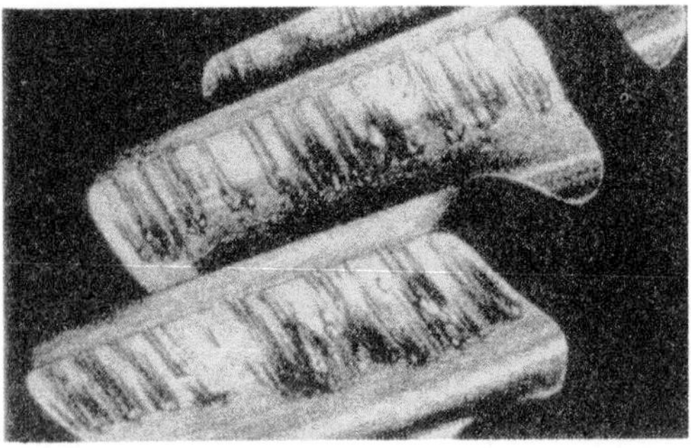

Figure 14.36 *Wear pattern caused by abrasives in lubricating oil*

Abnormal Wear

Figures 14.36 through 14.39 illustrate common abnormal wear patterns found in gear sets. Each of these wear patterns suggests one or more potential failure modes for the gearbox.

Abrasion

Abrasion creates unique wear patterns on the teeth. The pattern varies, depending on the type of abrasion and its specific forcing function. Figure 14.36 illustrates severe abrasive wear caused by particulates in the lubricating oil. Note the score marks that run from the root to the tip of the gear teeth.

Figure 14.37 *Pattern caused by corrosive attack on gear teeth*

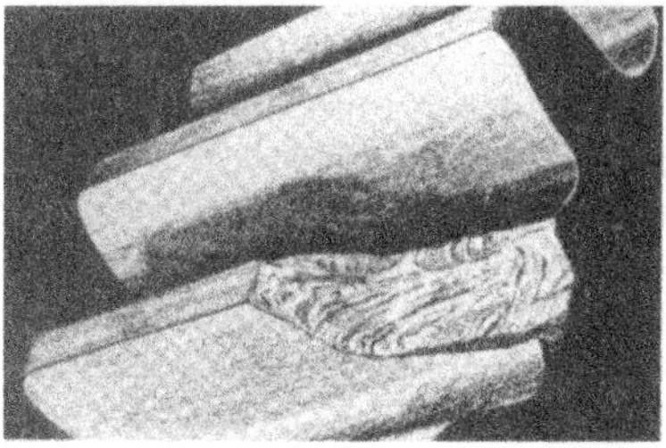

Figure 14.38 *Pitting caused by gear overloading*

Chemical Attack or Corrosion

Water and other foreign substances in the lubricating oil supply also cause gear degradation and premature failure. Figure 14.37 illustrates a typical wear pattern on gears caused by this failure mode.

Overloading

The wear patterns generated by excessive gear loading vary, but all share similar components. Figure 14.38 illustrates pitting caused by excessive torsional loading. The pits are created by the implosion of lubricating oil. Other wear patterns, such as spalling and burning, can also help to identify specific forcing functions or root causes of gear failure.

15 Hydraulics

"Only Permanent Repairs Made Here"

Hydraulic Knowledge

People say knowledge is power. This is true in hydraulic maintenance. Many maintenance organizations do not know what their maintenance personnel should know. We believe in an industrial maintenance organization where we should divide the hydraulic skill necessary into two groups. One is the hydraulic troubleshooter; they must be your experts in maintenance, and this should be as a rule of thumb 10% or less of your maintenance workforce. The other 90% plus would be your general hydraulic maintenance personnel. They are the personnel that provide the preventive maintenance expertise. The percentages we give you are based on a company developing a true preventive/proactive maintenance approach to its hydraulic systems. Let's talk about what the hydraulic troubleshooter knowledge and skills should be.

Hydraulic Troubleshooter

Knowledge:

- Mechanical principles (force, work, rate, simple machines)
- Math (basic math, complex math equations)
- Hydraulic components (application and function of all hydraulic system components)
- Hydraulic schematic symbols (understanding all symbols and their relationship to a hydraulic system)
- Calculating flow, pressure, and speed
- Calculating the system filtration necessary to achieve the system's proper ISO particulate code

Skill:

- Trace a hydraulic circuit to 100% proficiency
- Set the pressure on a pressure compensated pump
- Tune the voltage on an amplifier card
- Null a servo valve
- Troubleshoot a hydraulic system and utilize "Root Cause Failure Analysis"
- Replace any system component to manufacturer's specification.
- Develop a PM program for a hydraulic system.
- Flush a hydraulic system after a major component failure

General Maintenance Person

Knowledge:

- Filters (function, application, installation techniques)
- Reservoirs (function, application)
- Basic hydraulic system operation
- Cleaning of hydraulic systems
- Hydraulic lubrication principles
- Proper PM techniques for hydraulics

Skill:

- Change a hydraulic filter and other system components
- Clean a hydraulic reservoir
- Perform PM on a hydraulic system
- Change a strainer on a hydraulic pump
- Add filtered fluid to a hydraulic system
- Identify potential problems on a hydraulic system
- Change a hydraulic hose, fitting, or tubing

Best Maintenance Hydraulic Repair Practices

In order to maintain your hydraulic systems, you must have preventive maintenance procedures and you must have a good understanding and knowledge of "Best Maintenance Practices" for hydraulic systems. We will convey these practices to you. See Table 15.1.

Table 15.1 *Best maintenance repair practices: hydraulics*

Component	Component knowledge	Best practices	Frequency
Hydraulic fluid filter	There are two types of filters on a hydraulic system. 1. Pressure filter: Pressure filters come in collapsible and noncollapsible types. The preferred filter is the noncollapsible type. 2. Return filter: Typically has a bypass, which will allow contaminated oil to bypass the filter before indicating the filter needs to be changed.	Clean the filter cover or housing with a cleaning agent and clean rags. Remove the old filter with clean hands and install new filter into the filter housing or screw into place. CAUTION: NEVER allow your hand to touch a filter cartridge. Open the plastic bag and insert the filter without touching the filter with your hand.	Preferred: based on historical trending of oil samples. Least preferred: Based on equipment manufacture's recommendations.
Reservoir air breather	The typical screen breather should not be used in a contaminated environment. A filtered air breather with a rating of 10 micron is preferred because of the introduction of contaminants to a hydraulic system.	Remove and throw away the filter.	Preferred: Based on historical trending of oil samples. Least preferred: Based on equipment manufacturer's recommendations.

Table 15.1 *continued*

Component	Component knowledge	Best practices	Frequency
Hydraulic reservoir	A reservoir is used to: Remove contamination. Dissipate heat from the fluid. Store a volume of oil.	Clean the outside of the reservoir to include the area under and around the reservoir. Remove the oil by a filter pump into a clean container, which has not had other types of fluid in it before. Clean the insides of the reservoir by opening the reservoir and cleaning the reservoir with a lint-free rag. Afterward, spray clean hydraulic fluid into the reservoir and drain out of the system.	If any of the following conditions are met: A hydraulic pump fails. If the system has been opened for major work. If an oil analysis reveals excessive contamination.
Hydraulic pumps	A maintenance person needs to know the type of pump in the system and determine how it operates in the system. Example: What is the flow and pressure of the pump during a given operating cycle? This information allows a maintenance person to trend potential pump failure and troubleshoot a system problem quickly.	Check and record flow and pressure during specific operating cycles. Review graphs of pressure and flow. Check for excessive fluctuation of the hydraulic system. (Designate the fluctuation allowed.)	Pressure checks: Preferred: daily Least preferred: Weekly Flow & pressure checks: Preferred: two weeks Least preferred: monthly

Root Cause Failure Analysis

As in any proactive maintenance organization you must perform Root Cause Failure Analysis in order to eliminate future component failures. Most maintenance problems or failures will repeat themselves without someone identifying what caused the failure and proactively eliminating it. A preferred method is to inspect and analyze all component failures. Identify the following:

- Component name and model number
- Location of component at the time of failure
- Sequence or activity the system was operating at when the failure occurred
- What caused the failure?
- How will the failure be prevented from happening again?

Failures are not caused by an unknown factor like "bad luck," or "it just happened," or "the manufacturer made a bad part." We have found most failures can be analyzed and prevention taken to prevent their recurrence. Establishing teams to review each failure can pay off in major ways.

Preventive Maintenance

Preventive maintenance (PM) of a hydraulic system is very basic and simple and if followed properly can eliminate most hydraulic component failure. Preventive maintenance is a discipline and must be followed as such in order to obtain results. We must view PM programs as performance oriented and not activity oriented. Many organizations have good PM procedures but do not require maintenance personnel to follow them or hold personnel accountable for the proper execution of these procedures. In order to develop a preventive maintenance program for your system you must follow these steps:

First: Identify the system operating condition.

- Does the system operate 24 hours a day, 7 days a week?

- Does the system operate at maximum flow and pressure 70% or better during operation?
- Is the system located in a dirty or hot environment?

Second: What requirements does the equipment manufacturer state for preventive maintenance on the hydraulic system?

Third: What requirements and operating parameters does the component manufacturer state concerning the hydraulic fluid ISO particulate?

Fourth: What requirements and operating parameters does the filter company state concerning its filters' ability to meet this requirement?

Fifth: What equipment history is available to verify the above procedures for the hydraulic system?

As in all preventive maintenance programs, we must write procedures required for each PM task. Steps or procedures must be written for each task, and they must be accurate and understandable by all maintenance personnel from entry level to master.

Preventive maintenance procedures must be a part of the PM job plan, which includes (see Figure 15.1):

- Tools or special equipment required for performing the task;
- Parts or material required for performing the procedure with store room number;
- Safety precautions for this procedure;
- Environmental concerns or potential hazards.

A list of preventive maintenance tasks for a hydraulic system could be:

- Change the hydraulic filter (could be the return or pressure filter).
- Obtain a hydraulic fluid sample.
- Filter hydraulic fluid.
- Check hydraulic actuators.
- Clean the inside of a hydraulic reservoir.
- Clean the outside of a hydraulic reservoir.
- Check and record hydraulic pressures.

ABC COMPANY
PREVENTIVE MAINTENANCE PROCEDURE

TASK DESCRIPTION:	P.M. – Inspect hydraulic oil reserve tank level
EQUIPMENT NUMBER:	311111
FILE NUMBER:	09
FREQUENCY:	52
KEYWORD, QUALIFIER:	Unit, Hydraulic (Dynamic Press)
SKILL/CRAFT:	Production
PM TYPE:	Inspection
SHUTDOWN REQUIRED:	No

REFERENCE MANUAL/DWGS:
1. See operator manual F-378

REQUIRED TOOLS/MATERIALS:
1. Oil, Texaco Rando 68 SDK #400310
2. Flashlight
3. Oil Filter Pump

SAFETY PRECAUTIONS:
1. Observe plant and area specific safe work practices.

MAINTENANCE PROCEDURE:
1. Inspect hydraulic oil reserve tank level as follows:
   a) If equipped with sight glass, verify oil level at the full mark. Add oil as required.
   b) If not equipped with sight glass, remove fill plug/cap.
   c) Using flashlight, verify that oil is at proper level in tank. Add oil as required.
2. Record discrepancies or unacceptable conditions in comments.

PM Procedure Courtesy of Life Cycle Engineering, Inc.

Figure 15.1 *Sample preventive maintenance procedure*

- Check and record pump flow.
- Check hydraulic hoses, tubing, and fittings.
- Check and record voltage reading to proportional or servo valves.
- Check and record vacuum on the suction side of the pump.
- Check and record amperage on the main pump motor.
- Check machine cycle time and record.

Preventive maintenance is the core support that a hydraulic system must have in order to maximize component life and reduce system failure. Preventive maintenance procedures that are properly written and followed properly will allow equipment to operate to its full potential and life cycle. Preventive maintenance allows a maintenance department to control a hydraulic system rather than the system controlling the maintenance department. We must control a hydraulic system by telling it when we will perform maintenance on it and how much money we will spend on the maintenance for the system. Most companies allow hydraulic systems to control the maintenance on them at a much higher cost.

Measuring Success

In any program we must track success in order to have support from management and maintenance personnel. We must also understand that any action will have a reaction, negative or positive. We know successful maintenance programs will provide success, but we must have a checks and balances system to ensure we are on track.

In order to measure success of a hydraulic maintenance program we must have a way of tracking success but first we need to establish a benchmark. A benchmark is method by which we will establish certain key measurement tools that will tell you the current status of your hydraulic system and then tell you if you are succeeding in your maintenance program.

Before you begin the implementation of your new hydraulic maintenance program it would be helpful to identify and track the following information.

1 Track all downtime (in minutes) on the hydraulic system with these questions (tracked daily):

- What component failed?
- Cause of failure?
- Was the problem resolved?
- Could this failure have been prevented?
- Track all costs associated with the downtime (tracked daily).
- Parts and material cost?
- Labor cost?

- Production downtime cost?
- Any other cost you may know that can be associated with a hydraulic system failure?

2 Track hydraulic system fluid analysis. Track the following from the results (taking samples once a month):

- Copper content
- Silicon content
- H_2O
- Iron content
- ISO particulate count
- Fluid condition (viscosity, additives, and oxidation)

When the tracking process begins, you need to trend the information that can be trended. This allows management the ability to identify trends that can lead to positive or negative consequences. See Figure 15.2.

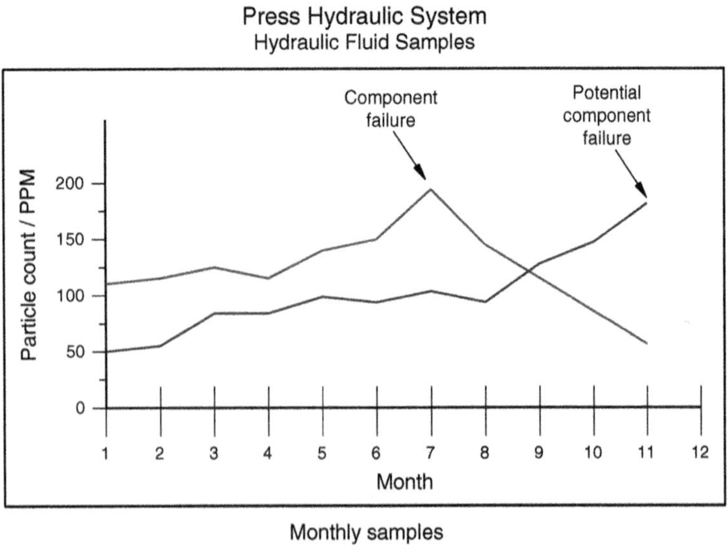

Figure 15.2 *Hydraulic fluid samples*

Fluid analysis will prove the need for better filtration. The addition of a 3-micron absolute return line filter to supplement the "kidney loop" filter can solve the problem.

Many organizations do no know where to find the method for tracking and trending the information you need accurately. A good computerized maintenance management system can track and trend most of this information for you.

Recommended Maintenance Modifications

Modifications to an existing hydraulic system need to be accomplished professionally. A modification to a hydraulic system in order to improve the maintenance efficiency is important to a company's goal of maximum equipment reliability and reduced maintenance cost.

First: Filtration pump with accessories

Objective: The objective of this pump and modification is to reduce contamination that is introduced into an existing hydraulic system through the addition of new fluid and the device used to add oil to the system.

Additional information: Hydraulic fluid from the distributor is usually not filtered to the requirements of an operating hydraulic system. Typically, this oil is strained to a mesh rating and not a micron rating. How clean is clean? Typically, hydraulic fluid must be filtered to 10 microns absolute or less for most hydraulic systems; 25 microns is the size of a white blood cell, and 40 microns is the lower limit of visibility with the unaided eye.

Many maintenance organizations add hydraulic fluid to a system through a contaminated funnel and may even, without cleaning it, use a bucket that has had other types of fluids and lubricants in it previously.

Recommended equipment and parts:

- Portable filter pump with a filter rating of 3 microns absolute.
- Quick disconnects that meet or exceed the flow rating of the portable filter pump.
- A $\frac{3}{4}$" pipe long enough to reach the bottom of the hydraulic container your fluids are delivered in from the distributor.

- A 2" reducer bushing to $\frac{3}{4}$" NPT to fit into the 55-gallon drum, if you receive your fluid by the drum. Otherwise, mount the filter buggy to the double wall "tote" tank supports if you receive larger quantities.
- Reservoir vent screens should be replaced with 3/10 micron filters, and openings around piping entering the reservoir sealed.

Show a double wall tote tank of about 300 gallons mounted on a frame for fork truck handling, with the pump mounted on the framework.

Also show pumping from a drum mounted on a frame for fork truck handling, sitting in a catch pan, for secondary containment, with the filter buggy attached.

Regulations require that you have secondary containment, so make everything "leak" into the pan. See Figure 15.3.

Second: Modify the Hydraulic Reservoir (See Figure 15.4)

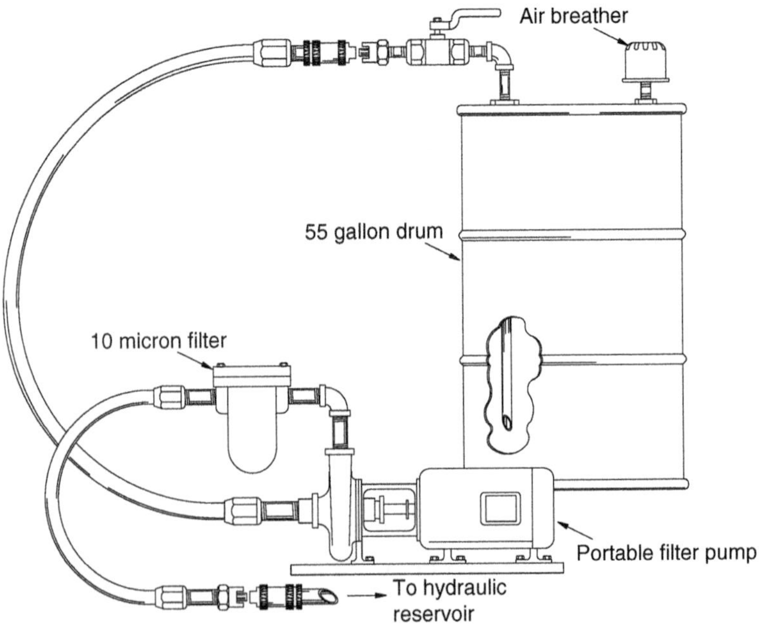

Figure 15.3 *Filter pumping unit*

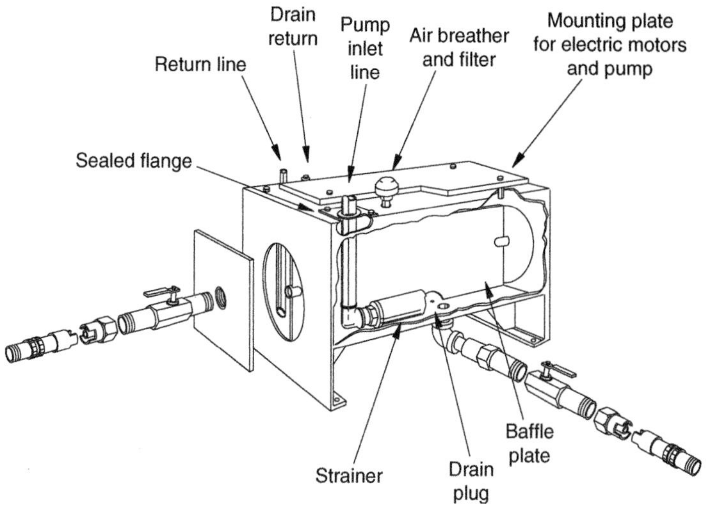

Figure 15.4 *Hydraulic reservoir modification*

Objective: The objective is to eliminate the introduction of contamination through oil being added to the system or contaminants being added through the air intake of the reservoir. A valve needs to be installed for oil sampling.

Additional Information: The air breather strainer should be replaced with a 10-micron filter if the hydraulic reservoir cycles. A quick disconnect should be installed on the bottom of the hydraulic unit and at the $\frac{3}{4}$ level point on the reservoir with valves to isolate the quick disconnects in case of failure. This allows the oil to add from a filter pump as previously discussed and would allow for external filtering of the hydraulic reservoir oil if needed. Install a petcock valve on the front of the reservoir, which will be used for consistent oil sampling.

Equipment and parts needed:

- Quick disconnects that meet or exceed the flow rating of the portable filter pump
- Two gate valves with pipe nipples.
- One 10-micron filter breather.

WARNING: Do not weld on a hydraulic reservoir to install the quick disconnects or air filter.

To summarize, maintenance of a hydraulic system is the first line of defense to prevent component failure and thus improve equipment reliability. As discussed earlier, discipline is the key to the success of any proactive maintenance program.

16 Lubrication

"The Foundation of Equipment Maintainability"

Lubrication Principles

Friction occurs when two surfaces in contact with each other attempt to move in opposing directions at the same time. It is also defined as the resistance to movement between two surfaces in contact with each other. If friction happens without the benefit of a lubricant, it is called a "solid" friction. Lubrication is defined as reducing friction to a minimum by replacing solid friction with fluid friction. Reducing the friction increases the equipment efficiency.

Kinds of Friction

Even the most carefully finished metal surface is not truly flat but is covered with microscopic irregularities, projections, and depressions. When two dry surfaces are rubbed together, the irregularities have a tendency to interlock and resist the sliding motion. Under conditions of extreme pressure the irregularities tend to weld together. Friction between moving surfaces is grouped into three main types: sliding, rolling, and fluid.

Sliding Friction

Sliding friction occurs when two surfaces slide over each other, as in a brake slowing down a rotating wheel on a vehicle, or a piston sliding in a cylinder. In sliding friction, because the contact pressure is usually spread over a large area, the pressure per square inch is relatively low.

Rolling Friction

Rolling friction takes place when a spherical or cylindrical body rolls over a surface. Common examples of rolling friction are ball and roller bearings. With ball or roller friction bearings the area of contact is quite small; however, the pressure loading, or pressure per square inch, is high. There is also a very small amount of sliding friction between the ball or roller and the separators because the components are rolling instead of sliding as in the piston example above.

With gears, both sliding and rolling conditions exist as the gears mesh and unmesh. They are grouped according to their contact area and action.

Fluid Friction

Fluid friction refers to air, water, or other types of fluid providing the resistance to movement between two objects. One example of fluid friction would be the resistance of air to an airplane flying. Another example would be the torque converter in an automatic transmission; the transmission fluid provides the power to drive the automobile through friction with the impeller blades.

Lubrication Theory

When lubricating oil is applied to each of the component surfaces, a thin film of oil is formed, filling up the depressions and covering the projections. Due to the film of oil between the two surfaces, sliding, not friction, will occur. This condition is called fluid lubrication. See Figure 16.1.

In theory, the oil forms in layers of globules, one layer adhering to each metal surface and any number of layers of globules in between. (See Figure 16.2.) In the illustrations, layer (1) adheres to the top surface, layer (9) in Figure 16.2(a) or layer (8) in Figure 16.2(b) adheres to the bottom surface, and the layers in between roll over each other when the bearing surfaces move. When these layers of oil roll over each other, the only friction present is

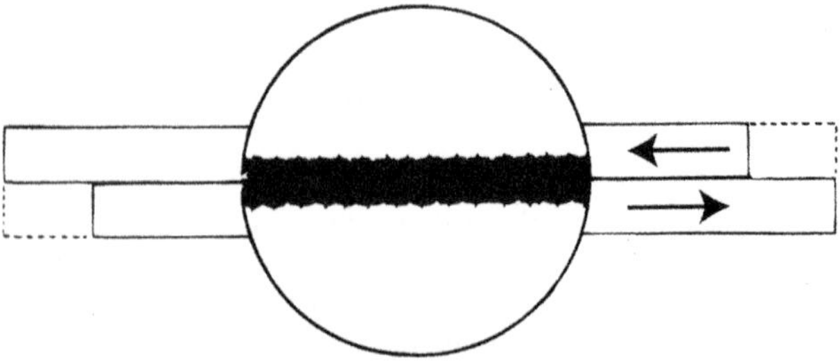

Figure 16.1 *Magnified bearing surface with a fluid film*

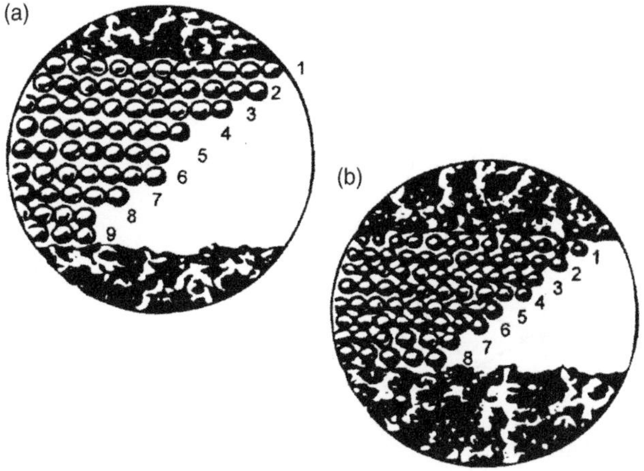

Figure 16.2 *Globules*

between the oil globules, forming what is called fluid friction. This state of fluid friction will be maintained as long as a suitable quantity of oil is supplied.

In plain bearings, the cohesion between the molecules of oil, plus the adhesion of the oil to the metal surfaces, causes the shaft to draw oil in under it as it revolves. This is known as "wedge action" and accounts for the presence of the lubricating film even in heavily loaded bearings.

When the shaft is at rest, most of the film of oil between it and the bearing is squeezed out, allowing some direct metal-to-metal contact. See Figure 16.3.

As the shaft starts to rotate, oil climbs up the bearing side in a direction opposite to the direction of rotation. The layer of oil on the slowly turning shaft clings to the surface and turns with it. As the oil is carried between the shaft and the bearing it separates the bearing surfaces with a continuous layer of oil. See Figure 16.4.

As the speed is increased, more oil is forced between the shaft and the bearing. The shaft then has a tendency to fall to the bottom of the bearing, but the layer of oil prevents metal-to-metal contact. See Figure 16.5.

At final speed the wedging action of the oil moves in the direction of rotation and becomes strong enough to lift the shaft into the location.

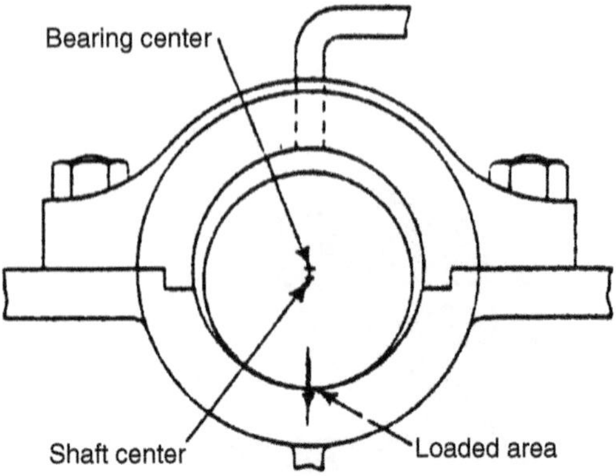

Figure 16.3 *Journal at rest*

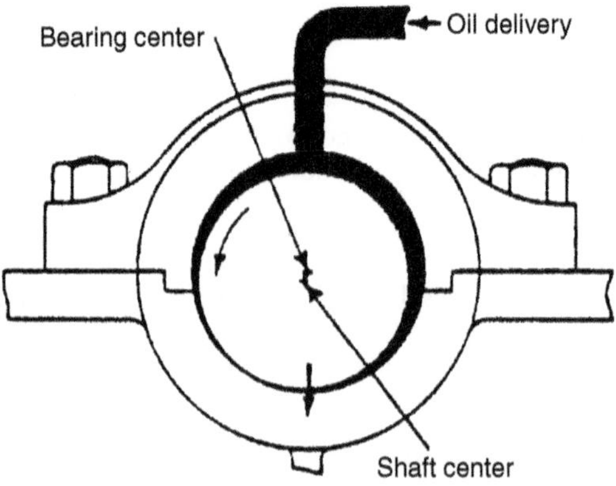

Figure 16.4 *Rolling action*

The turning shaft has been likened to a pump forcing oil between shaft and bearing, with hydraulic pressure creating an oil wedge to force the shaft against the opposite side.

It should be noted that this theory depends on a satisfactory supply of oil to form a continuous film. Lack of oil after the rotation begins means that a lubricating film and wedge cannot be established, and the metal-to-metal

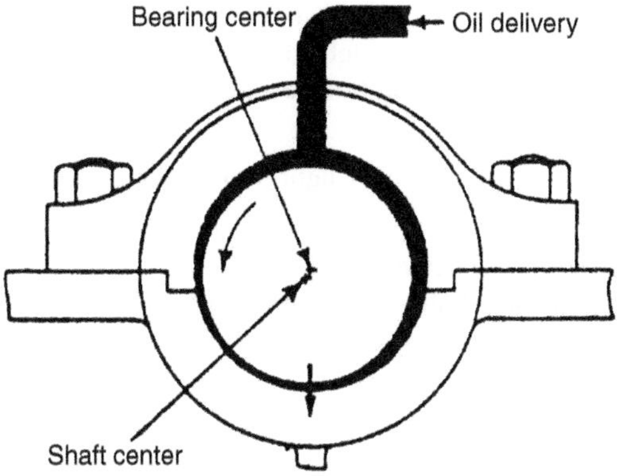

Figure 16.5 *Establishment of fluid film*

contact will be maintained, generating heat and eventually wearing out the bearing.

In Figure 16.6, the area marked C is the point of high pressure, and the oil film is thinnest in that area. Oil should come in from the top of low-pressure area, where it can be picked up by the shaft, and brought around to the high-pressure area.

When rotation starts, the coefficient of friction is quite high, but as soon as the shaft has made about half a turn, or enough to form a film of oil with the bearing, the coefficient of friction drops to a low level.

In an antifriction bearing there are two oil wedge formations due to the three-unit construction of the bearing.

Ball Bearing Oil Wedge Formation

- Outer race
- Ball
- Inner race

Operating Conditions

Viscosity is the most important property of lubrication oil (source: *Petroleum Handbook*). Viscosity is a measure of a fluid's resistance to flow, and it largely determines the suitability of an oil for any particular application. The best oil for a bearing is one with the right viscosity to maintain the "oil wedge" action efficiently, subject to conditions of speed, pressure, and heat.

Oils with low viscosity rating are quite thin or light; while oils with high viscosity rating flow very slowly.

The speed of a shaft and the clearance between shaft and bearing will determine the choice of oil. A slow-turning shaft with relatively wide clearance can use heavy or high-viscosity oil, while high-speed shafts with close tolerance bearings require a light or low viscosity oil.

Bearing load must be considered, as the oil must have enough body to maintain a good oil film under estimated maximum load. An oil that maintains a film under 300 lbs. load will not stand up under a 1,000-lb. load in the same bearing. Generally, a heavy load demands a heavier grade of oil than does a light load, bearing areas being equal.

Pressure in the oil film builds up from zero on the incoming side to a peak slightly past the centerline of the bearing, then drops to zero. The oil film pressure is directly proportional to the load on the bearing. Increase the load and the pressure increases; decrease the load and the pressure decreases. Regardless of the load, the pressure adjusts to provide sufficient pressure to carry the load. Speed does not have any effect on film pressure.

Oils become thinner when heated and thicker when cooled so that temperature will be a factor in determining viscosity. Heat should be considered in two ways: heat from operation, and heat or lack of heat from surroundings.

Heat from operation is usually in a very small range, but in some machines an allowable rise of 100°F is predicted. Heat from surroundings will vary, from an exposed bearing in winter to a bearing next to a large boiler. The temperature range could be as much as 150°F.

Properties of Oil

Viscosity

A lubricant for any machine must meet the requirements set by critical load, speed, and temperature. The correct lubricating oil is selected for its physical properties of viscosity and pour point, plus the extra qualities obtained by additives or special agents. Lubricating oil is used to minimize wear, heat rise, and power loss due to friction, to act as a cushion to absorb shock and vibration, and to act as a cleansing agent by washing away minute wear particles.

Viscosity ratings are obtained by a viscometer that measures the amount of time it takes for a measured amount of oil to flow through a measured opening at a definite temperature. (Saybolt Universal Viscosity [SUS] is the time in seconds for 60 cubic centimeters of a fluid to flow through the orifice of the Standard Saybolt Viscometer at a given temperature under specified conditions.) Temperatures taken are 100°F and 210°F (100°, 130°, 210°F—Shell Oil). For example, one sample of oil will take 60 seconds to flow through the opening, while another sample of oil of the same volume takes 600 seconds. The oil taking 60 seconds has a low viscosity rating and is called thin oil, while the oil taking 600 seconds has a high viscosity rating and is called heavier oil. The rate of flow of oil through the test hole will vary with the temperature, and the viscosity readings for the different temperatures give an index to the oil's ability to withstand temperature changes. This is called a Viscosity Index or V.I. A high V.I. means that the oil does not change viscosity through the temperature range as much as oil with a lower V.I.

Table 16.1 is for oils to lubricate journal bearings.

Note that for any speed, viscosity ratings increase with the heat, but that at any heat above freezing, viscosity decreases with the speed. Pour point of any oil is the lowest temperature at which the lubricant will flow. This is an important characteristic when selecting an oil to be used at below-freezing temperatures. A machine installed in a heated building will take one grade of oil all year, but he same machine exposed to weather conditions will take one grade of oil in the summer and a lighter grade in the winter. Any new

Table 16.1 *Oil to lubricate journals—in SUS ratings*

Surface speed ft./min.	Below 32°	32°–150°	150°–200°
Below 150	42	65	150
150–300	42	65	120
300–750	42	50	65
Over 750	42	50	55

machine will have lubrication recommendations for maximum temperature range, or any oil company will recommend a lubricant to suit.

Oiliness

Oiliness, or the ability to adhere to metal and bearing surfaces, is common to all petroleum oils. If a super degree of oiliness is required, an additive must be added to the oil.

Flash and Fire Points

Flash and fire point are considered only when the lubricant is exposed to high temperatures. Flash point is the temperature at which the oil gives off fumes that can be ignited by an open flame, usually over 300°F. Fire point is when the oil itself burns and is about 50°F higher than flash point. Flash point depends on a supply of oxygen.

Carbon Deposits

Carbon deposits are formed when oils are heated to 675°F and higher and occur mainly in the lubrication of steam and internal combustion engine cylinders.

Anticorrosives

New lubricating oils are noncorrosive to most metals used in machines, but through continuous use the oils slowly oxidize and form acids. Some oils contain an additive to prevent acid formation, thus extending the life of the oil. On aging in service, reactions sometimes take place that result in the formation of insoluble substances. All deposits that settle in lubricating systems are not the fault of oil deterioration but are usually from contamination. At high temperatures the insoluble substances may be deposited as a varnish. Detergent or dispersing agents are added to the oil to keep the deterioration products in a fine state so their separation as a sludge or varnish is prevented.

Additives

Extreme Pressure (EP) additives give the oils high film strength to support extreme loads and pressures met with hypoid gears. (This oil is not recommended for use on machinery with brass washers, bearings, or sleeves unless the EP agent is of a type that does not attack brass.)

Antifoaming additives are added to oils used in gearboxes and light oils used to lubricate roller bearings in high-speed applications to reduce the amount of foaming.

A circulating oil system will encounter water in the oil due to condensation of water vapor in the system, and sometimes by water working by the oil seals in wet locations. Oil for this type of system should have an additive to assure quick separation of oil from the water to prevent the formation of emulsions. This characteristic of separation is called "demulsibility."

Grease Characteristics and Types

Grease is made by adding a metallic soap to lubricating oil, effectively thickening it to the point that it turns into grease. The soap molecules in the grease cling together and have such a strong attraction with the oil molecules that it is very difficult to separate the soap and the oil. The soap molecules are "polar"—that is, they carry an electric charge that causes them to be attracted to any electric field extending out a few molecule lengths from most metallic bearing surfaces. This electrical attraction causes the formation of a minute layer of soap molecules on the metallic surfaces, and these soap molecules attract molecules of oil. This attraction anchors a very thin film of grease to the bearing surface.

Grease has a peculiar characteristic called directional fluidity. When moving in a bearing, the grease tends to "shear" into thin layers that move in the direction of rotation. As the shearing speed increases, the grease becomes easier to shear. This directional fluidity is encountered only in the direction of the shearing force, and the grease does not tend to run or be squeezed out of a bearing, even though it is acting like a liquid. Under shearing stress, the apparent viscosity of the grease falls rapidly until it approaches the viscosity of the oil used in its manufacture.

Classification of Grease

The penetration number, dropping point, metallic base, and the thickening agents used are all elements that are used to grade grease.

Penetration numbers indicate the consistency of a grease and are determined by the depth a rod with a definite surface area and weight will sink into the grease at a certain temperature during a given time. Soft grease has a high penetration number, while hard grease has a low penetration number. In general, the hardness of grease increases with an increase in the amount of thickening agent NLGI (National Lubricating Grease Institute).

Standards range from Grade 0 for the softest grade—which has the consistency of rendered lard at room temperature—to 6 for the stiffest, which approaches bar soap in hardness.

Drop point or melting point is a specification requirement. A sample of grease is heated at a given rate in a small cup with an opening in the bottom. The drop point is the temperature at which a drop of the sample falls from the cup. This in not an accurate way to measure the heat tolerance of grease, as many different greases flow from a bearing at a temperature far below their drop point.

Thickening Agents

Soap greases (calcium base grease is an example) are made by cooking a mixture of a suitable fatty acid and a portion of the petroleum oil with calcium hydroxide. When the saponification of the acid by the lime is complete, the water content is adjusted and the remainder of the oil incorporated. A fine mesh wire screen removes impurities and lumps before packaging (Imperial Oil).

Calcium base greases depend on a definite water content to stabilize the soap/oil structure. If the water is removed, the grease has a tendency to separate. For this reason calcium soap greases are not recommended for temperatures over 150°F due to water evaporation. Calcium soaps do not dissolve in water, and calcium base greases may be used in damp locations—damp or wet but not submerged in water.

Sodium base greases have very little water content and are suitable for use at higher temperatures. Soda soaps are water-soluble, and soda base greases are not suitable in wet conditions. Soda greases have greater stability than lime base greases and are more often used to lubricate higher speed antifriction bearings.

Mixtures of calcium and sodium soaps give greases that will stand a higher temperature than a straight lime base grease but are not as water resistant as a soda base grease.

Barium and lithium base greases are water resistant and will withstand temperatures of around 350°F. These greases have a long life expectancy and are suitable for bearings in hard-to-get-at places. Lithium greases are workable at −20°F.

Aluminum base greases vary from liquid grease to solid grease. A major use is car lubrication.

Aluminum greases are both fibrous and nonfibrous. Nonfibrous grades resemble lime greases and are used as such. Fibrous grades are quite stringy and are often used on slow turning shafts to cushion shock loading or in badly worn journal bearings to prolong bearing life.

Regardless of the composition of the grease, the basic shearing action is the same in use, one layer of grease slides over another.

The above facts for grease are approximate, and maker's trade or information sheets should be consulted to get actual ratings for any lubricant.

Grease and Oil: A Comparison

- Oil is easier to handle for draining, cleaning, and refilling bearings or gear cases.
- Oil is more suitable for wide temperature and speed variations.
- Oil can be used in a circulating system to act as a cooling agent and to wash away impurities.
- Oil can be used in a gravity flow system to lubricate a number of bearings from one location.
- Grease will stay in a bearing with less leaking than oil, and the seals can be quite simple.
- With a grease gun, grease can be forced to flow in any direction, but oil will only flow down unless a pressure pumping system is installed.
- In operating conditions near lubrication failure, grease is better than an oil of the same viscosity as the blended oil, due to the extra lubrication provided by the soap.
- Under many working conditions grease will carry a heavier load than the oil from which it is compounded, since the soaps impart superior lubricating ability.
- Greases are often more versatile than oils, and fewer grades are required for different speed and load conditions.

Other Oil Applicators

The hand oiler or squirt can is the oldest method of applying oil and is still in use. This method leads to extremes of over- or underlubrication.

Common oilers such as the bottle, wick, or drop feed are means of adding oil at a gradual rate to suit operating conditions. They can be used only above the bearing as the oil flow from them is by gravity. See Figure 16.1.

- The wick feeder oiler uses the capillary action of a strand or strands of wool to lift the oil out of the reservoir. The flow of oil varies according to the number of strands of wool and the height of oil in the reservoir. The flow will continue as long as there is a supply of oil. The capillary action of the wick tends to filter the oil, but after a time the wick will get dirty and the flow will decrease.
- A drip feed oiler offers a visual check and a means of controlling the flow of oil by adjusting the needle valve. It can be shut off when the machine is not used, avoiding a waste of oil. The oiler is filled through a small hole in the top, requiring care to avoid spilling oil and to keep foreign material from entering the system. Once contaminated, the needle valve is fouled easily by a small piece of dirt or waste.

Several types of lubricators for oiling a bearing or series of bearings (or drip oiling chains) can be made to suit local conditions. The basic style is a tank made from a short length of pipe with a metal removable or hinged lid covering a smaller opening for adding oil. The bottom has a $\frac{1}{2}$" pipe coupling welded on to connect to the drain line. The rate of flow is controlled by a valve and sight glass on the drain line, which can be either pipe or tubing. Tubing is preferred, as it can easily be led around obstructions and will withstand more vibration. See Figure 16.6.

With the increasing use of antifriction bearings this type of lubrication is being eliminated, or else is used only for lubricating chains.

Figure 16.6 *Lubricators*

Where oil is used over a period of time a highly stable oil with additives to prevent acid formation, rust formation, and formation or emulsions by any condensation in the oil reservoir is desirable.

The ring oiler is a mechanical way of oiling a slow speed shaft. The ring has a larger ID than the OD of the shaft. It rests on the top of the shaft with the bottom of the ring in the oil in the bottom of the housing. As the shaft turns, friction pulls the ring around with oil clinging to its surfaces. For long bearings, two or more rings can be used.

- The rings are usually of one-piece metal or two pieces hinged, or can be a flexible light ladder chain. A one-piece ring limits the bearing to two individual shells, but a two-piece ring or chain ring allows unit construction for the bottom bearing.

Enclosed System

A circulating system is used mainly when there are a large number of bearings on one machine or machines close together, all using the same oil. The other general application is where a bearing or bearings on a machine are expected to run at a high temperature, and cool oil is pumped from the reservoir over the shaft and bearing to control heat rise. The basic circulating system consists of pumped oil from the reservoir, piping to each bearing, and a drain from each bearing back to the reservoir. The refinements are individual flow control to each bearing and a visual check of each flow.

Grease Lubrication Methods

Grease used for friction bearings is usually applied by a handheld grease gun. Greasing with a gun has the advantage of not depending on gravity for flow conditions. The oiler can walk on the floor level and grease bearings at any level when they are piped to a suitable location. For fixed bearings the usual piping is $\frac{1}{8}$" tubing. For movable bearings or take-up bearings in hard-to-reach places, a loop of oil-resistant pressure hose between the bearing and the fixed tubing will allow greasing from a distance. Bearings should be checked at close range at frequent intervals in case the grease line breaks or works out of the bearing. When using two or more types of grease, it is good policy to have a gun for each.

Compression grease cups are used in hazardous areas and are screwed directly into the bearing or into a short pipe connection to the bearing.

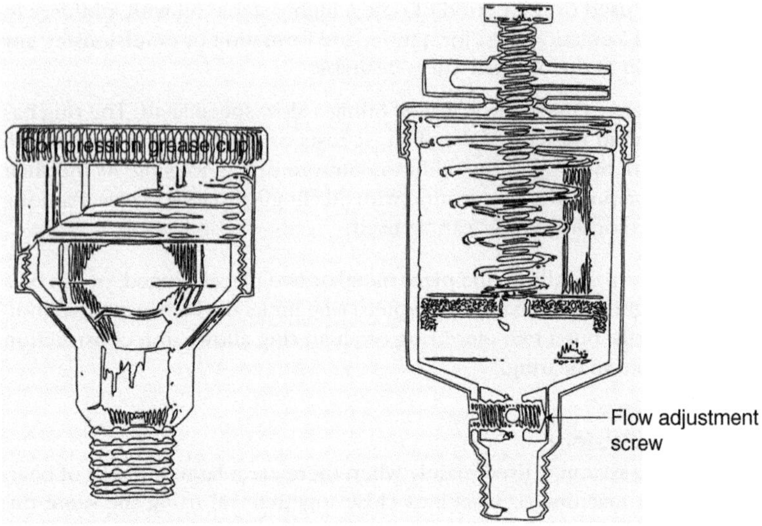

Compression grease cup

Spring compression grease cup

Figure 16.7 *Compression grease cups*

They present a safety hazard; if the oiler has to reach near moving machinery to screw down the top or take it off there is a possibility of dirt or foreign material getting into the grease. A spring compression grease cup will give a steady metered supply of grease for a period of time of up to four hours. See Figure 16.7.

Oil and Grease Lubrication: Special Applications

Enclosed Gears

The proper lubrication of gears depends on several factors, the first four of which are most important:

- Type of gear
- Load
- Speed

- Temperature
- Methods

Worm Gears

Wheel gears and hypoid gears generate high pressure on the contact line and greater friction. This will call for heavier oil or one with special additives.

The higher the load on a gear, the greater will be the tooth pressure. When the pressure is too high the oil film is broken, and metal-to-metal contact takes place. Heavier-bodied oil or one with EP additives is needed for gears operating under high loads.

Higher speeds call for lighter lubricants, while slower speeds allow heavier lubricants. With multiple reduction gear sets having two or more steps of reduction, the oil is selected to suit the low-speed pinion on the last reduction.

Temperature variations are influenced by the surroundings and the heat rise of operation. During operation the heat generated by friction and by churning of the oil will increase the temperature of the unit. Hypoid and worm gears operate with a permissible rise of 90° to 100°F, while other gear types run with an estimated permissible rise of 40° to 50°F. Surroundings will present a large temperature range. If the gear set is in a heated building, the temperature variation will be only a few degrees. If the gear set is in an outside location, the winter temperature may be so cold that the oil will not flow properly. When this is the case, the starting temperature must be considered when selecting the proper oil. For units located near a source of heat or in hot areas such as a steam plant, the final temperature will decide the choice of oil.

Splash lubrication is the most common way of lubrication in enclosed gear systems. In most units the larger gear picks up the oil and carries it to the mesh point, as well as splashing oil to a trough that drains to the bearings' worm-wheel units, with the worm on the bottom lubricated by the worm passing oil to the wheel. The oil must be kept high enough to ensure that the gear will pick up a sufficient quantity of oil.

Open Gears

Lubrication of open gears is by means of a grease or very heavy oil. Operating conditions have to be considered for:

- Temperature

- Method of application
- Surrounding conditions
- Gear material
- Choice of oil

When applied by either a brush or paddle, the lubricant must be sufficiently fluid to flow easily. During operations, the lubricant should be heavy-bodied, viscous, and tacky. Some oils and greases can be thinned enough for application by heating them and applying them hot. When heating is not practical, heavy-body diluted oils can be used. These oils are thinned with a nonflammable solvent that evaporates after exposure to air, leaving the heavy oil to cover the surface.

Oil can also be applied to gears by a drip cup or oil can. Very slow-moving gears can be lubricated from a bottom pan; the grease is picked up by the teeth of the larger gear and brought around the smaller gear or gears.

If surroundings are clean, either grease or oil can be used for good results, but if surroundings are wet or dirty, the chosen lubricant must be capable of providing good service to meet the equipment requirements regardless of the environment.

With the development of gears made from synthetics such as Bakelite, Nylon, Celoron, Micarta, and others, special lubricating methods may be needed. The ability of a nonmetallic gear to withstand petroleum lubricants should be identified and known before a selection is made.

Choice of Oil

When making a choice of what type of oil to use in a particular situation, several elements must be considered. Some of those items are:

- Viscosity must be suitable to form a lubricating film under expected maximum working conditions.
- Chemical stability is important as the oil is continually churned into contact with the air. Low stability oil will break down to form acids and sludge.
- Demulsibility or water separation is necessary as water is frequently formed from condensation inside the housing.
- Antirust additives are needed to minimize rust formation rising from water in the gear housing.

- High film strength is needed to sustain the oil film between the gear teeth under load conditions.
- EP additives are needed for use in hypoid gear trains. Note: An EP-additive oil should be checked with the equipment supplier, as some additives are not recommended for use with brass.

Any new equipment should have the manufacturer's lubrication recommendations. An alternative method of obtaining recommendations is to ask your oil supplier for assistance; they will normally recommend a suitable lubricant.

Bearings

A reduction unit with shafting in a vertical position has a pump to lubricate any bearings and any gears not touching the oil in the reservoir. This pump supplies the proper lubricant under pressure to the moving parts throughout the equipment. See Figure 16.8.

Proper lubrication results in:

- Reduced friction between bearing races, rolling elements, and the separator;
- Protection of the finished surfaces of the bearing from rust;
- Removal or dissipation of heat;
- Exclusion or isolation of foreign material.

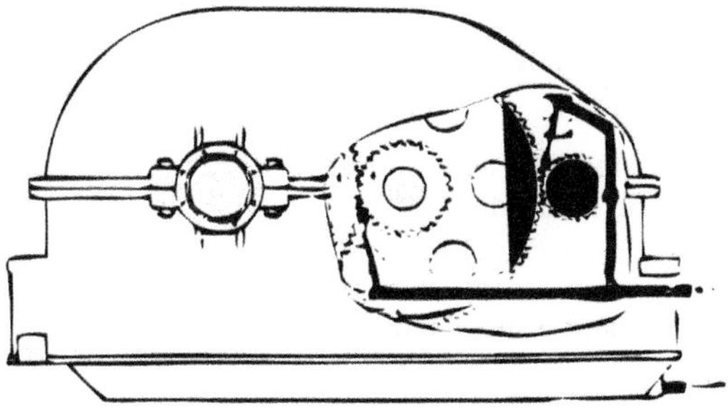

Figure 16.8 *Reduction unit*

In many cases either oil or grease can be used, but the choice of lubricant depends on the following conditions:

- Operating speed;
- Load;
- Temperature;
- Ease of access to bearing;
- Cleanliness required in surrounding area.

Grease is used in the large majority of bearing applications with slow to moderate speeds and moderate temperatures.

Grease Advantages

- Lubrication for longer periods of time without renewal or additions
- Requires fewer additions
- Permits the use of simple seals as its tenacity enables it to stay in place
- Provides a better antirust film during periods of downtime

Grease Disadvantages

- Does not easily dissipate heat
- Old grease is not easily removed
- Limited to low and moderate speeds

Oil Advantages

- Develops less fluid friction, which leads to a cooler running bearing
- Has a flushing action on bearing surfaces to wash off dirt and abrasive material and help dissipate heat
- Is easily removed from a housing with a minimum of down time, and oil can be changed while machine is running

Greases

- Sodium-based greases for moderate speeds, dry conditions, and temperatures up to 200°F.
- Calcium-based greases for moderate speeds, damp or wet conditions, and temperatures up to 150°F.

- Lithium-based greases for moderate speeds, damp or wet conditions, and temperatures from −30 degrees to 300°F.
- Special greases for extremes of temperature and loading conditions

Grease should be selected for its consistency at the operating temperature, when it should be fluid enough to gradually flow to the bearing. Very soft grease will have a tendency to churn and generate heat. Instead of being submerged in grease, the rolling elements operate in a channel and flake off the grease as they rotate.

The spaces between the rings should be filled with grease after assembly and the housing packed one-third full for high speeds and one-half full for slow speeds.

Overpacking will cause churning of the grease, higher temperatures, a reduction in the lubricating value of the grease, and a shorter life for the seals and for the equipment.

On low-speed applications with extreme conditions of moisture and dirt, the housing can be completely filled with grease. This will reduce the amount of foreign material passing the grease seal. Do not completely assemble the bearing without grease and then use a grease gun unless the housing has a relief fitting. Pressure will blow out the grease seals.

Oil must be selected to meet the temperature extremes and must have additives to prevent corrosion, foaming, and rusting. Oil is used for high speeds or temperatures below zero and above 200°F.

The oil bath is commonly used with the oil at a level to cover the bottom of the outer race but not higher than the center point of the lowest ball or roller. Overoiling produces churning and a rise in temperature. In high-speed conditions generating heat, oil can be circulated and cooled as a separate function, drawing heat away from the bearing and shaft.

A hot bearing is caused mainly by a lubricating fault or misalignment. No lubrication or not enough lubrication are the common causes of overheating, but dirty or wrong grades of oil are possible causes. Misalignment of the bearing concentrates the load on small areas at the edges of the bearing. Newly installed bearings can have too little side clearance between the shaft and the base, or too little clearance between the cap and the shaft. When the bearing cap is tightened down, there is severe binding between the shaft and the bearing. Pulling the bearing to one side when tightening down the base bolts frequently causes misalignment.

The main objective is to keep the bearing and the shaft within normal operating temperatures. Flood the bearing with the usual grade of oil to wash out any impurities or cuttings, then lubricate with a heavier grade of oil. The higher velocity oil will give better protection at a higher heat. Slack off cap bolts in a split bearing, or loosen base bolts if misalignment is suspected. If an air hose is available, direct a flow of air against the bearing and shaft, keeping the end of the hose a short distance from the bearing. Check the shaft for any signs of kinking or bending, as sometimes a chain will catch or break and wrap around the sprocket, pulling in the shaft. The hot bearing will not show until some time after the misalignment took place.

Storage

Oil drums should be stored under cover when possible. Drums exposed to the weather are best stood up with plastic lids to keep water from leaking past the plugs. Storage at a regular oil house at room temperature makes the oil easier to handle and pour. Without a separate and proper location to store the oil, a fire hazard will exist if oil is stored at random in the plant.

Small-quantity storage areas or stations in the plant, away from traffic and pedestrian ways and in a clean area, will do for day-to-day lubricating requirements. The storage bin or shelter should be made of metal or wood-covered metal and troughed to catch spills. Oils should be kept in covered containers with each grade of oil stored in a clearly marked container. Grease drums should be kept with the lids on.

For fire protection, a CO_2 or dry chemical extinguisher is needed at each storage area. Good housekeeping in wiping up oil spills will reduce fire and slipping hazards.

Best Maintenance Lubrication Practices

1 When using a grease hand gun: always clean the end of the grease gun and the grease fitting with a clean rag or towel.

2 When using a grease hand gun: always know the amount of grease required and the frequency. Ask a vendor's lubrication engineer to assist in this area. Check and mark your grease gun to ensure the amount of grease is known and can be visualized for each type of grease gun one uses.

3 Always take oil samples when changing oil in a gearbox.

4 When installing a new gearbox, replace the oil 24 hours after installation to remove any contamination that may have been washed from the gearbox cavity and gears.

5 Always add hydraulic fluid into a reservoir using a filter cart (see Chapter 17).

6 Never touch a hydraulic filter with your hand during installation. By touching the filter you will introduce contamination to the hydraulic system.

7 Never accept leaks on any type of lubrication line or bearing. Identify the true problem and make a permanent repair.

8 Ensure maintenance personnel performing lubrication practices score at least a 90% on the lubrication assessment in Chapter 3.

9 ALWAYS read and follow lubrication instructions from an equipment manufacturer. If you must change the instructions, contact the manufacturer first for comments.

In conclusion, lubrication can cause up to 80% of your equipment problems if not performed in a disciplined manner.

17 Machinery Installation

"Installed to Specification"

Introduction

Many people believe that machinery installation starts with the foundation. This is only true where this is a used piece of equipment. With new equipment, machinery installation actually starts after the selection is made and before the contract is drawn up. There are many items that need to be included in the contract before the new piece of machinery is even delivered. Remember, this is the time the manufacturer is trying to court you.

When drawing up the contract, make sure that these things are included. You may not get every point, but now is the time to try:

- Complete list of drawings (preferably on CAD). This list should include a complete set of components down to bearings, shaft size, etc.
- Installation recommendations including all electrical power needs.
- Spare parts recommendations. Be careful with this list because it is usually only intended to get you through the warranty period. Also make sure the list is broken down with manufacturer numbers, such as Dodge part #, etc.
- Extra copies of all manuals.
- Warranty to start once equipment is put into service, not when delivered to site.
- A list of preventive maintenance tasks with frequencies and lubrication requirements.

Foundation

The most important aspect of machinery installation is to provide a suitable base or support. The capacity of this foundation must have the ability to carry the machinery load without movement and maintaining placement.

Most foundations are constructed of concrete, but depending on the application, structural steel can be used. It's very important to look at the condition of the cement or steel foundation. Many times you are in an older facility where the concrete is already crumbling or the steel is rusted beyond use. If this is the case you may have to take out the old concrete and lay a new pad over the section or replace the structural steel. Other considerations should be adequate space for working/running the equipment and safety.

There are many types of anchor bolts in use in industry today, such as hooked in new concrete, compression, epoxy, etc. Make sure you check with the manufacturer for recommendations. When hold-down bolts are subject to extreme vibration and could possibly break at the thread section, a sleeve can be used extending to the pocket of the foundation to form what is commonly called a boxed anchor bolt. To allow for variations in casting or errors in layout, a space can be left around each bolt to allow for some change of bolt position. The use of short sections of pipe is commonly done. The pipe should be larger in diameter than the bolt and held firmly against the bottom of the template. Two or three times rod diameter is allowable for the pipe. The best and most common way of locating the anchor bolts is by the creation of a template with holes that match the mounting needs of the machine. This does not have to be anything too elaborate and can be made up of normal scrap parts. See Figure 17.1.

Due to the fact that over time most concrete slabs settle, whenever possible, do not mount the machinery directly to the concrete foundation. Best practices include the use of a bed plate, which tends to be bolted to the concrete by use of the anchor bolts. This is what the equipment is mounted and fastened to. This provides a firm, level surface allowing for shims to be used to accurately align the piece of equipment. By staying firmly in place during the leveling phase, the bed plate will help machinery to stay closely lined up. It is practical to use shims and grout. The grout will help support all the parts and hold them in position, and the shims are used for alignment and leveling. When constructing the foundation for grout and leveling, $\frac{3}{4}$" inches to $1\frac{1}{2}$" should be allowed. See Figure 17.2

Leveling and Elevation

The level is used to create a true horizontal plane. It is always a good idea to check the accuracy of your level on a known source before starting.

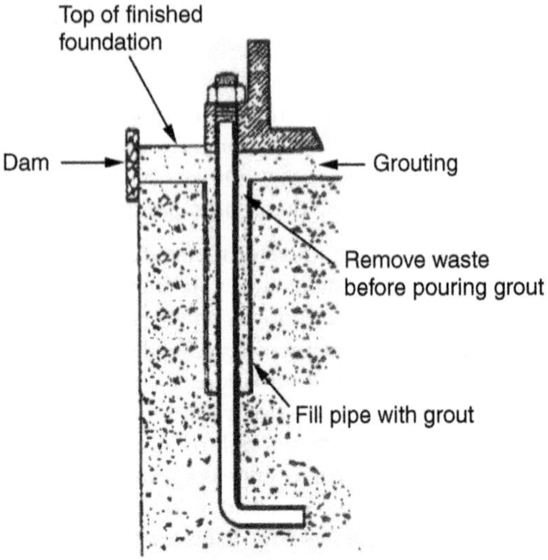

Figure 17.1 *Anchors*

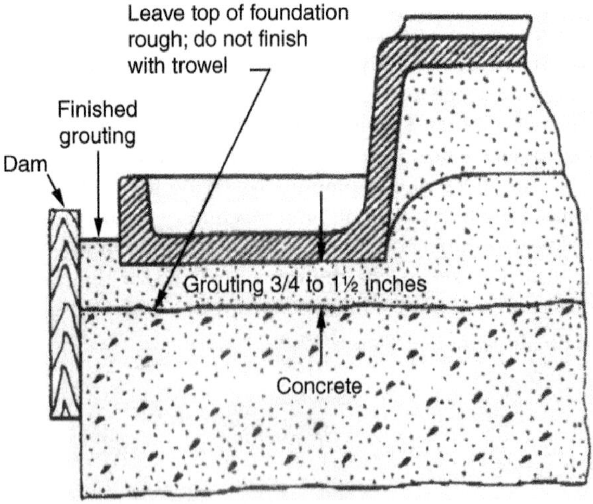

Figure 17.2 *Grouting and foundations*

Remember, level to the true plane may not be level to the continuous equipment you are attaching to. The level should be used with feeler gauges to measure the amount of drift to the true plane. The wedges of shims must bear the weight equally to prevent the frame from distorting when the bolts are tightened. Level on a machine is checked on two directions, both lengthwise and sideways with the use of a level on any major machine horizontal surface. For equipment with a shaft, a shaft level will give an easy means of checking the level in one direction. Once level has been established, tighten down all bolts and recheck.

Elevation and line of center is just as important as level. If this piece of machinery is used in a continuous line process, you need to have a true line of center with the process. The best and most accurate method to use is laser alignment. Once the piece of machinery is set into the process by the laser, make sure to mark the equipment in its position relative to the rest of the line at both ends. If for some reason the equipment gets moved, or you have a problem in the process, you can easily check to see if it is still in alignment.

Grout can be hand-mixed in buckets and is poured between the foundation and the plate. It is extremely important to grout fully around the inside of the bed plate. This will insure the stability of all shims and provide a supporting surface. Installations using pipe to allow for movement of the anchor bolt should have the pipe filled with grout for grouting under the base. Special care must be taken with boxed anchor bolts to keep grout from getting into the space between the bolt and the plate. Some people will actually get fancy with the grout and build wooden frames and tuck point edges. This is not necessary but does give a very professional look.

Before Mounting

One of the most common errors in machinery installation is not doing your homework. It is extremely important to check with the operators of that equipment to see how it functions and what requirements are needed. In some cases it may be getting to the reservoir to polish the hydraulic fluid or just having clear access underneath to keep it clean. Another point to consider is stub-ups for hydraulics or electricity, etc. You want to make sure that they come up in the most opportune spot to allow for easy excess, cleaning, and operation.

Development of your preventive maintenance tasks is highly recommended before mounting. This will help you to determine any special needs or considerations to accomplish these PM tasks. There are many times that maintenance is not considered until well after equipment is put into operation and it is too late to make changes to the equipment.

Deciding on the type of mounts to be used is critical. In today's modular factory settings equipment sometimes needs to have the flexibility to be pulled out or modified quickly. Also, certain types of equipment that require large amounts of rebuild may need to have quick disconnecting capabilities so they can be taken to the shop. Redundant equipment also needs this type of flexibility.

Machinery Mounts

There are many ways of mounting machinery that are not just rigidly bolting to the bed plate. Some equipment requires vibration dampening, automatic leveling, or mobility. These devices need to have the capability of quick adjustment for alignment and leveling needs. The industry standard term for this device is "Machinery Mounts," since we no longer require the use of anchor bolts. In essence the machine is now freestanding. The built-in leveling and alignment capabilities usually allow the machine to be set in a minimum amount of time to exacting limits. Since the primary anchor is the machine weight itself, it is important that the weight be solidly placed to the floor. As we stated earlier, floors are very seldom level; hence the need for the leveling devices to be built in.

Making sure to put all the recommendations of the manufacturer, operations, and maintenance engineering together is extremely important. The key to a successful installation and start-up is the involvement of all parties throughout the installation process. One other key point is to have a start-up plan. This should include prechecks, slow ramp-up, postchecks, and lookouts positioned at full ramp-up. Once the machine has been fully checked out, be sure to go back and check for alignment, tightness of bolts, and level.

18 Mixers and Agitators

Mixers are devices that blend combinations of liquids and solids into a homogenous product. They come in a variety of sizes and configurations designed for specific applications. Agitators provide the mechanical action required to keep dissolved or suspended solids in solution.

Both operate on basically the same principles, but variations in design, operating speed, and applications divide the actual function of these devices. Agitators generally work just as hard as mixers, and the terms are often used interchangeably.

Configuration

There are two primary types of mixers: propeller/paddle and screw. Screw mixers can be further divided into two types: batch and mixer-extruder.

Propeller/Paddle

Propeller/paddle mixers are used to blend or agitate liquid mixtures in tanks, pipelines, or vessels. Figure 18.1 illustrates a typical top-entering propeller/paddle mixer. This unit consists of an electric motor, a mounting bracket, an extended shaft, and one or more impeller(s) or propeller(s). Materials of construction range from bronze to stainless steel and are selected based on the particular requirements of the application.

The propeller/paddle mixer is also available in a side-entering configuration, which is shown in Figure 18.2. This configuration is typically used to agitate liquids in large vessels or pipelines. The side-entering mixer is essentially the same as the top-entering version except for the mounting configuration.

Both the top-entering and side-entering mixers may use either propellers, as shown in the preceding figures, or paddles, as illustrated by part b of Figure 18.3. Generally, propellers are used for medium- to high-speed applications where the viscosity is relatively low. Paddles are used in low-speed, high-viscosity applications.

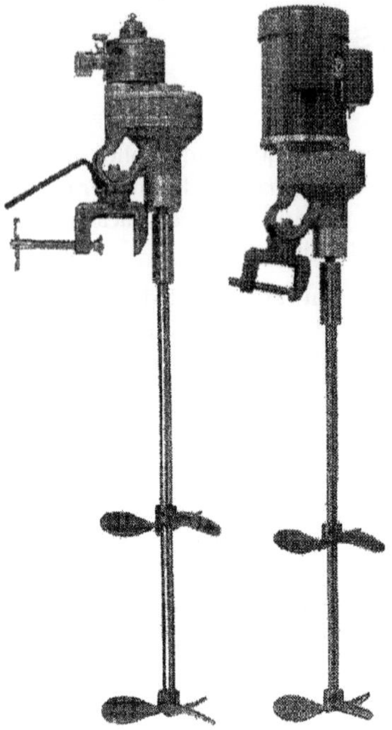

Figure 18.1 *Top-entering propeller-type mixer*

Screw

The screw mixer uses a single- or dual-screw arrangement to mix liquids, solids, or a combination of both. It comes in two basic configurations: batch and combination mixer-extruder.

Batch

Figure 18.4 illustrates a typical batch-type screw mixer. This unit consists of a mixing drum or cylinder, a single- or dual-screw mixer, and a power supply.

The screw configuration is normally either a ribbon-type helical screw or a series of paddles mounted on a common shaft. Materials of construction are selected based on the specific application and materials to be mixed.

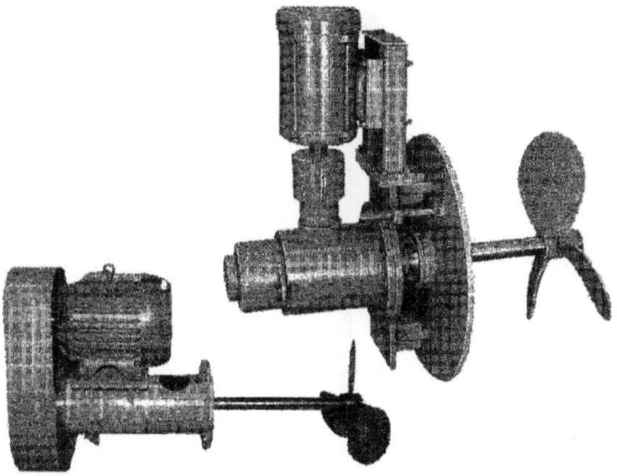

Figure 18.2 *Side-entering propeller-type mixer*

Typically, the screws are either steel or stainless steel, but other materials are available.

Combination Mixer-Extruder

The mixer-extruder combination unit shown in Figure 18.5 combines the functions of a mixer and screw conveyor. This type of mixer is used for mixing viscous products.

Performance

Unlike for centrifugal pumps and compressors, for mixers there are few criteria that can be used directly to determine effectiveness and efficiency. However, product quality and brake horsepower are indices that can be used to indirectly gauge performance.

Product Quality

The primary indicator of acceptable performance is the quality of the product delivered by the mixer. Although there is no direct way to measure this indicator, feedback from the quality assurance group should be used to verify that acceptable performance levels are attained.

Figure 18.3 *Mixer can use either propellers (a) or paddles (b) to provide agitation*

Figure 18.4 *Batch-type mixer uses single or dual screws to mix product*

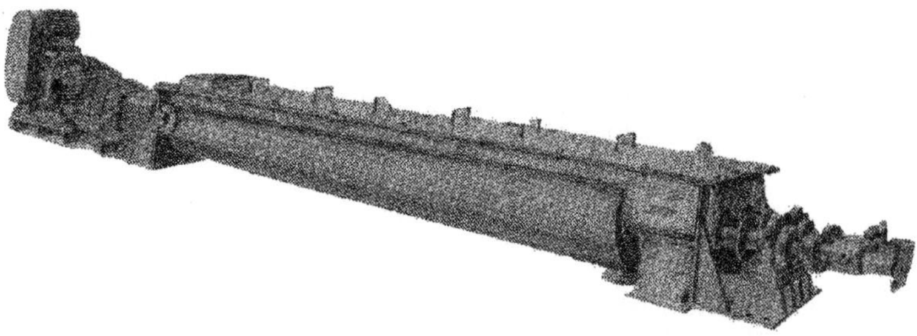

Figure 18.5 *Combination mixer-extruder*

Brake Horsepower

Variation in the actual brake horsepower required to operate a mixer is the primary indicator of its performance envelope. Mixer design, whether propeller- or screw-type, is based on the viscosity of both the incoming and finished product. These variables determine the brake horsepower required to drive the mixer, which will follow variations in the viscosity of the products being mixed. As the viscosity increases, so will the brake horsepower demand. Conversely, as the viscosity decreases, so will the horsepower require driving the mixer.

Installation

Installation of propeller-type mixers varies greatly, depending on the specific application. Top-entering mixers utilize either a clamp- or flange-type mounting. It is important that the mixer be installed so the propeller or paddle placement is at a point within the tank, vessel, or piping that assures proper mixing. Vendor recommendations found in O&M manuals should be followed to ensure proper operation of the mixer.

Mixers should be mounted on a rigid base that assures level alignment and prevents lateral movement of the mixer and its drive train. While most mixers can be bolted directly to a base, care must be taken to ensure that it is rigid and has the structural capacity to stabilize the mixer.

Operating Methods

There are only three major operating concerns for mixers: setup, incoming-feed rate, and product viscosity.

Mixer Setup

Both propeller and screw mixers have specific setup requirements. In the case of *propeller/paddle-type mixers*, the primary factor is the position of the propellers or paddles within the tank or vessel. Vendor recommendations should be followed to assure proper operation of the mixer.

If the propellers or paddles are too close to the liquid level, the mixer will create a vortex that will entrain air and prevent adequate blending or mixing. If the propellers are set too low, compress vortexing may occur. When this happens, the mixer will create a stagnant zone in the area under the rotating assembly. As a result, some of the product will settle in this zone, and proper mixing cannot occur. Setting the mixer too close to a corner or the side of the mixing vessel can also create a stagnant zone that will prevent proper blending or mixing of the product.

For *screw-type mixers*, proper clearance between the rotating element and the mixer housing must be maintained to vendor specifications. If the clearance is improperly set, the mixer will bind (i.e., not enough clearance) or fail to blend properly.

Feed Rate

Mixers are designed to handle a relatively narrow band of incoming product flow rate. Therefore, care must be exercised to ensure that the actual feed rate is maintained within acceptable limits. The O&M manuals provided by the vendor will provide the feed-rate limitations for various products. Normally, these rates must be adjusted for viscosity and temperature variations.

Viscosity

Variations in viscosity of both the incoming and finished products have a dramatic effect on mixer performance. Standard operating procedures should include specific operating guidelines for the range of variation that is acceptable for each application. The recommended range should include adjustments for temperature, flow rates, mixing speeds, and other factors that directly or indirectly affect viscosity.

Troubleshooting

Table 18.1 identifies common failure modes and their causes for mixers and agitators. Most of the problems that affect performance and reliability are caused by improper installation or variations in the product's physical properties.

Table 18.1 *Common failure modes of mixers and agitators*

	THE PROBLEM						
THE CAUSES	Surface vortex visible	Incomplete mixing of product	Excessive vibration	Excessive wear	Motor overheats	Excessive power demand	Excessive bearing failures
Abrasives in product				•			
Mixer/agitator setting too close to side or corner		•	•	•		•	•
Mixer/agitator setting too high	•	•					
Mixer/agitator setting too low		•		•			
Mixer/agitator shaft too long							•
Product temperature too low		•			•	•	
Rotating element imbalanced or damaged		•	•		•	•	•
Speed too high	•		•	•			
Speed too low		•					
Viscosity/specific gravity too high		•			•	•	
Wrong direction of rotation		•			•		•

Proper installation of mixers and agitators is critical. The physical location of the vanes or propellers within the vessel is the dominant factor to consider. If the vanes are set too close to the side, corner, or bottom of the vessel, a stagnant zone will develop that causes both loss of mixing quality and premature damage to the equipment. If the vanes are set too close to the liquid level, vortexing can develop. This will also cause a loss of efficiency and accelerated component wear.

Variations in the product's physical properties, such as viscosity, also will cause loss of mixing efficiency and premature wear of mixer components. Although the initial selection of the mixer or agitator may have addressed the full range of physical properties that will be encountered, applications sometimes change. Such a change may result in the use of improper equipment for a particular application.

19 Packing and Seals

All machines such as pumps and compressors that handle liquids or gases must include a reliable means of sealing around their shafts so that the fluid being pumped or compressed does not leak. To accomplish this, the machine design must include a seal located at various points to prevent leakage between the shaft and housing. In order to provide a full understanding of seal and packing use and performance, this manual discusses fundamentals, seal design, and installation practices.

Fundamentals

Shaft seal requirements and two common types of seals, packed stuffing boxes and simple mechanical seals, are described and discussed in this section. A packed box typically is used on slow- to moderate-speed machinery where a slight amount of leakage is permissible. A mechanical seal is used on centrifugal pumps or other type of fluid handling equipment where shaft sealing is critical.

Shaft Seal Requirements

Figure 19.1 shows the cross-section of a typical end-suction centrifugal pump where the fluid to be pumped enters the suction inlet at the eye of the impeller. Due to the relatively high speed of rotation, the fluid collected within the impeller vanes is held captive because of the close tolerance between the front face of the impeller and the pump housing.

With no other available escape route, the fluid is passed to the outside of the impeller by centrifugal force and into the volute, where its kinetic energy is converted into pressure. At the point of discharge (i.e., discharge nozzle), the fluid is highly pressurized compared to its pressure at the inlet nozzle of the pump. This pressure drives the fluid from the pump and allows a centrifugal pump to move fluids to considerable heights above the centerline of the pump.

This highly pressurized fluid also flows around the impeller to a lower pressure zone where, without an adequate seal, the fluid will leak along the drive

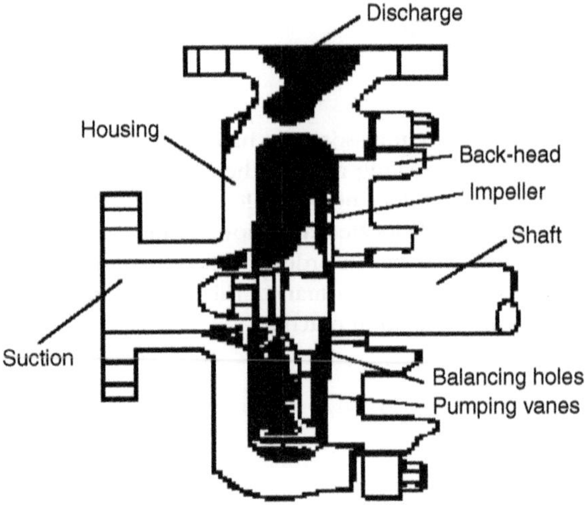

Figure 19.1 *Cross-section of a typical end-suction centrifugal pump*

shaft to the outside of the pump housing. The lower pressure results from a pump design intended to minimize the pressure behind the impeller. Note that this design element is specifically aimed at making drive shaft sealing easier.

Reducing the pressure acting on the fluid behind the impeller can be accomplished by two different methods, or a combination of both, on an open-impeller unit. One method is where small pumping vanes are cast on the backside of the impeller. The other method is for balance holes to be drilled through the impeller to the suction eye.

In addition to reducing the driving force behind shaft leakage, decreasing the pressure differential between the front and rear of the impeller using one or both of the methods described above greatly decreases the axial thrust on the drive shaft. This decreased pressure prolongs the thrust bearing life significantly.

Sealing Devices

Two sealing devices are described and discussed in this section: packed stuffing boxes and simple mechanical seals.

Packed Stuffing Boxes

Before the development of mechanical seals, a soft pliable material or packing placed in a box and compressed into rings encircling the drive shaft was used to prevent leakage. Compressed packing rings between the pump housing and the drive shaft, accomplished by tightening the gland-stuffing follower, formed an effective seal.

Figure 19.2 shows a typical packed box that seals with rings of compressed packing. Note that if this packing is allowed to operate against the shaft without adequate lubrication and cooling, frictional heat eventually builds up to

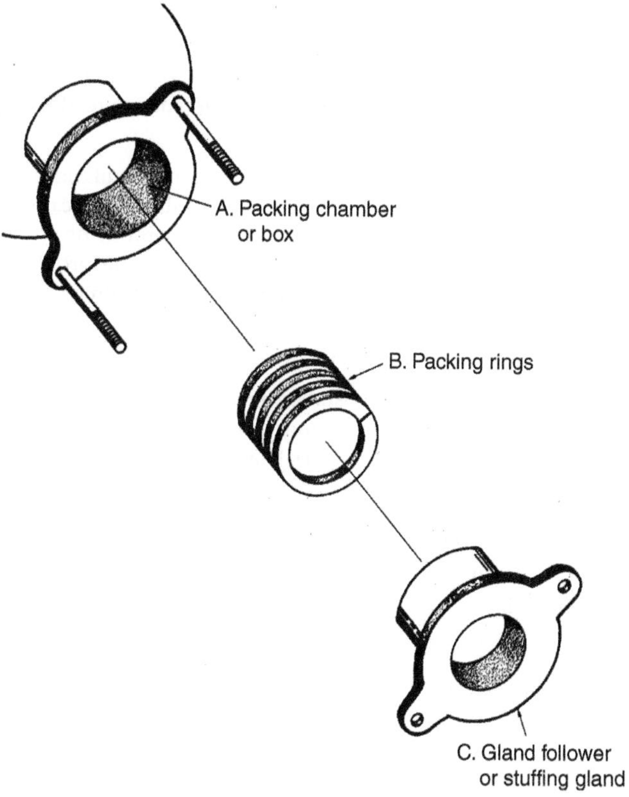

Figure 19.2 *Typical packed stuffing box*

the point of total destruction of the packing and damage to the drive shaft. Therefore, all packed boxes must have a means of lubrication and cooling.

Lubrication and cooling can be accomplished by allowing a small amount of leakage of fluid from the machine or by providing an external source of fluid. When leakage from the machine is used, leaking fluid is captured in collection basins that are built into the machine housing or baseplate. Note that periodic maintenance to recompress the packing must be carried out when leakage becomes excessive.

Packed boxes must be protected against ingress of dirt and air, which can result in loss of resilience and lubricity. When this occurs, packing will act like a grinding stone, effectively destroying the shaft's sacrificial sleeve, and cause the gland to leak excessively. When the sacrificial sleeve on the drive shaft becomes ridged and worn, it should be replaced as soon as possible. In effect, this is a continuing maintenance program that can readily be measured in terms of dollars and time.

Uneven pressures can be exerted on the drive shaft due to irregularities in the packing rings, resulting in irregular contact with the shaft. This causes uneven distribution of lubrication flow at certain locations, producing acute wear and packed-box leakages. The only effective solution to this problem is to replace the shaft sleeve or drive shaft at the earliest opportunity.

Simple Mechanical Seal

Mechanical seals, which are typically installed in applications where no leakage can be tolerated, are described and discussed in this section. Toxic chemicals and other hazardous materials are primary examples of applications where mechanical seals are used.

Components and Assembly

Figure 19.3 shows the components of a simple mechanical seal, which is made up of the following:

- Coil spring
- O-ring shaft packing
- Seal ring

The seal ring fits over the shaft and rotates with it. The spring must be made from a material that is compatible with the fluid being pumped so that it will withstand corrosion. Likewise, the same care must be taken in

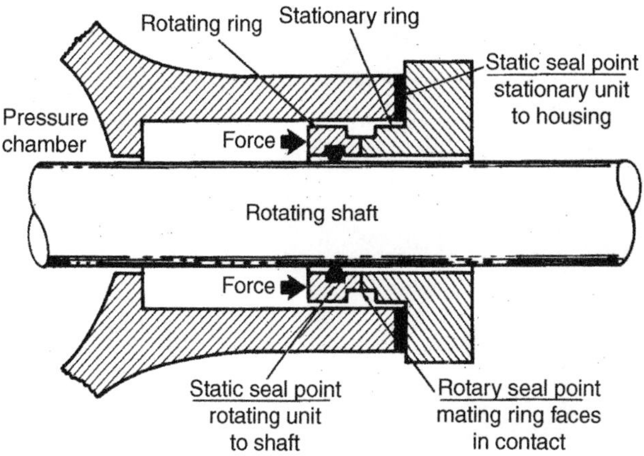

Figure 19.3 *Simple mechanical seal*

material selection of the O-ring and seal materials. The insert and insert O-ring mounting are installed in the bore cavity provided in the gland ring. This assembly is installed in a pump-stuffing box, which remains stationary when the pump shaft rotates.

A carbon graphite insertion ring provides a good bearing surface for the seal ring to rotate against. It is also resistive to attack by corrosive chemicals over a wide range of temperatures.

Figure 19.4 depicts a simple seal that has been installed in the pump's stuffing box. Note how the coil spring sits against the back of the pump's impeller, pushing the packing O-ring against the seal ring. By doing so, it remains in constant contact with the stationary insert ring.

As the pump shaft rotates, the shaft packing rotates with it due to friction. (In more complex mechanical seals, the shaft-packing element is secured to the rotating shaft by Allen screws.) There is also friction between the spring, the impeller, and the compressed O-ring. Thus, the whole assembly rotates together when the pump shaft rotates. The stationary insert ring is located within the gland bore. The gland itself is bolted to the face of the stuffing box. This part is held stationary due to the friction between the O-ring insert mounting and the inside diameter (ID) of the gland bore as the shaft rotates within the bore of the insert.

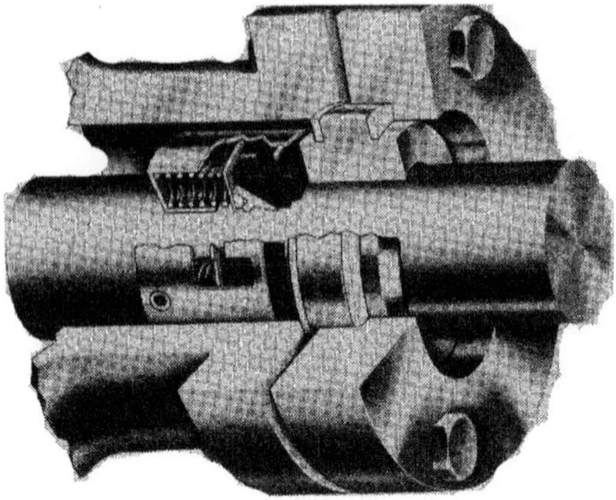

Figure 19.4 *Pump stuffing box seal*

How It Prevents Leakage

Having discussed how a simple mechanical seal is assembled in the stuffing box, we must now consider how the pumped fluid is stopped from leaking out to the atmosphere.

In Figure 19.4, the O-ring shaft packing blocks the path of the fluid along the drive shaft. Any fluid attempting to pass through the seal ring is stopped by the O-ring shaft packing. Any further attempt by the fluid to pass through the seal ring to the atmospheric side of the pump is prevented by the gland gasket and the O-ring insert. The only other place where fluid can potentially escape is the joint surface, which is between the rotating carbon ring and the stationary insert. (Note: The surface areas of both rings must be machined-lapped perfectly flat, measured in light bands with tolerances of one-millionth of an inch.)

Sealing Area and Lubrication

The efficiency of all mechanical seals is dependent upon the condition of the sealing area surfaces. The surfaces remain in contact with each other for the effective working life of the seal and are friction-bearing surfaces.

As in the compressed packing gland, lubrication also must be provided in mechanical seals. The sealing area surfaces should be lubricated and cooled with pumped fluid (if it is clean enough) or an outside source of clean fluid. However, much less lubrication is required with this type of seal because the frictional surface area is smaller than that of a compressed packing gland, and the contact pressure is equally distributed throughout the interface. As a result, a smaller amount of lubrication passes between the seal faces to exit as leakage.

In most packing glands there is a measurable flow of lubrication fluid between the packing rings and the shaft. With mechanical seals, the faces ride on a microscopic film of fluid that migrates between them, resulting in leakage. However, leakage is so slight that if the temperature of the fluid is above its saturation point at atmospheric pressure, it flashes off to vapor before it can be visually detected.

Advantages and Disadvantages

Mechanical seals offer a more reliable seal than compressed packing seals. Because the spring in a mechanical seal exerts a constant pressure on the seal ring, it automatically adjusts for wear at the faces. Thus, the need for manual adjustment is eliminated. Additionally, because the bearing surface is between the rotating and stationary components of the seal, the shaft or shaft sleeve does not become worn. Although the seal will eventually wear out and need replacing, the shaft will not experience wear.

However, much more precision and attention to detail must be given to the installation of mechanical seals compared to conventional packing. Nevertheless, it is not unusual for mechanical seals to remain in service for many thousands of operational hours if they have been properly installed and maintained.

Mechanical Seal Designs

Mechanical seal designs are referred to as friction drives, or single-coil spring seals, and positive drives.

Single-Coil Spring Seal

The seal shown back in Figure 19.4 depicts a typical friction drive or single-coil spring seal unit. This design is limited in its use because the seal relies on friction to turn the rotary unit. Because of this, its use is limited to liquids such as water or other nonlubricating fluids. If this type of seal is to be used with liquids that have natural lubricating properties, it must be mechanically locked to the drive shaft.

Although this simple seal performs its function satisfactorily, there are two drawbacks that must be considered. Both drawbacks are related to the use of a coil spring that is fitted over the drive shaft:

- One drawback of the spring is the need for relatively low shaft speeds because of a natural tendency of the components to distort at high surface speeds. This makes the spring push harder on one side of the seal than the other, resulting in an uneven liquid film between the faces. These cause excessive leakage and wear at the seal.
- The other drawback is simply one of economics. Because pumps come in a variety of shaft sizes and speeds, the use of this type of seal requires inventorying several sizes of spare springs, which ties up capital.

Nevertheless, the simple and reliable coil spring seal has proven itself in the pumping industry and is often selected for use despite its drawbacks. In regulated industries, this type of seal design far exceeds the capabilities of a compressed packing ring seal.

Positive Drive

There are two methods of converting a simple seal to positive drive. Both methods, which use collars secured to the drive shaft by setscrews, are shown in Figure 19.5. In this figure, the end tabs of the spring are bent at

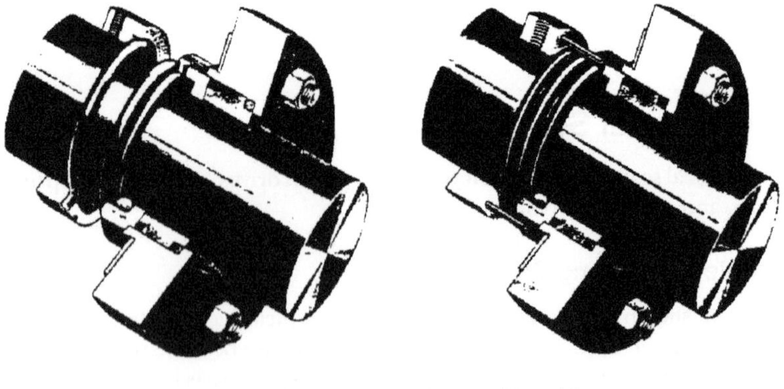

Figure 19.5 *Conversion of a simple seal to positive drive*

90 degrees to the natural curve of the spring. These end tabs fit into notches in both the collar and the seal ring. This design transmits rotational drive from the collar to the seal ring by the spring. Figure 19.5 also shows two horizontally mounted pins that extend over the spring from the collar to the seal ring.

Installation Procedures

This section describes the installation procedures for packed stuffing boxes and mechanical seals.

Packed Stuffing Box

This procedure provides detailed instructions on how to repack centrifugal pump packed stuffing boxes or glands. The methodology described here is applicable to other gland sealed units such as valves and reciprocating machinery.

Tool List

The following is a list of the tools needed to repack a centrifugal pump gland:

- Approved packing for specific equipment
- Mandrel sized to shaft diameter
- Packing ring extractor tool
- Packing board
- Sharp knife
- Approved cleaning solvent
- Lint-free cleaning rags

Precautions

The following precautions should be taken in repacking a packed stuffing box:

- Ensure coordination with operations control.
- Observe site and area safety precautions at all times.

- Ensure equipment has been electrically isolated and suitably locked out and tagged.
- Ensure machine is isolated and depressurized, with suction and discharge valves chained and locked shut.

Installation

The following are the steps to follow in installing a gland:

1 Loosen and remove nuts from the gland bolts.

2 Examine threads on bolts and nuts for stretching or damage. Replace if defective.

3 Remove the gland follower from the stuffing box and slide it along the shaft to provide access to the packing area.

4 Use packing extraction tool to carefully remove packing from the gland.

5 Keep the packing rings in the order they are extracted from the gland box. This is important in evaluating wear characteristics. Look for rub marks and any other unusual markings that would identify operational problems.

6 Carefully remove the lantern ring. This is a grooved, bobbin-like spool piece that is situated exactly on the centerline of the seal water inlet connection to the gland (Figure 19.6).

NOTE: It is most important to place the lantern ring under the seal water inlet connection to ensure the water is properly distributed within the gland to perform its cooling and lubricating functions.

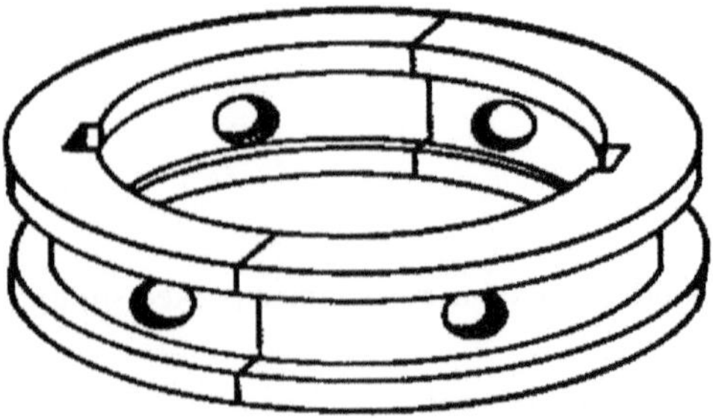

Figure 19.6 *Lantern ring or seal cage*

7 Examine the lantern ring for scoring and possible signs of crushing. Make sure the lantern ring's outside diameter (OD) provides a sliding fit with the gland box's internal dimension. Check that the lantern ring's ID is a free fit along the pump's shaft sleeve. If the lantern ring does not meet this simple criterion, replace it with a new one.

8 Continue to remove the rest of the packing rings as previously described. Retain each ring in the sequence that it was removed for examination.

9 Do not discard packing rings until they have been thoroughly examined for potential problems.

10 Turn on the gland seal cooling water slightly to ensure there is no blockage in the line. Shut the valve when good flow conditions are established.

11 Repeat Steps 1 through 10 with the other gland box.

12 Carefully clean out the gland stuffing boxes with a solvent-soaked rag to ensure that no debris is left behind.

13 Examine the shaft sleeve in both gland areas for excessive wear caused by poorly lubricated or overtightened packing.

NOTE: If the shaft sleeve is ridged or badly scratched in any way, the split housing of the pump may have to be split and the impeller removed for the sleeve to be replaced. Badly installed and maintained packing causes this.

14 Check total indicated runout (TIR) of the pump shaft by placing a magnetic base-mounted dial indicator on the pump housing and a dial stem on the shaft. Zero the dial and rotate the pump shaft one full turn. Record reading (Figure 19.7).

NOTE: If the TIR is greater than +/−0. 002 inches, the pump shaft should be straightened.

15 Determine the correct packing size before installing using the following method (Figure 19.8):
Measure the ID of the stuffing box, which is the OD at the packing (*B*), and the diameter of the shaft (*A*). With this data, the packing cross-section size is calculated by:

$$\text{Packing Cross Section} = \frac{B - A}{2}$$

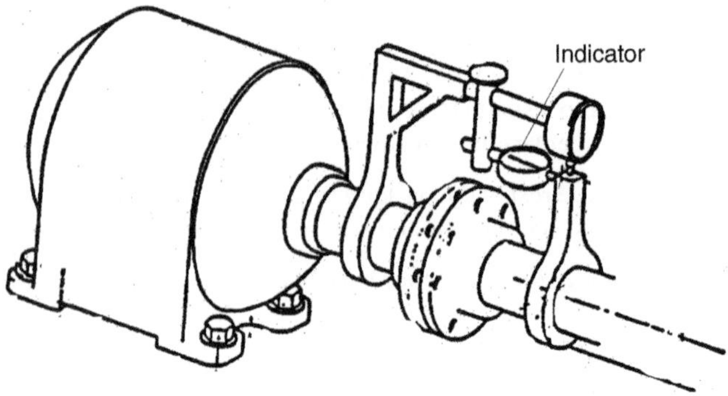

Figure 19.7 *Dial indicator check for run-out*

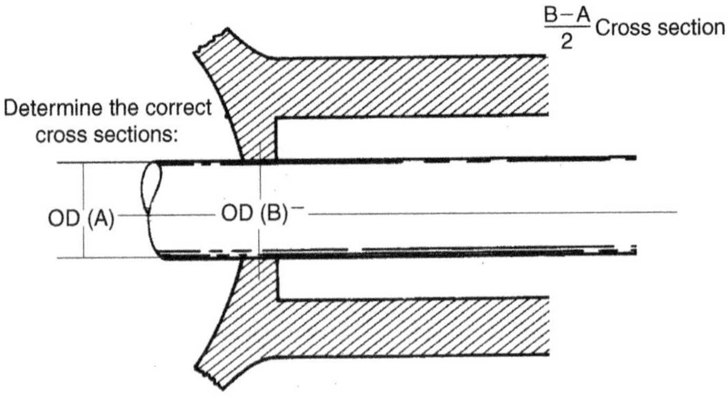

Figure 19.8 *Selecting correct packing size*

The packing length is determined by calculating the circumference of the packing within the stuffing box. The centerline diameter is calculated by adding the diameter of the shaft to the packing cross-section that was calculated in the preceding formula. For example, a stuffing box with a 4" ID and a shaft with a 2" diameter will require a packing cross-section of 1". The centerline of the packing would then be 3".

Therefore, the approximate length of each piece of packing would be:

$$\text{Packing Length} = \text{Centerline Diameter} \times 3.1416$$
$$= 3.0 \times 3.1416 = 9.43 \text{ inches}$$

The packing should be cut approximately $\frac{1}{4}$" longer than the calculated length so that the end can be bevel cut.

16 Controlled leakage rates easily can be achieved with the correct size packing.

17 Cut the packing rings to size on a wooden mandrel that is the same diameter as the pump shaft. Rings can be cut either square (butt cut) or diagonally (approximately 30 degrees). NOTE: Leave at least a $\frac{1}{6}$" gap between the butts regardless of the type of cut used. This permits the packing rings to move under compression or temperature without binding on the shaft surface.

18 Ensure that the gland area is perfectly clean and is not scratched in any way before installing the packing rings.

19 Lubricate each ring lightly before installing in the stuffing box. NOTE: When putting packing rings around the shaft, use an "S" twist. *Do not bend open*. See Figure 19.9.

20 Use a split bushing to install each ring, ensuring that the ring bottoms out inside the stuffing box. An offset tamping stick may be used if a split bushing is not available. *Do not use a screwdriver.*

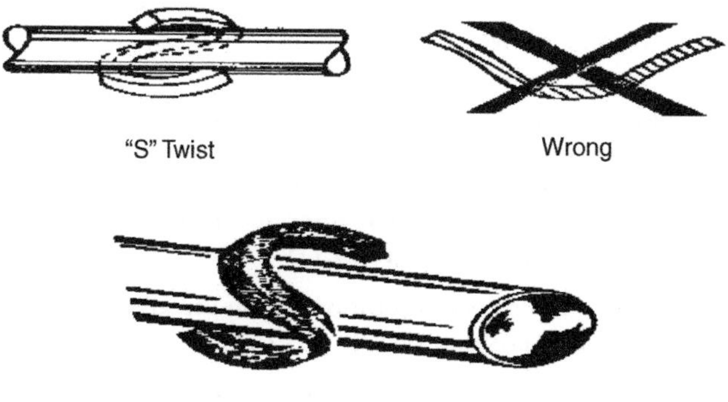

Figure 19.9 *Proper and improper installation of packing*

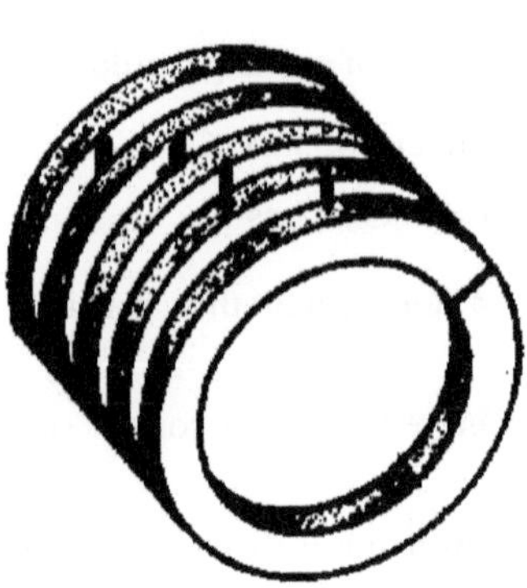

Figure 19.10 *Stagger butt joints*

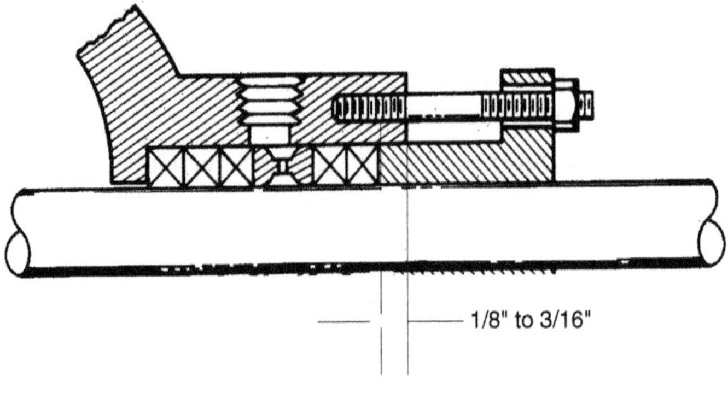

Figure 19.11 *Proper gland follower clearance*

21 Stagger the butt joints, placing the first ring butt at 12 o'clock; the second at 6 o'clock; the third at 3 o'clock; the fourth at 9 o'clock; etc., until the packing box is filled (Figure 19.10). NOTE: When the last ring has been installed, there should be enough room to insert the gland follower $\frac{1}{8}$ to 3/16 inches into the stuffing box (Figure 19.11).

22 Install the lantern ring in its correct location within the gland. Do not force the lantern ring into position (Figure 19.12).

23 Tighten up the gland bolts with a wrench to seat and form the packing to the stuffing box and shaft.

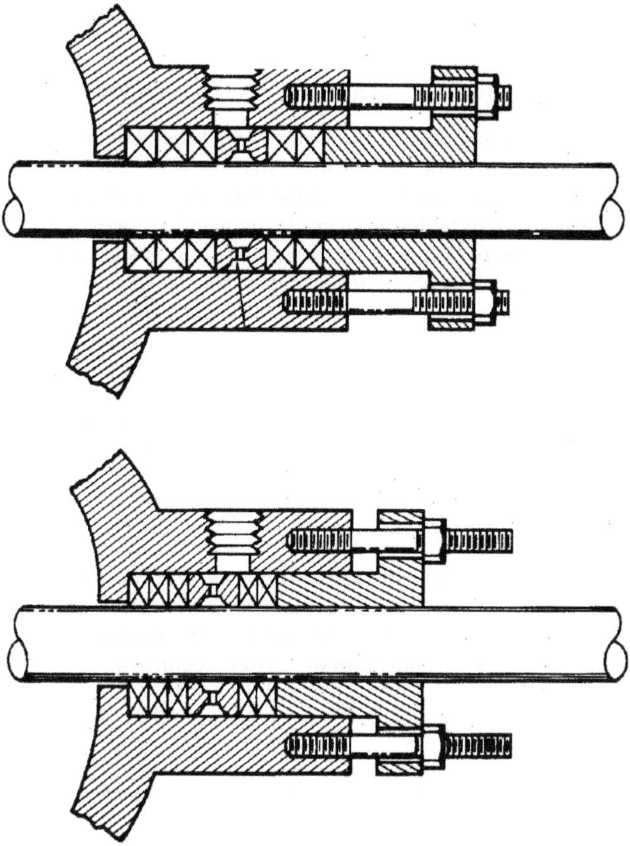

Figure 19.12 *Proper lantern ring installation*

24 Loosen the gland nuts one complete turn and rotate the shaft by hand to get running clearance.

25 Retighten the nuts finger tight only. Again, rotate the shaft by hand to make sure the packing is not too tight.

26 Contact operations to start the pump and allow the stuffing box to leak freely. Tighten the gland bolts one flat at a time until the desired leakage is obtained and the pump runs cool.

27 Clean up the work area and account for all tools before returning them to the tool crib.

28 Inform operations of project status and complete all paperwork.

29 After the pump is in operation, periodically inspect the gland to determine its performance. If it tends to leak more than the allowable amount, tighten by turning the nuts one flat at a time. Give the packing enough time to adjust before tightening it more. If the gland is tightened too much at one time, the packing can be excessively compressed, causing unnecessary friction and subsequent burnout of the packing.

Mechanical Seals

A mechanical seal's performance depends on the operating condition of the equipment where it is installed. Therefore, inspection of the equipment before seal installation can potentially prevent seal failure and reduce overall maintenance expenses.

Equipment Checkpoints

The pre-installation equipment inspection should include the following: stuffing box space, lateral or axial shaft movement (end play), radial shaft movement (whip or deflection), shaft runout (bent shaft), stuffing box face squareness, stuffing box bore concentricity, driver alignment, and pipe strain.

Stuffing Box Space

To properly receive the seal, the radial space and depth of the stuffing box must be the same as the dimensions shown on the seal assembly drawing.

Lateral or Axial Shaft Movement (Endplay)

Install a dial indicator with the stem against the shoulder of the shaft. Use a soft hammer or mallet to lightly tap the shaft on one end and then on the other. Total indicated endplay should be between 0.001 and 0.004 inches. A mechanical seal cannot work properly with a large amount of endplay or lateral movement. If the hydraulic condition changes (as frequently happens), the shaft could "float," resulting in sealing problems. Minimum endplay is a desirable condition for the following reasons:

- Excessive endplay can cause pitting, fretting, or wear at the point of contact between the shaft packing in the mechanical seal and the shaft or sleeve O.D. As the mechanical seal-driving element is locked to the shaft or sleeve, any excessive endplay will result in either overloading or underloading of the springs, causing excessive wear or leaks.

- Excessive endplay as a result of defective thrust bearings can reduce seal performance by disturbing both the established wear pattern and the lubricating film.
- A floating shaft can cause chattering, which results in chipping of the seal faces, especially the carbon element. Ideal mechanical seal performance requires a uniform wear pattern and a liquid film between the mating contact faces.

Radial Shaft Movement (Whip or Deflection)

Install the dial indicator as close to the radial bearing as possible. Lift the shaft or exert light pressure at the impeller end. If more than 0.002 to 0.003 inches of radial movement occurs, investigate bearings and bearing fits (especially the bore) for the radial bearing fit. An oversized radial bearing bore caused by wear, improper machining, or corrosion will cause excessive radial shaft movement resulting in shaft whip and deflection. Minimum radial shaft movement is important for the following reasons:

- Excessive radial movement can cause wear, fretting, or pitting of the shaft packing or secondary sealing element at the point of contact between the shaft packing and the shaft or sleeve OD.
- Extreme wear at the mating contact faces will occur when excessive shaft whip or deflection is present due to defective radial bearings or bearing fits. The contact area of the mating faces will be increased, resulting in increased wear and the elimination or reduction of the lubricating film between the faces, further shortening seal life.

Shaft Runout (Bent Shaft)

A bent shaft can lead to poor sealing and cause vibration. Bearing life is greatly reduced, and the operating conditions of both radial and thrust bearings can be affected.

Clamp the dial indicator to the pump housing and measure the shaft runout at two or more points on the OD of the impeller end of the shaft. Also measure the shaft runout at the coupling end of the shaft. If the runout exceeds 0.002 inches, remove the shaft and straighten or replace it.

Square Stuffing Box Face

With the pump stuffing box cover bolted down, clamp the dial indicator to the shaft with the stem against the face of the stuffing box. The total indicator runout should not exceed 0.003 inches.

When the face of the stuffing box is "out-of-square," or not perpendicular to the shaft axis, the result can be serious malfunction of a mechanical seal for the following reasons:

- The stationary gland plate that holds the stationary insert or seat in position is bolted to the face of the stuffing box. Misalignment will cause the gland to cock, resulting in cocking of the stationary element. This results in seal wobble or operation in an elliptical pattern. This condition is a major factor in fretting, pitting, and wearing of the mechanical seal shaft packing at the point of contact with the shaft or sleeve.
- A seal that is wobbling on the shaft can also cause wear on the drive pins. Erratic wear on the face contact causes poor seal performance.

Stuffing Box Bore Concentricity

With the dial indicator set up as described above, place the indicator stem well into the bore of the stuffing box. The stuffing box should be concentric to the shaft axis to within a 0.005-inch total indicator reading.

Eccentricity alters the hydraulic loading of the seal faces, reducing seal life and performance. If the shaft is eccentric to the box bore, check the slop, or looseness, in the pump bracket fits. Rust, atmospheric corrosion, or corrosion from leaking gaskets can cause damage to these fits, making it impossible to ensure a stuffing box that is concentric with the shaft. A possible remedy for this condition is welding the corroded area and remachining to proper dimensions.

Driver Alignment and Pipe Strain

Driver alignment is extremely important, and periodic checks should be performed. Pipe strain can also damage pumps, bearings, and seals.

In most plants, it is customary to blind the suction and discharge flanges of inactive pumps. These blinds should be removed before the pump driver alignment is made, or the alignment job is incomplete.

After the blinds have been removed and as the flanges on the suction and discharge are being connected to the piping, check the dial indicator reading on the outside diameter (OD) of the coupling half and observe movement of the indicator dial as the flanges are being secured. Deviation indicates pipe strain. If severe strain exists, corrective measures should be taken, or damage to the pump and unsatisfactory seal service can result.

Seal Checkpoints

The following are important seal checkpoints:

- Ensure that all parts are kept clean, especially the running faces of the seal ring and insert.
- Check the seal rotary unit, and make sure the drive pins and spring pins are free in the pinholes or slots.
- Check the setscrews in the rotary unit collar to see that they are free in the threads. Setscrews should be replaced after each use.
- Check the thickness of all gaskets against the dimensions shown on the assembly drawing. Improper gasket thickness will affect the seal setting and the spring load imposed on the seal.
- Check the fit of the gland ring to the equipment. Make sure there is no interference or binding on the studs or bolts or other obstructions. Be sure the gland ring pilot, if any, enters the bore with a reasonable guiding fit for proper seal alignment.
- Make sure all rotary unit parts of the seal fit over the shaft freely.
- Check both running faces of the seal (seal ring and insert) and be sure there are no nicks or scratches. Imperfections of any kind on either of these faces will cause leaks.

Installing the Seal

The following steps should be taken when installing a seal:

- Instruction booklets and a copy of the assembly drawing are shipped with each seal. Be sure each is available, and read the instructions before starting installation.
- Remove all burrs and sharp edges from the shaft or shaft sleeve, including sharp edges of keyways and threads. Worn shafts or sleeves should be replaced.
- Check the stuffing box bore and face to ensure they are clean and free of burrs.

- The shaft or sleeve should be lightly oiled before the seal is assembled to allow the seal parts to move freely over it. This is especially desirable when assembling the seal collar because the bore of the collar usually has only a few thousandths of an inch clearance. Care should be taken to avoid getting the collar cocked.
- Install the rotary unit parts on the shaft or sleeve in the proper order.
- Be careful when passing the seal gland ring and insert over the shaft. Do not bring the insert against the shaft because it might chip away small pieces from the edge of the running face.
- Wipe the seal faces clean and apply a clean oil film before completing the equipment assembly. A clean finger, which is not apt to leave lint, will do the best job when giving the seal faces the final wiping.
- Complete the equipment assembly, taking care when compressing the seal into the stuffing box.
- Seat the gland ring and gland ring gasket to the face of the stuffing box by tightening the nuts or bolts evenly and firmly. Be sure the gland ring is not cocked. Tighten the nuts or bolts only enough to affect a seal at the gland ring gasket, usually finger tight and $\frac{1}{2}$ to $\frac{3}{4}$ of a turn with a wrench. Excessively tightening the gland ring nut or bolt will cause distortion that will be transmitted to the running face, resulting in leaks.

If the seal assembly drawing is not available, the proper seal setting dimension for inside seals can be determined as follows:

- Establish a reference mark on the shaft or sleeve flush with the face of the stuffing box.
- Determine how far the face of the insert will extend into the stuffing box bore. This dimension is taken from the face of the gasket.
- Determine the compressed length of the rotary unit by compressing the rotary unit to the proper spring gap.
- This dimension added to the distance the insert extends into the stuffing box will give the seal setting dimension from the reference mark on the shaft or sleeve to the back of the seal collar.
- Outside seals are set with the spring gap equal to the dimension stamped on the seal collar.

- Cartridge seals are set at the factory and installed as complete assemblies. These assemblies contain spacers that must be removed after the seal assembly is bolted into position and the sleeve collar is in place.

Installation of Environmental Controls

Mechanical seals are often chosen and designed to operate with environmental controls. If this is the case, check the seal assembly drawing or equipment drawing to ensure that all environmental control piping is properly installed. Before equipment startup, all cooling and heating lines should be operating and remain so for at least a short period after equipment shutdown.

Before startup, all systems should be properly vented. This is especially important on vertical installations where the stuffing box is the uppermost portion of the pressure-containing part of the equipment. The stuffing box area must be properly vented to avoid a vapor lock in the seal area that would cause the seal to run dry.

On double seal installations, be sure the sealing liquid lines are connected, the pressure control valves are properly adjusted, and the sealing liquid system is operating before starting the equipment.

Seal Startup Procedures

When starting equipment with mechanical seals, make sure the seal faces are immersed in liquid from the beginning so they will not be damaged from dry operation. The following recommendations for seal startup apply to most types of seal installations and will improve seal life if followed:

- Caution the electrician not to run the equipment dry while checking motor rotation. A slight turnover will not hurt the seal, but operating full speed for several minutes under dry conditions will destroy or severely damage the rubbing faces.
- The stuffing box of the equipment, especially centrifugal pumps, should always be vented before startup. Even though the pump has a flooded suction, it is still possible that air may be trapped in the top of the stuffing box after the initial liquid purge of the pump.
- Check installation for need of priming. Priming might be necessary in applications with a low or negative suction head.
- Where cooling or bypass recirculation taps are incorporated in the seal gland, piping must be connected to or from these taps before startup.

These specific environmental control features must be used to protect the organic materials in the seal and to ensure its proper performance. Cooling lines should be left open at all times or whenever possible. This is especially true when a hot product might be passing through standby equipment while it is not online. Many systems provide for product to pass through the standby equipment, so the need for additional product volume or an equipment change is only a matter of pushing a button.

- With hot operational equipment that is shut down at the end of each day, it is best to leave the cooling water on at least long enough for the seal area to cool below the temperature limits of the organic materials in the seal.
- Face lubricated-type seals must be connected from the source of lubrication to the tap openings in the seal gland before startup. This is another predetermined environmental control feature that is mandatory for proper seal function. Where double seals are to be operated, it is necessary that the lubrication feed lines be connected to the proper ports for both circulatory or dead-end systems before equipment startup. This is very important because all types of double seals depend on the controlled pressure and flow of the sealing fluid to function properly. Even before the shaft is rotated, the sealing liquid pressure must exceed the product pressure opposing the seal. Be sure a vapor trap does not prevent the lubricant from reaching the seal face promptly.
- Thorough warm-up procedures include a check of all steam piping arrangements to be sure that all are connected and functioning, as products that will solidify must be fully melted before startup. It is advisable to leave all heat sources on during shutdown to ensure a liquid condition of the product at all times. Leaving the heat on at all times further facilitates quick startups and equipment switchovers that may be necessary during a production cycle.
- Thorough chilling procedures are necessary on some installations, especially liquefied petroleum gases (LPG) applications. LPG must always be kept in a liquid state in the seal area, and startup is usually the most critical time. Even during operation, the re-circulation line piped to the stuffing box might have to be run through a cooler in order to overcome frictional heat generated at the seal faces. LPG requires a stuffing box pressure that is greater than the vapor pressure of the product at pumping temperature (25 to 50 psi differential is desired).

Troubleshooting

Failure modes that affect shaft seals are normally limited to excessive leakage and premature failure of the mechanical seal or packing. Table 19.1 lists the common failure modes for both mechanical seals and packed boxes. As the table indicates, most of these failure modes can be directly attributed to misapplication, improper installation, or poor maintenance practices.

Mechanical Seals

By design, mechanical seals are the weakest link in a machine train. If there is any misalignment or eccentric shaft rotation, the probability of a mechanical seal failure is extremely high. Most seal tolerances are limited to no more than 0.002 inches of total shaft deflection or misalignment. Any deviation outside of this limited range will cause catastrophic seal failure.

Physical misalignment of a shaft will either cause seal damage, permitting some leakage through the seal, or it will result in total seal failure. Therefore, it is imperative that good alignment practices be followed for all shafts that have an installed mechanical seal.

Process and machine-induced shaft instability also create seal problems. Primary causes for this failure mode include: aerodynamic or hydraulic instability, critical speeds, mechanical imbalance, process load changes, or radical speed changes. These can cause the shaft to deviate from its true centerline enough to result in seal damage.

Chemical attack (i.e., corrosion or chemical reaction with the liquid being sealed) is another primary source of mechanical seal problems. Generally, two primary factors cause chemical attack: misapplication or improper flushing of the seal.

Misapplication is another major cause of premature seal failure. Little attention is generally given to the selection of mechanical seals. Most plants rely on the vendor to provide a seal that is compatible with the application. Too often there is a serious breakdown in communications between the end user and the vendor on this subject. Either the procurement specification does not provide the vendor with appropriate information, or the vendor does not offer the option of custom ordering the seals. Regardless of the reason, mechanical seals are often improperly selected and used in inappropriate applications.

Table 19.1 *Common failure modes of packing and mechanical seals*

		THE CAUSES	THE PROBLEM: Excessive leakage	Continuous stream of liquid	No leakage	Shaft hard to turn	Shaft damage under packing	Frequent replacement required	Bellows spring failure	Seal face failure
Packed box	Nonrotating	Cut ends of packing not staggered	•	•				•		
		Line pressure too high	•							
		Not packed properly				•	•	•		
		Packed box too loose	•	•						
		Packing gland too loose	•	•						
		Packing gland too tight	•	•		•	•	•		
	Rotating	Cut end of packing not staggered		•						
		Line pressure too high	•							
		Mechanical damage (seals, seat)	•	•	•			•		
		Noncompatible packing	•	•			•			
		Packing gland too loose	•							
		Packing gland too tight			•		•	•		
Mechanical seal	Internal flush	Flush flow/pressure too low							•	•
		Flush pressure too high	•	•					•	•
		Improperly installed	•						•	•
		Induced Misalignment	•							
		Internal flush line plugged							•	•
		Line pressure too high							•	•
		Physical shaft misalignment	•							
		Seal not compatible with application	•							
	External flush	Contamination in flush liquid	•							•
		External flush line plugged							•	•
		Flush flow/pressure too low							•	•
		flush pressure too high	•	•					•	•
		Improperly installed	•							•
		Induced misalignment	•						•	•
		Line pressure too high							•	•
		Physical shaft misalignment	•						•	•
		Seal not compatible with application	•							•

Source: Integrated Systems Inc.

Seal Flushing

When installed in corrosive chemical applications, mechanical seals must have a clear water flush system to prevent chemical attack. The flushing system must provide a positive flow of clean liquid to the seal and also provide an enclosed drain line that removes the flushing liquid. The flow rate and pressure of the flushing liquid will vary depending on the specific type of seal but must be enough to assure complete, continuous flushing.

Packed Boxes

Packing is used to seal shafts in a variety of applications. In equipment where the shaft is not continuously rotating (e.g., valves), packed boxes can be used successfully without any leakage around the shaft. In rotating applications, such as pump shafts, the application must be able to tolerate some leakage around the shaft.

Nonrotating Applications

In nonrotating applications, packing can be installed tightly enough to prevent leakage around the shaft. As long as the packing is properly installed and the stuffing-box gland is properly tightened, there is very little probability that seal failure will occur. This type of application does require periodic maintenance to ensure that the stuffing-box gland is properly tightened or that the packing is replaced when required.

Rotating Applications

In applications where a shaft continuously rotates, packing cannot be tight enough to prevent leakage. In fact, some leakage is required to provide both flushing and cooling of the packing. Properly installed and maintained packed boxes should not fail or contribute to equipment reliability problems. Proper installation is relatively easy, and routine maintenance is limited to periodic tightening of the stuffing-box gland.

20 Precision Measurement

Introduction

Precision measurement is an important part of any maintenance procedure. Without micrometers, telescopic gauges, dial calipers, edge finders, and other precision measuring tools, the job cannot be done correctly. The areas covered in this chapter are:

1 The proper use of an outside micrometer

2 The proper use of an inside micrometer

3 The proper use of telescopic gauges

4 The proper use of dial calipers

Micrometers

Precision measurement is an important part of the correct installation of equipment. One of the most important precision measurement tools available to the technician is the micrometer.

A difference of 0.001" may not seem important for most purposes, but some parts of equipment or tools must fit even more closely than that, even as close as .0001".

The most common type of micrometer is operated by a screw that has 40 threads to the inch. Each revolution of the screw moves the measuring spindle 0.025". A scale revolving with the screw is divided into 25 parts and indicates, therefore, the fractions of a turn in units of 0.001".

Outside Micrometer

A Vernier scale micrometer can measure objects to .001" or .0001". Measurements for the outside micrometer are taken on the outside of an object like a shaft (see Figure 20.1).

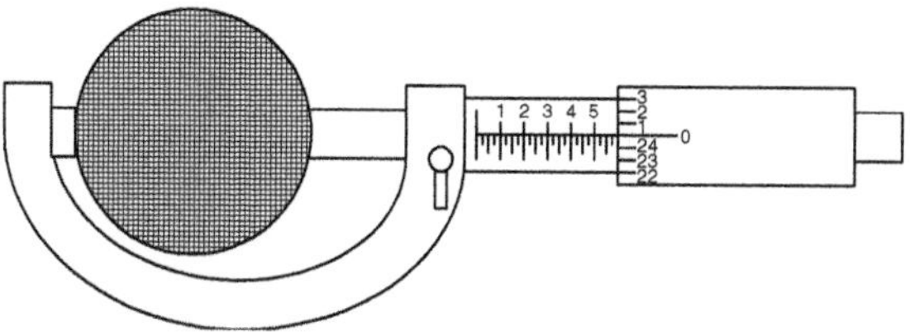

Figure 20.1 *Outside micrometer*

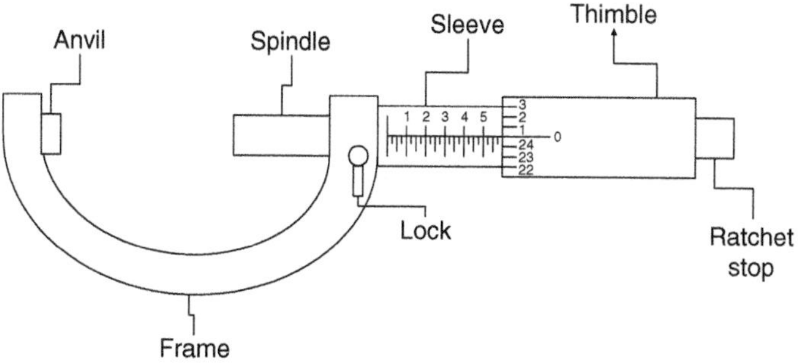

Figure 20.2 *Defining parts of a micrometer*

Standards

Standards are used to check the accuracy of the micrometers. These are precision blocks that are cut to an exact measurement. The micrometer is then used to measure the standard. The measurement on the micrometer must match that of the standard. If there is any variation then the micrometer must be adjusted.

Let's take a look at the names for the specific parts of the micrometer (see Figure 20.2).

The scale on the sleeve is graduated in .025". The scale on the thimble is graduated in .001". See Figures 20.3 and 20.4.

Now let's see if we can put the two parts together and come up with a measurement. Write down the measurement for the following drawing. See Figure 20.5.

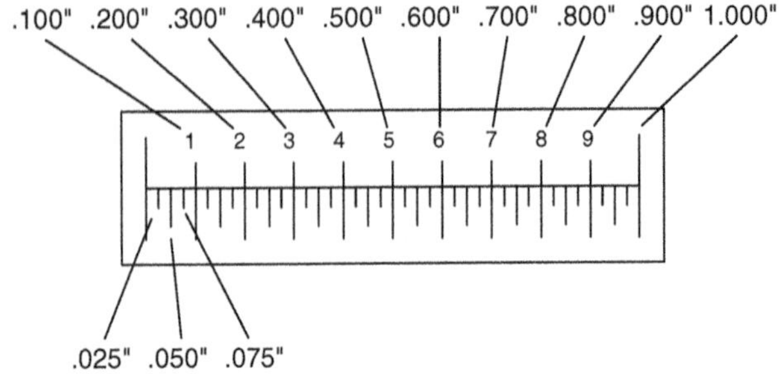

Figure 20.3 *Vernier scale*

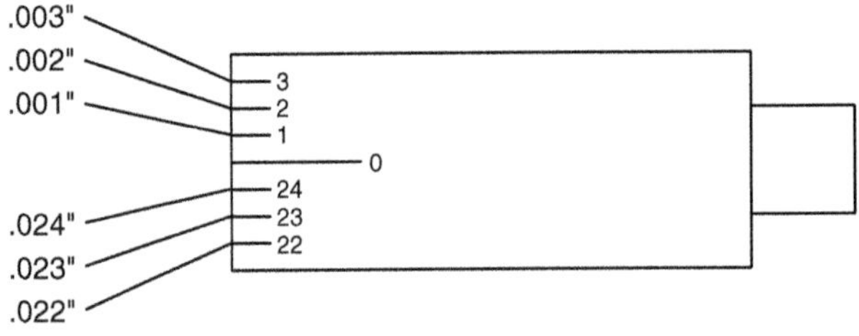

Figure 20.4 *Micrometer scale*

To find the measurement we start at the sleeve and add .600" + .075" + .000" = .675"

Vernier Scale

When using a micrometer, there may be a need to take measurements that are closer than .001". When this is necessary, a micrometer with a Vernier scale is used.

A Vernier scale will make measurements to within .0001 (one ten-thousandth) of an inch. The Vernier scale is located on top of the sleeve and is read by lining up the lines on the sleeve with those on the thimble.

In the Figure 20.6, we can see that the reading is .350", but we know that in order to take the reading to within .0001" we must also line up the lines

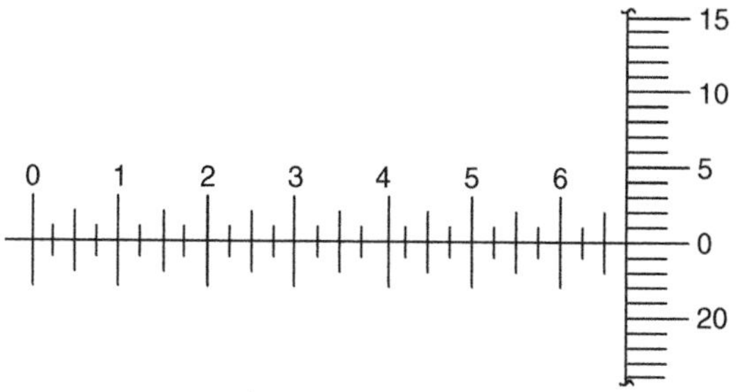

Figure 20.5 *Micrometer*

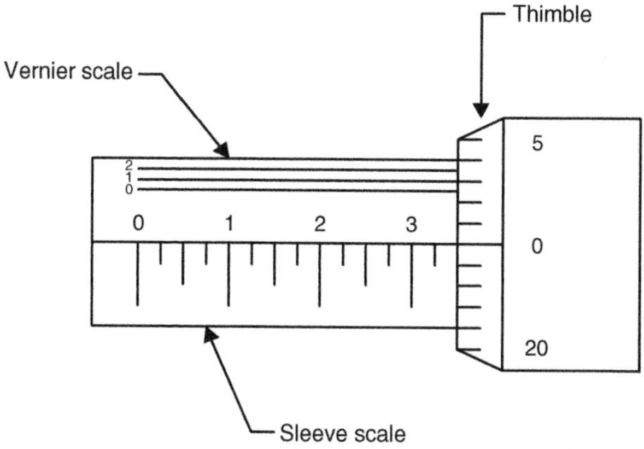

Figure 20.6 *Micrometer readings*

on the thimble with the lines on the sleeve. So, our actual reading would be .3501".

Inside Micrometer

In addition to outside micrometers, you must also become familiar with inside micrometers. Inside micrometers work the same as outside micrometers, except that they measure inside dimensions. See Figure 20.7.

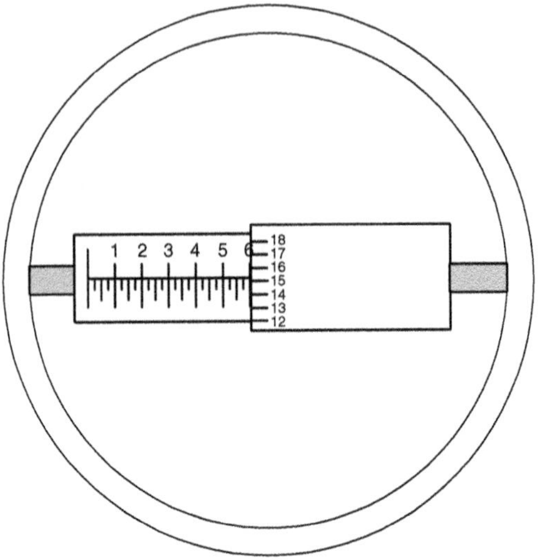

Figure 20.7 *Inside micrometer*

Telescopic Gauges

Another tool used for precision measurement is the snap, or telescopic, gauge. The telescopic gauge measures inside dimensions by adjusting to the correct bore size, then measuring it with a micrometer. See Figure 20.8.

Dial Caliper

Another tool that is widely used is the dial caliper. The dial caliper can take inside, outside, and depth measurements. The only drawback is that it is not as accurate as the micrometer (measurements to within .001"). See Figure 20.9.

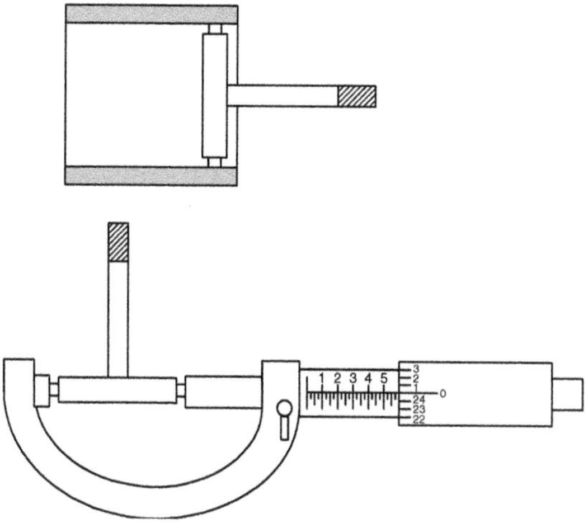

Figure 20.8 *Telescoping gauges measuring inside diameter*

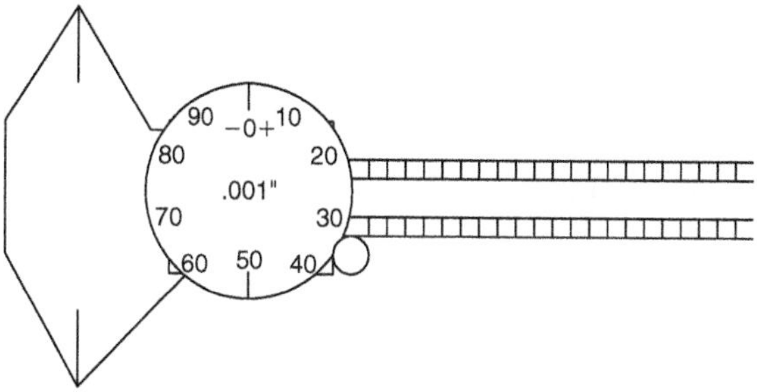

Figure 20.9 *Dial caliper*

Performance Exercises

Outside Micrometer

Now let's do some performance exercises with the outside micrometer. You will need a box of drill bits, drill rods, feeler gauges, or calibration standards that have the size written in thousandths of an inch (.001") and

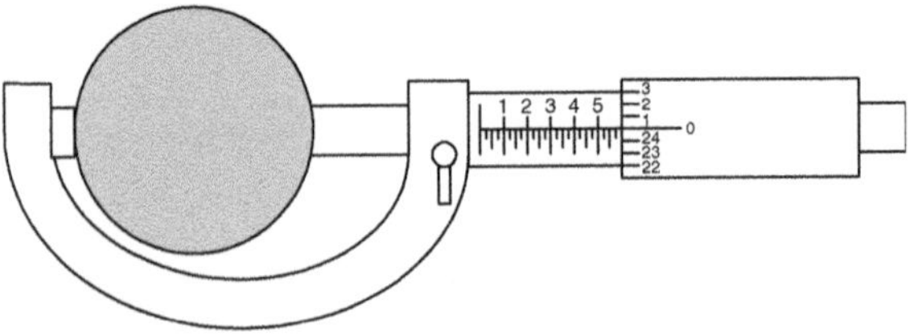

Figure 20.10 *Exercise 1*

an outside micrometer. Measure each drill bit and compare the reading that you get to the size on the drill index. Write down the answers that you get and keep them, because you will need them in exercises that follow. See Figure 20.10.

Inside Micrometer

Now let's try some inside measurements with the inside micrometer. Remember, this is similar to the outside micrometer, only you are measuring inside dimensions. You will need an assortment of pillow-block bearings to perform these exercises.

First note the dimension that is stamped on the outside of the bearing, then convert this from fractions to decimals. To do this simply divide the top number (numerator) by the bottom number (denominator). The problem will look like this:

Let's say that the bearing size is $\frac{3}{4}$". Just divide the top number by the bottom number, which will give you the decimal equivalent.

$$\frac{3}{4} = .75$$

Your measurement should be .75 on the micrometer scale. Remember to write down your answers, as you will need them in another exercise. See Figure 20.11.

Telescopic Gauges

Let's see how using a set of snap gauges compares to using an inside micrometer.

Using the same pillow-block bearings, turn the end of the handle counter-clockwise, squeeze together the snap gauge, and turn the handle back clockwise (this will lock the gauge). Then insert the snap gauges inside the

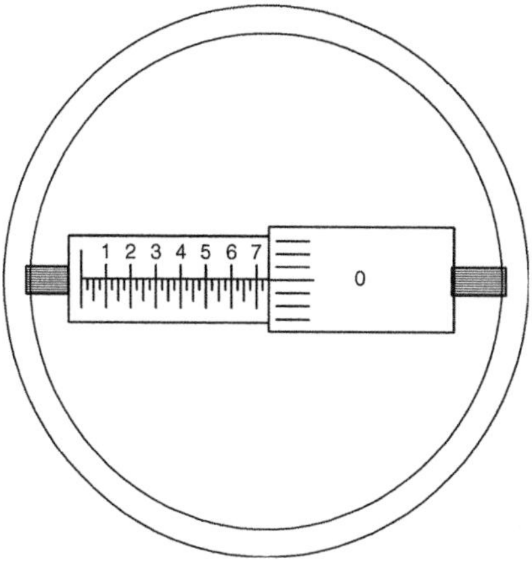

Figure 20.11 *Exercise 2*

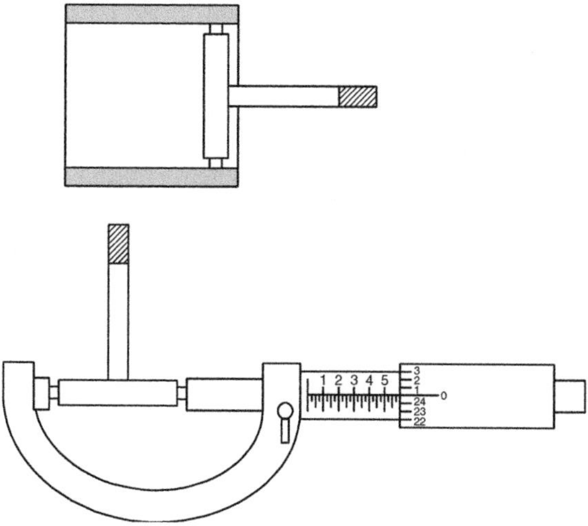

Figure 20.12 *Exercise 3*

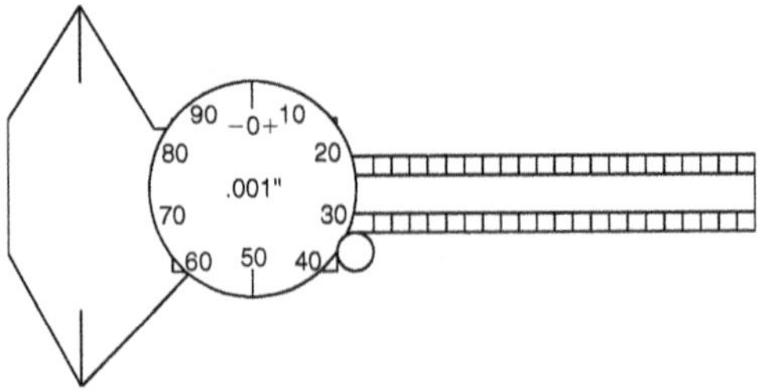

Figure 20.13 *Exercise 4*

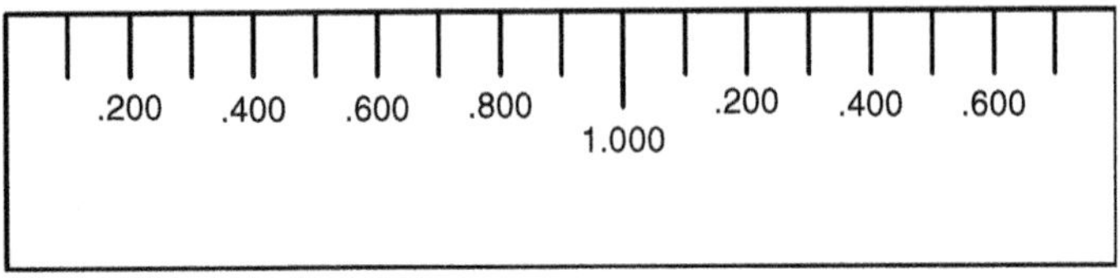

Figure 20.14 *Exercise 5*

bearing bore just as you did with the inside micrometer. Turn the handle counterclockwise to unlock the gauge. Holding the gauge perpendicular to the bearing, turn the handle clockwise to lock. Now using an outside micrometer, measure the dimension of the gauge. This is the inside diameter of the bearing bore. See Figure 20.12.

Dial Caliper

A dial caliper is similar to an inside and outside micrometer, but it can take both inside and outside measurements with just one device. A dial caliper has two measurement scales: the scale on the long flat body is graduated in .100 of an inch, and the round dial is graduated in .001 of an inch. See Figures 20.13 and 20.14.

Wrap-Up Exercise

Using the same pillow-block bearings and drill bits that were used in the previous exercises, measure the objects and compare the measurement that you get with the dial caliper to that of the micrometers. If you did not write down your answers from the previous exercise, then repeat the other exercises along with this one.

21 Pumps

Centrifugal Pumps

Centrifugal pumps basically consist of a stationary pump casing and an impeller mounted on a rotating shaft. The pump casing provides a pressure boundary for the pump and contains channels to properly direct the suction and discharge flow. The pump casing has suction and discharge penetrations for the main flow path of the pump and normally has a small drain and vent fittings to remove gases trapped in the pump casing or to drain the pump casing for maintenance.

Figure 21.1 is a simplified diagram of a typical centrifugal pump that shows the relative locations of the pump suction, impeller, volute, and discharge. The pump casing guides the liquid from the suction connection to the center, or eye, of the impeller. The vanes of the rotating *impeller* impart a radial and rotary motion to the liquid, forcing it to the outer periphery of the pump casing, where it is collected in the outer part of the pump casing called the volute.

The *volute* is a region that expands in cross-sectional areas as it wraps around the pump casing. The purpose of the volute is to collect the liquid discharged

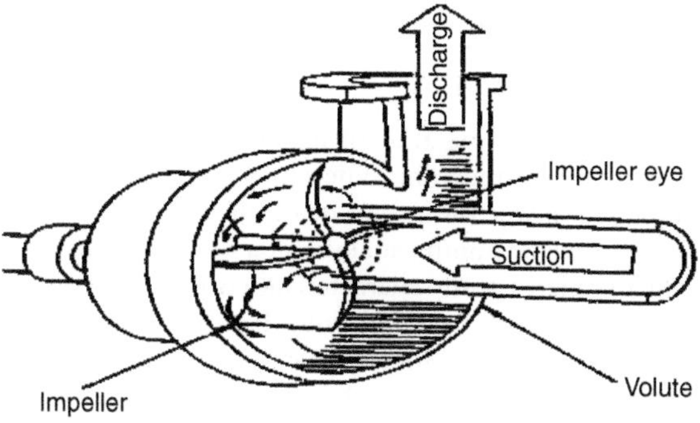

Figure 21.1 *Centrifugal pump*

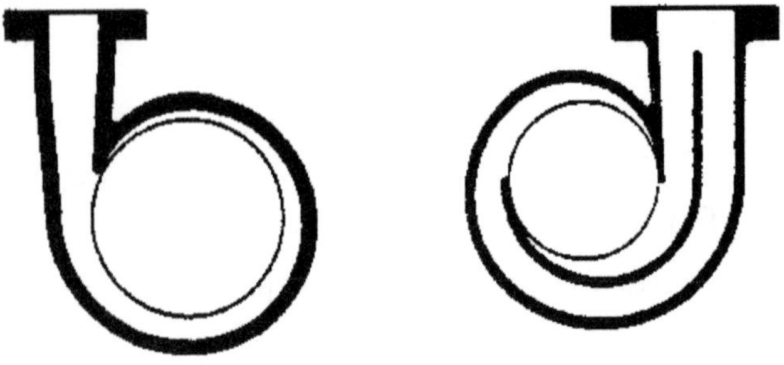

Figure 21.2 *Single and double volute*

from the periphery of the impeller at high velocity and gradually cause a reduction in fluid velocity by increasing the flow area. This converts the velocity head to static pressure. The fluid is then discharged from the pump through the discharge connection. Figure 21.2 illustrates the two types of volutes.

Centrifugal pumps can also be constructed in a manner that results in two distinct volutes, each receiving the liquid that is discharged from a 180 degrees region of the impeller at any given time. Pumps of this type are called double volute pumps. In some applications the double volute minimizes radial forces imparted to the shaft and bearings due to imbalances in the pressure around the impeller.

Characteristics Curve

For a given centrifugal pump operating at a constant speed, the flow rate through the pump is dependent upon the differential pressure or head developed by the pump. The lower the pump head, the higher the flow rate. A vendor manual for a specific pump usually contains a curve of pump flow rate versus pump head called a pump characteristic curve. After a pump is installed in a system, it is usually tested to ensure that the flow rate and head of the pump are within the required specifications. A typical centrifugal pump characteristic curve is shown in Figure 21.3.

There are several terms associated with the pump characteristic curve that must be defined. *Shutoff head* is the maximum head that can be developed

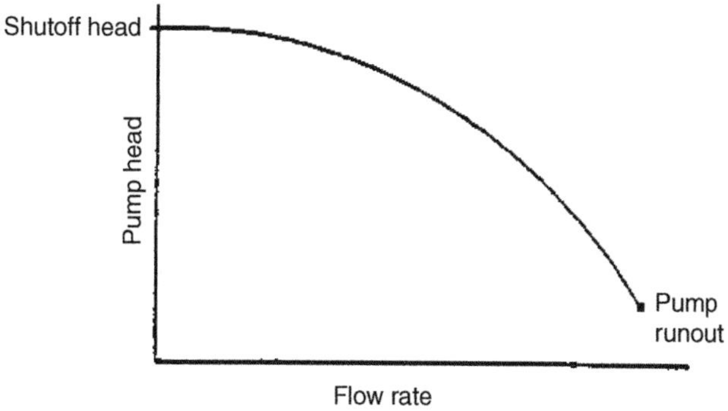

Figure 21.3 *Centrifugal pump characteristics curve*

by a centrifugal pump operating at a set speed. ***Pump run-out*** is the point where a centrifugal pump can develop the maximum flow without damaging the pump. Centrifugal pumps must be designed to be protected from the conditions of pump run-out or operating at shutoff head.

Protection

A centrifugal pump is deadheaded when it is operated with a closed discharge valve or against a seated check valve. If the discharge valve is closed and there is no other flow path available to the pump, the impeller will churn the same volume of water as it rotates in the pump casing. This will increase the temperature of the liquid in the pump casing to the point that it will flash to vapor. If the pump is run in this condition for a significant amount of time, it will become damaged.

When a centrifugal pump is installed in a system in such a way that it may be subjected to periodic shutoff head conditions, it is necessary to provide some means of pump protection. One method for protecting the pump from running deadheaded is to provide a recirculation line from the pump discharge line upstream of the discharge valve, back to the pump's supply source. The recirculation line should be sized to allow enough flow through the pump to prevent overheating and damage to the pump. Protection may also be accomplished by use of an automatic flow control device.

Centrifugal pumps must also be protected from runout. One method for ensuring that there is always adequate flow resistance at the pump discharge to prevent excessive flow through the pump is to place an orifice or a throttle valve immediately downstream of the pump discharge.

Gas Binding

Gas binding of a centrifugal pump is a condition in which the pump casing is filled with gases or vapors to the point where the impeller is no longer able to contact enough fluid to function correctly. The impeller spins in the gas bubble but is unable to force liquid through the pump.

Centrifugal pumps are designed so that their pump casings are completely filled with liquid during pump operation. Most centrifugal pumps can still operate when a small amount of gas accumulates in the pump casing, but pumps in systems containing dissolved gases that are not designed to be self-venting should be periodically vented manually to ensure that gases do not build up in the pump casing.

Priming

Most centrifugal pumps are not self-priming. In other words, the pump casing must be filled with liquid before the pump is started, or the pump will not be able to function. If the pump casing becomes filled with vapors or gases, the pump impeller becomes gas-bound and incapable of pumping. To ensure that a centrifugal pump remains primed and does not become gas-bound, most centrifugal pumps are located below the level of the source from which the pump is to take its suction. The same effect can be gained by supplying liquid to the pump suction under pressure supplied by another pump placed in the suction line.

Classification by Flow

Centrifugal pumps can be classified based on the manner in which fluid flows through the pump. The manner in which fluid flows through the pump is determined by the design of the pump casing and the impeller. The three types of flow through a centrifugal pump are radial flow, axial flow, and mixed flow.

Radial Flow

In a radial flow pump, the liquid enters at the center of the impeller and is directed out along the impeller blades in a direction at right angles to

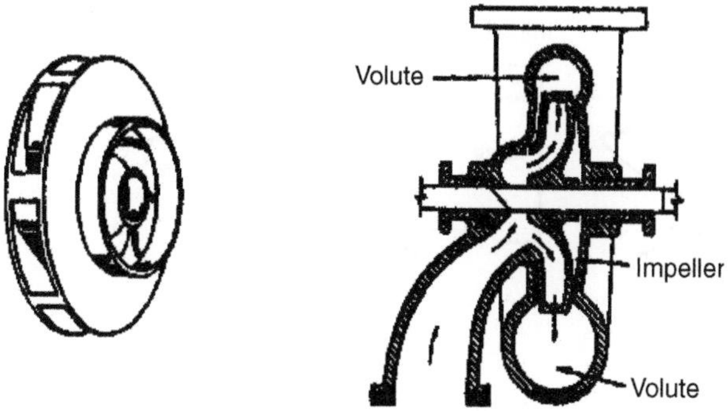

Figure 21.4 *Radial flow centrifugal pump*

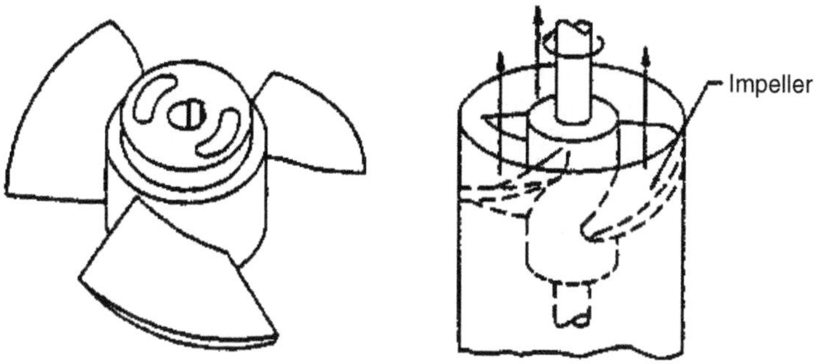

Figure 21.5 *Typical axial flow centrifugal pump*

the pump shaft. The impeller of a typical radial flow pump and the flow is illustrated in Figure 21.4.

Axial Flow

In an axial flow pump, the impeller pushes the liquid in a direction parallel to the pump shaft. Axial flow pumps are sometimes called propeller pumps because they operate essentially the same as the propeller of a boat. The impeller of a typical axial flow pump and the flow through a radial flow pump are shown in Figure 21.5.

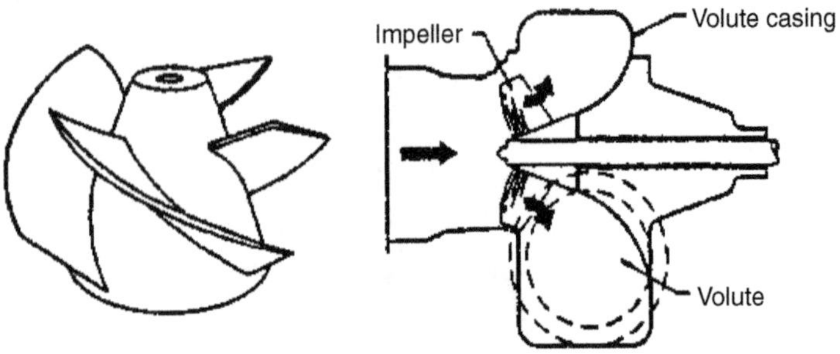

Figure 21.6 *Typical mixed flow pump*

Mixed Flow

Mixed flow pumps borrow characteristics from both radial flow and axial flow pumps. As liquid flows through the impeller of a mixed flow pump, the impeller blades push the liquid out away from the pump shaft and to the pump suction at an angle greater than 90 degrees. The impeller of a typical mixed flow pump and the flow through a mixed flow pump are shown in Figure 21.6.

Multistage Pumps

A centrifugal pump with a single impeller that can develop a differential pressure of more than 150 psid between the suction and the discharge is difficult and costly to design and construct. A more economical approach to developing high pressures with a single centrifugal pump is to include multiple impellers on a common shaft within the same pump casing. Internal channels in the pump casing route the discharge of one impeller to the suction of another impeller. Figure 21.7 shows a diagram of the arrangement of the impellers of a four-stage pump. The water enters the pump from the top left and passes through each of the four impellers, going from left to right. The water goes from the volute surrounding the discharge of one impeller to the suction of the next impeller.

A *pump stage* is defined as that portion of a centrifugal pump consisting of one impeller and its associated components. Most centrifugal pumps are single-stage pumps, containing only one impeller. A pump containing seven impellers within a single casing would be referred to as a seven-stage pump, or generally as a multistage pump.

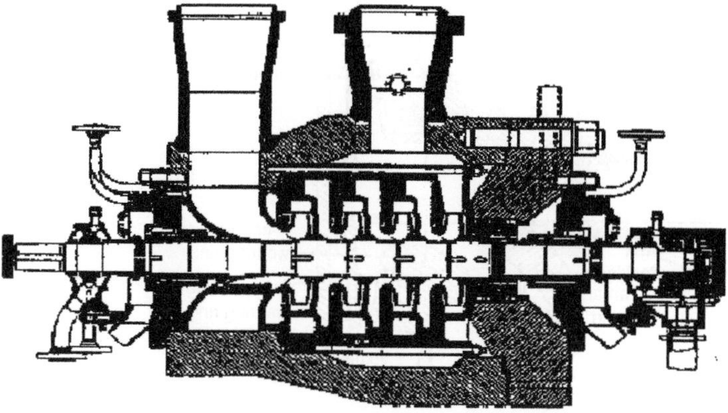

Figure 21.7 *Multistage centrifugal pump*

Components

Centrifugal pumps vary in design and construction from simple pumps with relatively few parts to extremely complicated pumps with hundreds of individual parts. Some of the most common components found in centrifugal pumps are wearing rings, stuffing boxes, packing, and lantern rings. These components are shown in Figure 21.8 and are described on the following pages.

Impellers

Impellers of pumps are classified based on the number of points at which the liquid can enter the impeller and also on the amount of webbing between the impeller blades.

Impellers can be either single-suction or double-suction. A single-suction impeller allows liquid to enter the center of the blades from only one direction. A double-suction impeller allows liquid to enter the center of the impeller blades from both sides simultaneously. Figure 21.9 shows simplified diagrams of single- and double-suction impellers.

Impellers can be open, semi-open, or enclosed. The open impeller consists only of blades attached to a hub. The semi-open impeller is constructed with a circular plate (the web) attached to one side of the blade. The enclosed impeller has circular plates attached to both sides of the blades.

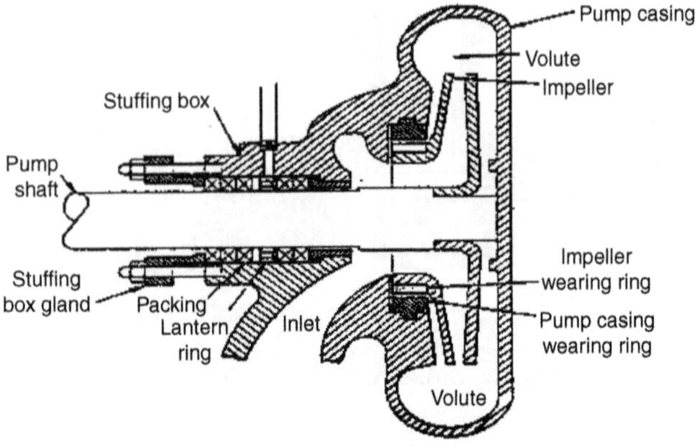

Figure 21.8 *Components of a centrifugal pump*

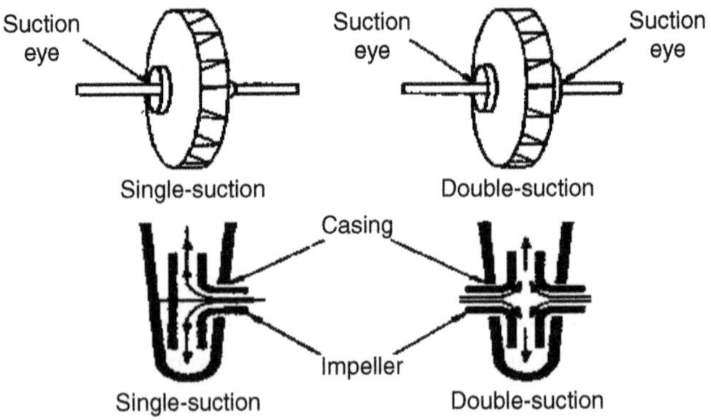

Figure 21.9 *Single-suction and double-suction impellers*

Enclosed impellers are also referred to as shrouded impellers. Figure 21.10 illustrates examples of open, semi-open, and enclosed impellers.

The impeller sometimes contains balancing holes that connect the space around the hub to the suction side of the impeller. The balancing holes have a total cross-sectional area that is considerably greater than the

Figure 21.10 *Open, semi-open, and enclosed impellers*

cross-sectional area of the annular space between the wearing ring and the hub. The result is suction pressure on both sides of the impeller hub, which maintains a hydraulic balance of axial thrust.

Diffuser

Some centrifugal pumps contain diffusers. A *diffuser* is a set of stationary vanes that surround the impeller. The purpose of the diffuser is to increase the efficiency of the centrifugal pump by allowing a more gradual expansion and less turbulent area for the liquid to reduce in velocity. The diffuser vanes are designed in a manner that the liquid exiting the impeller will encounter an ever increasing flow area as it passes through the diffuser. This increase in flow area causes a reduction in flow velocity, converting kinetic energy into flow energy. The increase in flow energy can be observed as an increase in the pressure of an incompressible fluid. Figure 21.11 shows a centrifugal pump diffuser.

Wearing Rings

Centrifugal pumps contain rotating impellers within stationary pump casings. To allow the impeller to rotate freely within the pump casing, a small clearance is maintained between the impeller and the pump casing. To maximize the efficiency of a centrifugal pump, it is necessary to minimize the amount of liquid leaking through this clearance from the high pressure side or discharge side of the pump back to the low pressure or suction side.

It is unavoidable that some wear will occur at the point where the impeller and the pump casing nearly come into contact. This wear is due to the

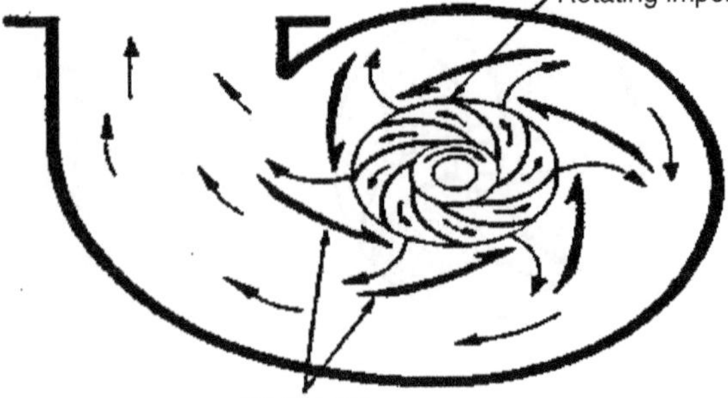

Figure 21.11 *Centrifugal pump diffuser*

erosion caused by liquid leaking through this tight clearance and other causes. Eventually, the leakage could become unacceptably large and maintenance would be required on the pump.

To minimize the cost of pump maintenance, many centrifugal pumps are designed with wearing rings. *Wearing rings* are replaceable rings that are attached to the impeller and/or the pump casing to allow a small running clearance between the impeller and pump casing without causing wear of the actual impeller or pump casing material.

Stuffing Box

In almost all centrifugal pumps, the rotating shaft that drives the impeller penetrates the pressure boundary of the pump casing. It is important that the pump is designed properly to control the amount of liquid that leaks along the shaft at the point that the shaft penetrates the pump casing. Factors considered when choosing a method include the pressure and temperature of the fluid being pumped, the size of the pump, and the chemical and physical characteristics of the fluid being pumped.

One of the simplest types of shaft seal is the stuffing box. The *stuffing box* is a cylindrical space in the pump casing surrounding the shaft. Rings of packing material are placed in this space. *Packing* is material in the form of rings or strands that is placed in the stuffing box to form a seal to control the rate of leakage along the shaft. The packing rings are held in place by a gland. The gland is, in turn, held in place by studs with adjusting nuts. As the adjusting nuts are tightened, they move the gland in and compress the packing. This axial compression causes the packing to expand radially, forming a tight seal between the rotating shaft and the inside wall of the stuffing box.

The high-speed rotation of the shaft generates a significant amount of heat as it rubs against the packing rings. If no lubrication and cooling are provided to the packing, the temperature of the packing increases to the point where damage occurs to the packing, the pump shaft, and possibly the nearby pump bearing. Stuffing boxes are normally designed to allow a small amount of controlled leakage along the shaft to provide lubrication and cooling to the packing. Tightening and loosening the packing gland can adjust the leakage rate.

Lantern Ring

It is not always possible to use a standard stuffing box to seal the shaft of a centrifugal pump. The pump suction may be under a vacuum so that outward leakage is impossible, or the fluid may be too hot to provide adequate cooling of the packing. These conditions require a modification to the standard stuffing box.

One method of adequately cooling the packing under these conditions is to include a lantern ring. A *lantern ring* is a perforated hollow ring located near the center of the packing box that receives relatively cool, clean liquid from either the discharge of the pump or from an external source and distributes the liquid uniformly around the shaft to provide lubrication and cooling. The fluid entering the lantern ring can cool the shaft and packing, lubricate the packing, or seal the joint between the shaft and packing against leakage of air into the pump in the event the pump suction pressure is less than that of the atmosphere.

Mechanical Seals

In some situations, packing material is not adequate for sealing the shaft. One common alternative method for sealing the shaft is with mechanical seals. Mechanical seals consist of two basic parts, a rotating element attached to the pump shaft and a stationary element attached to the pump casing. Each of these elements has a highly polished sealing surface. The polished faces of the rotating and stationary elements come into contact with each other to form a seal that prevents leakage along the shaft.

Summary

The important information is summarized below.

- Centrifugal pumps contain components with distinct purposes. The impeller contains rotating vanes that impart a radial and rotary motion to the liquid.
- The volute collects the liquid discharged from the impeller at high velocity and gradually causes a reduction in fluid velocity by increasing the flow area, converting the velocity head to a static head.
- A diffuser increases the efficiency of a centrifugal pump by allowing a more gradual expansion and less turbulent area for the liquid to slow as the flow area expands.
- Packing material provides a seal in the area where the pump shaft penetrates the pump casing.
- Wearing rings are replaceable rings that are attached to the impeller and/or the pump casing to allow a small running clearance between the impeller and pump casing without causing wear of the actual impeller or pump casing material.
- The lantern ring is inserted between rings of packing in the stuffing box to receive relatively cool, clean liquid and distribute the liquid uniformly around the shaft to provide lubrication and cooling to the packing.
- There are three indications that a centrifugal pump is cavitating:

  1 Noise

  2 Fluctuating discharge pressure and flow

  3 Fluctuating pump motor current

- Steps that can be taken to stop pump cavitation include:

  1 Increasing the pressure at the suction of the pump

  2 Reducing the temperature of the liquid being pumped

  3 Reducing head losses in the pump suction piping

4 Reducing the flow rate through the pump

5 Reducing the speed of the pump impeller

- Three effects of pump cavitation are:

  1 Degrading pump performance

  2 Excessive pump vibration

  3 Damage to pump impeller, bearing, wearing rings, and seals

- To avoid pump cavitation, the net positive suction head available must be greater than the net positive suction head required.
- Net positive suction head available is the difference between the pump suction pressure and the saturation pressure for the liquid being pumped.
- Cavitation is the process of the formation and subsequent collapse of vapor bubbles in a pump.
- Gas binding of a centrifugal pump is a condition where the pump casing is filled with gases or vapors to the point where the impeller is no longer able to contact enough fluid to function correctly.
- Shutoff head is the maximum head that can be developed by a centrifugal pump operating at a set speed.
- Pump run-out is the maximum flow that can be developed by a centrifugal pump without damaging the pump.
- The greater the head against which a centrifugal pump operates, the lower the flow rate through the pump. The relationship between pump flow rate and head is illustrated by the characteristic curve for the pump.
- Centrifugal pumps are protected from deadheading by providing a recirculation from the pump discharge back to the supply source of the pump.
- Centrifugal pumps are protected from run-out by placing an orifice or throttle valve immediately downstream of the pump discharge.

Positive Displacement Pumps

A positive displacement pump is one in which a definite volume of liquid is delivered for each cycle of pump operation. This volume is constant regardless of the resistance to flow offered by the system the pump is in, provided the capacity of the power unit driving the pump is not exceeded. The positive displacement pump delivers liquid in separate volumes with no delivery in between, although a pump having several chambers may have an overlapping delivery among individual chambers, which minimizes this effect. The positive displacement pump differs from other types of pumps that deliver a continuous even flow for any given pump speed and discharge.

Positive displacement pumps can be grouped into three basic categories based on their design and operation: reciprocating pumps, rotary pumps, and diaphragm pumps.

Principles of Operation

All positive displacement pumps operate on the same basic principle. This principle can be most easily demonstrated by considering a reciprocating positive displacement pump consisting of a single reciprocating piston in a cylinder with a single suction port and a single discharge port, as shown in Figure 21.12.

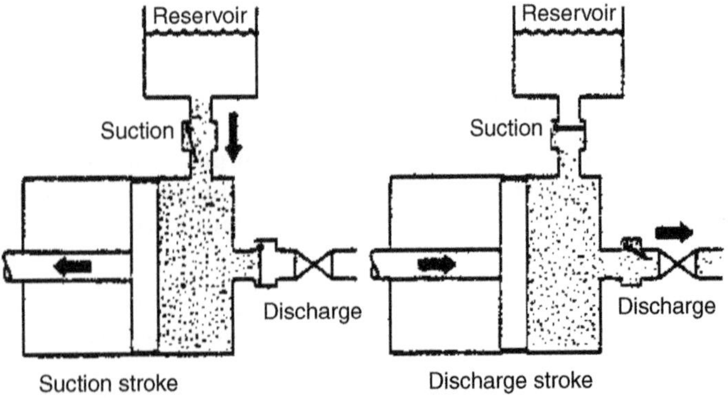

Figure 21.12 *Reciprocating positive displacement pump operation*

During the suction stroke, the piston moves to the left, causing the check valve in the suction line between the reservoir and the pump cylinder to open and admit water from the reservoir. During the discharge stroke, the piston moves to the right, seating the check valve in the suction line and opening the check valve in the discharge line. The volume of liquid moved by the pump in one cycle (one suction stroke and one discharge stroke) is equal to the change in the liquid volume of the cylinder as the piston moves from its farthest left position to its farthest right position.

Reciprocating Pumps

Reciprocating positive displacement pumps are generally categorized in four ways: direct-acting or indirect-acting; simplex or duplex; single-acting or double-acting; and power pumps.

Direct-Acting and Indirect-Acting

Some reciprocating pumps are powered by prime movers that also have reciprocating motion, such as a reciprocating pump powered by a reciprocating steam piston. The piston rod of the steam piston may be directly connected to the liquid piston of the pump, or it may be indirectly connected with a beam or linkage. Direct-acting pumps have a plunger on the liquid (pump) end that is directly driven by the pump rod (also the piston rod or extension thereof) and that carries the piston of the power end. Indirect-acting pumps are driven by means of a beam or linkage connected to and actuated by the power piston rod of a separate reciprocating engine.

Simplex and Duplex

A simplex pump, sometimes referred to as a single pump, is a pump having a single liquid (pump) cylinder. A duplex pump is the equivalent of two simplex pumps placed side by side on the same foundation.

The driving of the pistons of a duplex pump is arranged in such a manner that when one piston is on its upstroke, the other piston is on its downstroke and vice versa. This arrangement doubles the capacity of the duplex pump compared to a simplex pump of comparable design.

Single-Acting and Double-Acting

A single-acting pump is one that takes a suction, filling the pump cylinder on the stroke in only one direction, called the suction stroke, and then forces

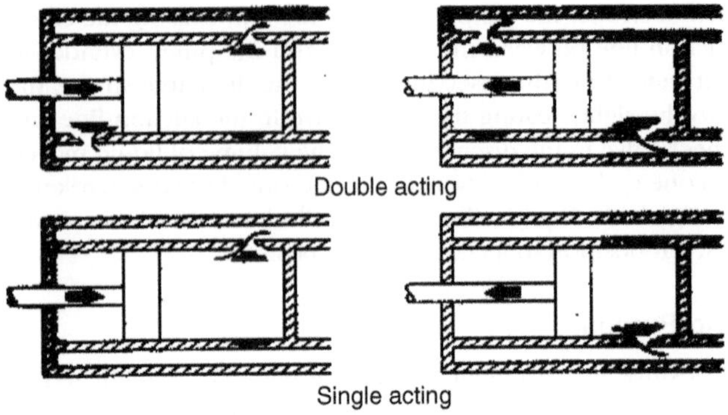

Figure 21.13 *Single-acting and double-acting pumps*

the liquid out of the cylinder on the return stroke, called the discharge stroke. A double-acting pump is one that, as it fills one end of the liquid cylinder, is discharging liquid from the other end of the cylinder. On the return stroke, the end of the cylinder just emptied is filled, and the end just filled is emptied. One possible arrangement for single-acting and double-acting pumps is shown in Figure 21.13.

Power

Power pumps convert rotary motion to low-speed reciprocating motion by reduction gearing, a crankshaft, connecting rods, and cross heads. Plungers or pistons are driven by the crosshead drives. The liquid ends of the low-pressure, higher-capacity units use rod and piston construction, similar to duplex double-acting steam pumps. The higher-pressure units are normally single-action plungers and usually employ three (triplex) plungers. Three or more plungers substantially reduce flow pulsations relative to simplex and even duplex pumps.

Power pumps typically have high efficiency and are capable of developing very high pressures. Either electric motors or turbines can drive them. They are relatively expensive pumps and can rarely be justified on the basis of efficiency over centrifugal pumps. However, they are frequently justified over steam reciprocating pumps where continuous duty service is needed due to the high steam requirements of direct acting steam pumps.

In general, the effective flow rate of reciprocating pumps decreases as the viscosity of the fluid being pumped increases, because the speed of the pump must be reduced. In contrast to centrifugal pumps, the differential pressure generated by reciprocating pumps is independent of fluid density. It is dependent entirely on the amount of force exerted on the piston.

Rotary

Rotary pumps operate on the principle that a rotating vane, screw, or gear traps the liquid in the suction side of the pump casing and forces it to the discharge side of the casing. These pumps are essentially self-priming due to their capability of removing air from suction lines and producing a high suction lift. In pumps designed for systems requiring high suction lift and self-priming features, it is essential that all clearances between rotating parts, and between rotating and stationary parts, be kept to a minimum in order to reduce slippage. Slippage is leakage of fluid from the discharge of the pump back to its suction.

Due to the close clearances in rotary pumps, it is necessary to operate these pumps at relatively low speed in order to secure reliable operation and maintain pump capacity over an extended period of time. Otherwise, the erosive action due to the high velocities of the liquid passing through the narrow clearance spaces would soon cause excessive wear and increased clearance, resulting in slippage.

There are many types of positive displacement rotary pumps, and they are normally grouped into three basic categories: gear pumps, screw pumps, and moving vane pumps.

Rotary Moving Vane

The rotary moving vane pump shown in Figure 21.14 is another type of positive displacement pump used in pumping viscous fluids. The pump consists of a cylindrically bored housing with a suction inlet on one side and a discharge outlet on the other. A cylindrically shaped rotor, with a diameter smaller than the cylinder, is driven about an axis place above the centerline of the cylinder. The clearance, between rotor and cylinder at the top, is small but increases at the bottom. The rotor carries vanes that move in and out as it rotates to maintain sealed space between the rotor and the cylinder wall. The vanes trap liquid on the suction side and carry it to the discharge side, where contraction of the space expels it through the discharge line. The vanes may swing on pivots, or they may slide in slots in the rotor.

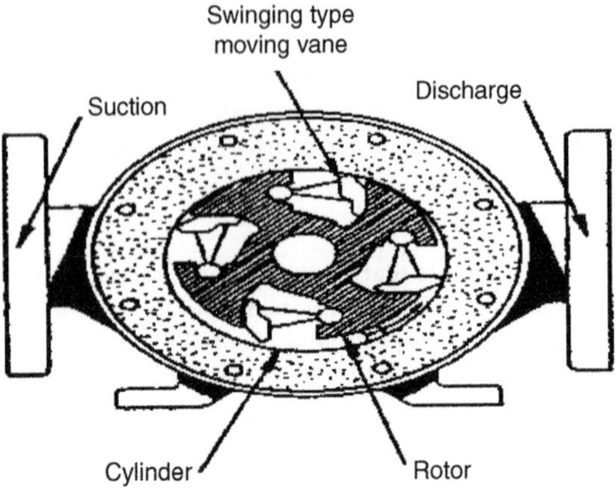

Figure 21.14 *Rotary moving vane pump*

Screw-Type, Positive Displacement Rotary

There are many variations in the design of the screw-type positive displacement rotary pump. The primary differences consist of the number of intermeshing screws involved, the pitch of the screws, and the general direction of fluid flow. Two designs include a two-screw, low-pitch double-flow pump, and a three-screw, high-pitch double-flow pump.

Two-Screw, Low-Pitch Screw Pump

The two-screw, low-pitch screw pump consists of two screws that mesh with close clearances, mounted on two parallel shafts. One screw has a right-handed thread, and the other screw has a left-handed thread. One shaft is the driving shaft and drives the other through a set of herringbone timing gears. The gears serve to maintain clearances between the screws as they turn and to promote quiet operation. The screws rotate in closely fitting duplex cylinders that have overlapping bores. All clearances are small, but there is no actual contact between the two screws or between the screws and the cylinder walls. The complete assembly and the usual path of flow are shown in Figure 21.15.

Liquid is trapped at the outer end of each pair of screws. As the first space between the screw threads rotated away from the opposite screw, a one-turn, spiral-shaped quantity of liquid is enclosed when the end of the screw

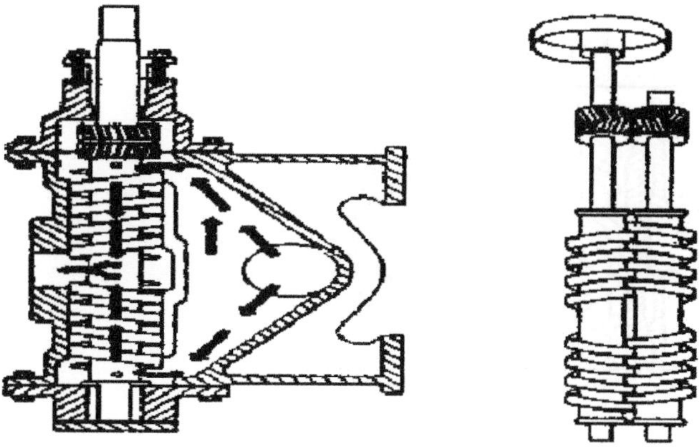

Figure 21.15 *Two-screw, low-pitch screw pump*

again meshes with the opposite screw. As the screw continues to rotate, the entrapped spiral turns of liquid slide along the cylinder toward the center discharge space while the next slug is being entrapped. Each screw functions similarly, and each pair of screws discharges an equal quantity of liquid in opposed streams toward the center, thus eliminating hydraulic thrust. The removal of liquid from the suction end by the screws produces a reduction in pressure, which draws liquid through the suction line.

Three-Screw, High-Pitch Screw Pump

The three-screw, high-pitch screw pump shown in Figure 21.16 has many of the same elements as the two-screw, low-pitch screw pump, and their operations are similar. Three screws, oppositely threaded on each end, are employed. They rotate in a triple cylinder, the two outer bores of which overlap the center bore. The pitch of the screws is much higher than in the low-pitch screw pump; therefore, the center screw, or power rotor, is used to drive the two outer idler rotors directly without external timing gears. Pedestal bearings at the base support the weight of the rotors and maintain their axial position and the liquid being pumped enters the suction opening, flows through passages around the rotor housing, and through the screws from each end, in opposed streams, toward the center discharge. This eliminates unbalanced hydraulic thrust. The screw

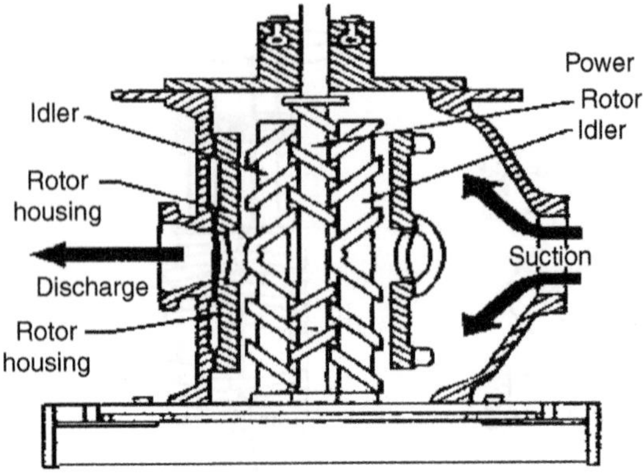

Figure 21.16 *Three-screw, high-pitch screw pump*

pump is used for pumping viscous fluids, usually lubricating, hydraulic, or fuel oil.

Diaphragm or Positive Displacement

Diaphragm pumps are also classified as positive displacement pumps because the diaphragm acts as a limited displacement piston. The pump will function when a diaphragm is forced into reciprocating motion by mechanical linkage, compressed air, or fluid from a pulsating, external source. The pump construction eliminates any contact between the liquid being pumped and the source of energy. This eliminates the possibility of leakage, which is important when handling toxic or very expensive liquids. Disadvantages include limited head and capacity range and the necessity of check valves in the suction and discharge nozzles. An example of a diaphragm pump is shown in Figure 21.17.

Characteristics Curve

Positive displacement pumps deliver a definite volume of liquid for each cycle of pump operation. Therefore, the only factor that affects flow rate in an ideal positive displacement is the speed at which it operates. The flow resistance of the system in which the pump is operating will not affect the flow rate through the pump. Figure 21.18 shows the characteristic curve for a positive displacement pump.

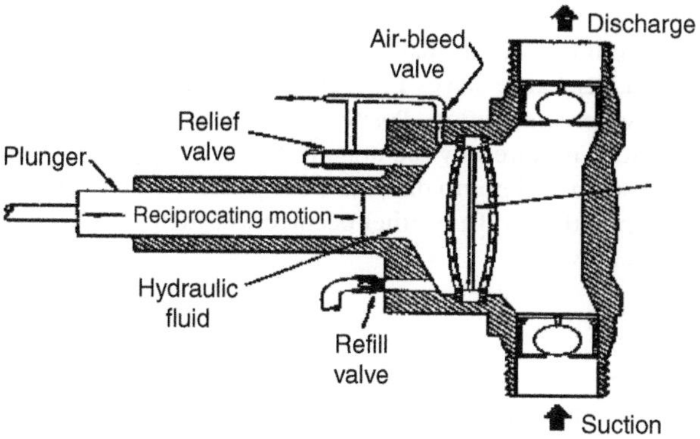

Figure 21.17 *Diaphragm or positive displacement pump*

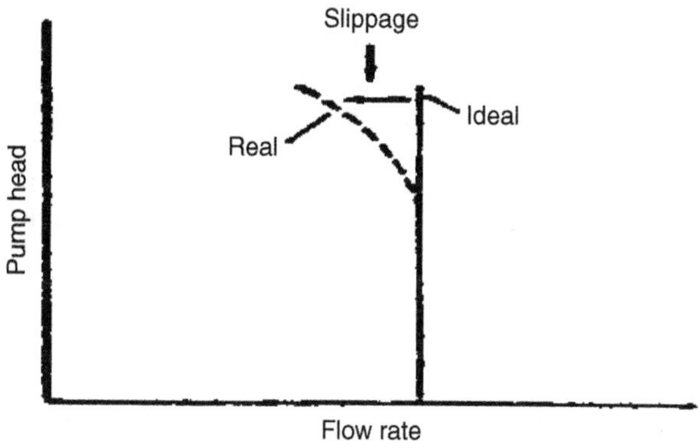

Figure 21.18 *Positive displacement pump characteristic curve*

The dashed line in Figure 21.18 shows actual positive displacement pump performance. This line reflects the fact that as the discharge pressure of the pump increases, some amount of liquid will leak from the discharge of the pump back to the pump suction, reducing the effective flow rate of the pump. The rate at which liquid leaks from the pump discharge to its suction is called slippage.

Protection

Positive displacement pumps are normally fitted with relief valves on the upstream side of their discharge valves to protect the pump and its discharge piping from overpressurization. Positive displacement pumps will discharge at the pressure required by the system they are supplying. The relief valve prevents system and pump damage if the pump discharge valve is shut during pump operation or if any other occurrence, such as a clogged strainer, blocks system flow.

Gear Pumps

Simple Gear Pumps

There are several variations of gear pumps. The simple gear pump shown in Figure 21.19 consists of two spur gears meshing together and revolving in opposite directions within a casing. Only a few thousandths of an inch of clearance exists between the case and the gear faces and teeth extremities. Any liquid that fills the space bounded by two successive gear teeth and the case must follow along with the teeth as they revolve. When the gear teeth mesh with the teeth of the other gear, the space between the teeth is reduced, and the entrapped liquid is forced out of the pump discharge pipe. As the gears revolve and the teeth disengage, the space again opens on the suction side of the pump, trapping new quantities of liquid and carrying it around the pump case to the discharge. As liquid is carried away from the suction side, a lower pressure is created, which draws liquid in through the suction line.

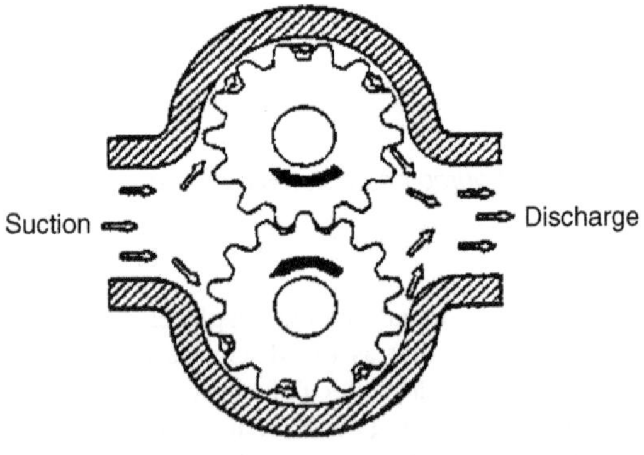

Figure 21.19 *Simple gear pump*

With the large number of teeth usually employed on the gears, the discharge is relatively smooth and continuous, with small quantities of liquid being delivered to the discharge line in rapid succession. If designed with fewer teeth, the space between the teeth is greater and the capacity increases for a given speed; however, the tendency toward a pulsating discharge increases. In all simple gear pumps, power is applied to the shaft of one of the gears, which transmits power to the driven gear through their meshing teeth.

There are no valves in the gear pump to cause friction losses as in the reciprocating pump. The high impeller velocities, with resultant friction losses, are not required as in the centrifugal pump. Therefore, the gear pump is well suited for handling viscous fluids such as fuel and lubricating oils.

Other Gear Pumps

There are two types of gears used in gear pumps in addition to the simple spur gear. One type is the helical gear. A helix is the curve produced when a straight line moves up or down the surface of a cylinder. The other type is the herringbone gear. A herringbone gear is composed of two helixes spiraling in different directions from the center of the gear. Spur, helical, and herringbone gears are shown in Figure 21.20.

The helical gear pump has advantages over the simple spur gear. In a spur gear, the entire length of the gear tooth engages at the same time. In a helical gear, the point of engagement moves along the length of the gear tooth as the gear rotates. This makes the helical gear operate with a steadier discharge pressure and fewer pulsations than a spur gear pump.

The herringbone gear pump is also a modification of the simple gear pump. Its principal difference in operation from the simple gear pump is that the pointed center section of the space between two teeth begins discharging

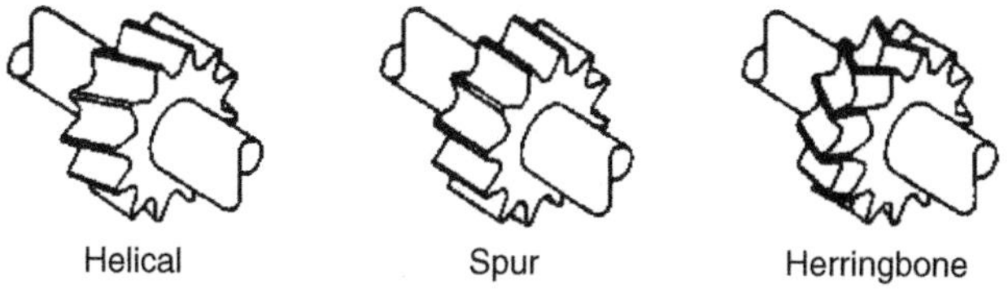

Figure 21.20 *Types of gears used in pumps*

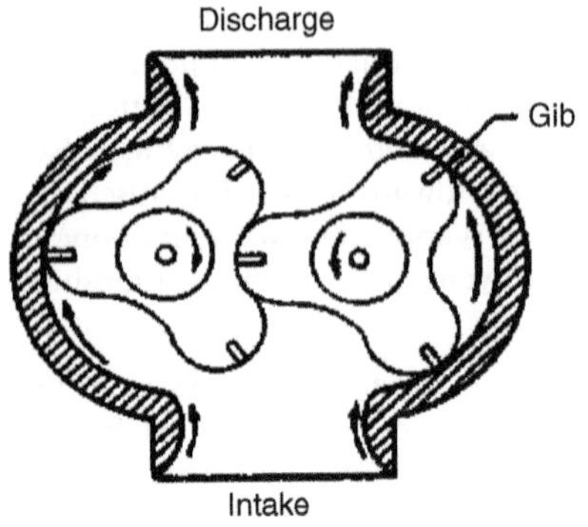

Figure 21.21 *Lobe-type pump*

before the divergent outer ends of the preceding space complete discharging. This overlapping tends to provide a steadier discharge pressure. The power transmission from the driving to the driven gear is also smoother and quieter.

Lobe-Type Pump

The lobe-type pump shown in Figure 21.21 is another variation of the simple gear pump. It is considered a simple gear pump having only two or three teeth per rotor; otherwise, its operation or the explanation of the function of its parts is no different. Some designs of lobe pumps are fitted with replaceable gibs, that is, thin plates carried in grooves at the extremity of each lobe where they make contact with the casing. The gibs promote tightness and absorb radial wear.

Summary

The important information is summarized below.

- The flow delivered by a centrifugal pump during one revolution of the impeller depends upon the head against which the pump is operating. The positive displacement pump delivers a fixed volume of fluid for each cycle of pump operation regardless of the head against which the pump is operating.

- Positive displacement pumps may be classified in the following ways:

  1 Reciprocating piston pump

  2 Gear-type rotary pump

  3 Lobe-type rotary pump

  4 Screw-type rotary pump

  5 Moving vane pump

  6 Diaphragm pump

- As the viscosity of a liquid increases, the maximum speed at which a reciprocating positive displacement pump can properly operate decreases. Therefore, as viscosity increases, the maximum flow rate through the pump decreases.
- Slippage is the rate at which liquid leaks from the discharge of the pump back to the pump suction.
- Positive displacement pumps are protected from overpressurization by a relief valve on the upstream side of the pump discharge valve.

Cavitation

Many centrifugal pumps are designed in a manner that allows the pump to operate continuously for months or even years. These centrifugal pumps often rely on the liquid that they are pumping to provide cooling and lubrication to the pump bearings and other internal components of the pump. If flow through the pump is stopped while the pump is still operating, the pump will no longer be adequately cooled, and the pump can quickly become damaged. Pump damage can also result from pumping a liquid that is close to saturated conditions. This phenomenon is referred to as *cavitation*. *Most centrifugal pumps are not designed to withstand cavitation.*

The flow area at the eye of the impeller is usually smaller than either the flow area of the pump suction piping or the flow area through the impeller vanes.

When the liquid being pumped enters the eye of a centrifugal pump, the decrease in flow area results in an increase in flow velocity accompanied by a decrease in pressure. The greater the pump flow rate, the greater the pressure drop between the pump suction and the eye of the impeller. If the pressure drop is large enough, or if the temperature is high enough, the pressure drop may be sufficient to cause the liquid to flash to vapor when the local pressure falls below the saturation pressure for the fluid being pumped. Any vapor bubbles formed by the pressure drop at the eye of the impeller are swept along the impeller vanes by the flow of the fluid. When the bubbles enter a region where local pressure is greater than saturation pressure farther out the impeller vane, the vapor bubbles abruptly collapse. This process of the formation and subsequent collapse of vapor bubbles in a pump is called cavitation.

Cavitation in a centrifugal pump has a significant effect on performance. It degrades the performance of a pump, resulting in a degraded, fluctuating flow rate and discharge pressure. Cavitation can also be destructive to pump internals. The formation and collapse of the vapor bubble can create small pits on the impeller vanes. Each individual pit is microscopic in size, but the cumulative effect of millions of these pits formed over a period of hours or days can literally destroy a pump impeller. Cavitation can also cause excessive pump vibration, which could damage pump bearings, wearing rings, and seals.

A small number of centrifugal pumps are designed to operate under conditions where cavitation is unavoidable. These pumps must be specially designed and maintained to withstand the small amount of cavitation that occurs during their operation.

Noise is one of the indications that a centrifugal pump is cavitating. A cavitating pump can sound like a can of marbles being shaken. Other indications that can be observed from a remote operating station are fluctuating discharge pressure, flow rate, and pump motor current.

Recirculation

When the discharge flow of a centrifugal pump is throttled by closing the discharge valve slightly, or by installing an orifice plate, the fluid flow through the pump is altered from its original design. This reduces the fluid's velocity as it exits the tips of the impeller vanes; therefore, the fluid does not flow

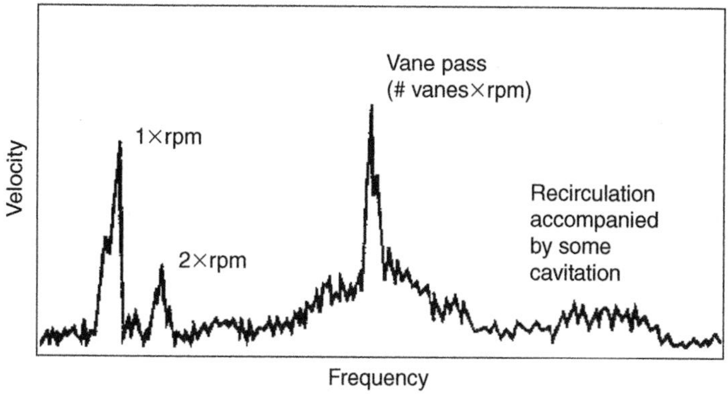

Figure 21.22 *Vane pass frequency*

as smoothly into the volute and discharge nozzle. This causes the fluid to impinge upon the "cutwater" and creates a vibration at a frequency equal to the vane pass × rpm. The resulting amplitude quite often exceeds alert set-point values, particularly when accompanied by resonance.

Random, low amplitude wide frequency vibration is often associated with vane pass frequency, resulting in vibrations similar to cavitation and turbulence, but it is usually found at lower frequencies. This can lead to misdiagnosis. Many pump impellers show metal reduction and pitting on the general area at the exit tips of the vanes. This has often been misdiagnosed as cavitation.

It is very important to note that recirculation is found to happen on the discharge side of the pump, whereas cavitation is found to happen on the suction side of the pump.

To prevent recirculation in pumps, pumps should be operated close to their operational rated capacity, and excessive throttling should be avoided.

When a permanent reduction in capacity is desired, the outside diameter of the pump impeller can be reduced slightly to increase the gap between the impeller tips and the cutwater.

Net Positive Suction Head

To avoid cavitation in centrifugal pumps, the pressure of the fluid at all points within the pump must remain above saturation pressure.

The quantity used to determine if the pressure of the liquid being pumped is adequate to avoid cavitation is the net positive suction head (NPSH). The *net positive suction head available* ($NPSH_A$) is the difference between the pressure at the suction of the pump and the saturation pressure for the liquid being pumped. The net positive suction head required ($NPSH_R$) is the minimum net positive suction head necessary to avoid cavitation.

The condition that must exist to avoid cavitation is that the net positive suction head available must be greater than or equal to the net positive suction head required. This requirement can be stated mathematically as shown below.

$$NPSH_A \geqslant NPSH_R$$

A formula for $NPSH_A$ can be stated as the following equation:

$$NPSH_A = P_{suction} - P_{saturation}$$

When a centrifugal pump is taking suction from a tank or other reservoir, the pressure at the suction of the pump is the sum of the absolute pressure at the surface of the liquid in the tank, plus the pressure due to the elevation difference between the surface of liquid in the tank, and the pump suction less the head losses due to friction in the suction line from the tank to the pump.

$$NPSH_A = P_a = P_{st} - h_f - P_{sat}$$

Where:

$NPSH_A$ = Net positive suction head available

P_a = Absolute pressure on the surface of the liquid

P_{st} = Pressure due to elevation between liquid surface and pump suction

h_f = Head losses in the pump suction piping

P_{sat} = Saturation pressure of the liquid being pumped

Preventing Cavitation

If a centrifugal pump is cavitating, several changes in the system design or operation may be necessary to increase the $NPSH_A$ above the $NPSH_R$ and stop the cavitation. One method for increasing the $NPSH_A$ is to increase the pressure at the suction of the pump. If a pump is taking suction from an enclosed tank, either raising the level of the liquid in the tank or increasing the pressure in the gas space above the liquid increases suction pressure.

It is also possible to increase the $NPSH_A$ by decreasing the temperature of the liquid being pumped. Decreasing the temperature of the liquid decreases the saturation pressure, causing $NPSH_A$ to increase.

If the head losses in the pump suction piping can be reduced, the $NPSH_A$ will be increased. Various methods for reducing head losses include increasing the pipe diameter, reducing the number of elbows, valves, and fittings in the pipe, and decreasing the length of the pipe.

It may also be possible to stop cavitation by reducing the $NPSH_R$ for the pump. The $NPSH_R$ is not a constant for a given pump under all conditions, but depends on certain factors. Typically, the $NPSH_R$ of a pump increases significantly as flow rate through the pump increases. Therefore, reducing the flow rate through a pump by throttling a discharge valve decreases $NPSH_R$. $NPSH_R$ is also dependent upon pump speed. The faster the impeller of a pump rotates, the greater the $NPSH_R$. Therefore, if the speed of a variable speed centrifugal pump is reduced, the $NPSH_R$ of the pump decreases.

The net positive suction head required to prevent cavitation is determined through testing by the pump manufacturer and depends upon factors including type of impeller inlet, impeller design, pump flow rate, impeller rotational speed, and the type of liquid being pumped. The manufacturer typically supplies curves of $NPSH_R$ as a function of pump flow rate for a particular liquid (usually water) in the vendor manual for the pump.

Troubleshooting

Design, installation, and operation are the dominant factors that affect a pump's mode of failure. This section identifies common failures for centrifugal and positive-displacement pumps.

Centrifugal

Centrifugal pumps are especially sensitive to: (1) variations in liquid condition (i.e., viscosity, specific gravity, and temperature); (2) suction variations, such as pressure and availability of a continuous volume of fluid; and (3) variations in demand. Table 21.1 lists common failure modes for centrifugal pumps and their causes.

Table 21.1 *Common failure modes of centrifugal pumps*

THE CAUSES	THE PROBLEM: Insufficient discharge pressure	Intermittent operation	Insufficient capacity	No liquid delivery	High bearing temperatures	Short bearing life	Short mechanical seal life	High vibration	High noise levels	Power demand excessive	Motor trips	Elevated motor temperature	Elevated liquid temperature
Bent shaft					•	•	•	•		•			
Casing distorted from excessive pipe strain					•	•	•	•		•		•	
Cavitation	•	•	•	•	•		•	•	•				•
Clogged impeller	•		•	•				•		•			
Driver imbalance						•	•	•					
Electrical problems (driver)						•	•	•		•	•	•	
Entrained air (suction or seal leaks)	•	•	•					•	•			•	
Hydraulic instability					•	•	•	•	•				
Impeller installed backward (double-suction only)	•		•							•			
Improper mechanical seal							•						
Inlet strainer partially clogged	•		•					•	•				•
Insufficient flow through pump													•
Insufficient suction pressure (NPSH)	•	•	•	•				•	•				
Insufficient suction volume	•	•	•	•	•			•	•				•
Internal wear	•		•					•		•			
Leakage in piping, valves, vessels	•		•	•									
Mechanical defects, worn, rusted, defective bearings					•		•			•			
Misalignment					•	•	•	•		•		•	
Misalignment (pump and driver)								•		•	•		•
Mismatched pumps in series	•		•			•		•		•			
Noncondensables in liquid	•	•	•					•	•			•	
Obstructions in lines or pump housing	•		•	•				•				•	•

Table 21.1 *continued*

THE CAUSES	THE PROBLEM												
	Insufficient discharge pressure	Intermittent operation	Insufficient capacity	No liquid delivery	High bearing temperatures	Short bearing life	Short mechanical seal life	High vibration	High noise levels	Power demand excessive	Motor trips	Elevated motor temperature	Elevated liquid temperature
Rotor imbalance						•	•	•					
Specific gravity too high	•									•		•	
Speed too high										•	•		
Speed too low	•		•	•								•	
Total system head higher than design	•	•	•	•	•		•					•	•
Total system head lower than design					•		•	•	•	•			•
Unsuitable pumps in parallel operation	•		•	•	•			•	•		•		•
Viscosity too high	•		•							•		•	
Wrong rotation	•			•						•		•	

Source: Integrated Systems Inc.

Mechanical failures may occur for a number of reasons. Some are induced by cavitation, hydraulic instability, or other system-related problems. Others are the direct result of improper maintenance. Maintenance-related problems include improper lubrication, misalignment, imbalance, seal leakage, and a variety of others that periodically affect machine reliability.

Cavitation

Cavitation in a centrifugal pump, which has a significant, negative effect on performance, is the most common failure mode. Cavitation not only degrades a pump's performance, but also greatly accelerates the wear rate of its internal components.

Causes

There are three causes of cavitation in centrifugal pumps: change of phase, entrained air or gas, and turbulent flow.

Change of Phase

The formation or collapse of vapor bubbles in either the suction piping or inside the pump is one cause of cavitation. This failure mode normally occurs in applications such as boiler feed, where the incoming liquid is at a temperature near its saturation point. In this situation, a slight change in suction pressure can cause the liquid to flash into its gaseous state. In the boiler-feed example, the water flashes into steam. The reverse process also can occur. A slight increase in suction pressure can force the entrained vapor to change phase to a liquid.

Cavitation due to phase change seriously damages the pump's internal components. Visual evidence of operation with phase-change cavitation is an impeller surface finish like an orange peel. Prolonged operation causes small pits or holes on both the impeller shroud and vanes.

Entrained Air/Gas

Pumps are designed to handle gas-free liquids. If a centrifugal pump's suction supply contains any appreciable quantity of gas, the pump will cavitate. In the example of cavitation due to entrainment, the liquid is reasonably stable, unlike with the change of phase described in the preceding section. Nevertheless, the entrained gas has a negative effect on pump performance. While this form of cavitation does not seriously affect the pump's internal components, it severely restricts its output and efficiency.

The primary causes of cavitation due to entrained gas include: two-phase suction supply, inadequate available net positive suction head ($NPSH_A$), and leakage in the suction-supply system. In some applications, the incoming liquid may contain moderate to high concentrations of air or gas. This may result from aeration or mixing of the liquid prior to reaching the pump or inadequate liquid levels in the supply reservoir. Regardless of the reason, the pump is forced to handle two-phase flow, which was not intended in its design.

Turbulent Flow

The effects of turbulent flow (not a true form of cavitation) on pump performance are almost identical to those described for entrained air or gas in the preceding section. Pumps are not designed to handle incoming liquids that do not have stable, laminar flow patterns. Therefore, if the flow is unstable, or turbulent, the symptoms are the same as for cavitation.

Symptoms

Noise (e.g., like a can of marbles being shaken) is one indication that a centrifugal pump is cavitating. Other indications are fluctuations of the pressure gauges, flow rate, and motor current, as well as changes in the vibration profile.

How to Eliminate

Several design or operational changes may be necessary to stop centrifugal-pump cavitation. Increasing the available net positive suction head ($NPSH_A$) above that required ($NPSH_R$) is one way to stop it. The NPSH required to prevent cavitation is determined through testing by the pump manufacturer. It depends upon several factors, including: type of impeller inlet, impeller design, impeller rotational speed, pump flow rate, and the type of liquid being pumped. The manufacturer typically supplies curves of $NPSH_R$ as a function of flow rate for a particular liquid (usually water) in the pump's manual.

One way to increase the $NPSH_A$ is to increase the pump's suction pressure. If a pump is fed from an enclosed tank, either raising the level of the liquid in the tank or increasing the pressure in the gas space above the liquid can increase suction pressure.

It also is possible to increase the $NPSH_A$ by decreasing the temperature of the liquid being pumped. This decreases the saturation pressure, which increases $NPSH_A$.

If the head losses in the suction piping can be reduced, the $NPSH_A$ will be increased. Methods for reducing head losses include: increasing the pipe diameter; reducing the number of elbows, valves, and fittings in the pipe; and decreasing the pipe length.

It also may be possible to stop cavitation by reducing the pump's $NPSH_R$, which is not a constant for a given pump under all conditions. Typically, the $NPSH_R$ increases significantly as the pump's flow rate increases. Therefore, reducing the flow rate by throttling a discharge valve decreases $NPSH_R$. In addition to flow rate, $NPSH_R$ depends on pump speed. The faster the pump's impeller rotates, the greater the $NPSH_R$. Therefore, if the speed of a variable-speed centrifugal pump is reduced, the $NPSH_R$ of the pump is decreased.

Variations in Total System Head

Centrifugal-pump performance follows its hydraulic curve (i.e., head versus flow rate). Therefore, any variation in the total backpressure of the system causes a change in the pump's flow or output. Because pumps are designed to operate at their Best Efficiency Point (BEP), they become more and more unstable as they are forced to operate at any other point because of changes in total system pressure, or head (TSH). This instability has a direct impact on centrifugal-pump performance, reliability, operating costs, and required maintenance.

Symptoms of Changed Conditions

The symptoms of failure due to variations in TSH include changes in motor speed and flow rate.

Motor Speed

The brake horsepower of the motor that drives a pump is load dependent. As the pump's operating point deviates from BEP, the amount of horsepower required also changes. This causes a change in the pump's rotating speed, which either increases or decreases depending on the amount of work that the pump must perform.

Flow Rate

The volume of liquid delivered by the pump varies with changes in TSH. An increase in the total system back-pressure results in decreased flow, while a back-pressure reduction increases the pump's output.

Correcting Problems

The best solution to problems caused by TSH variations is to prevent the variations. While it is not possible to completely eliminate them, the operating practices for centrifugal pumps should limit operation to an acceptable range of system demand for flow and pressure. If system demand exceeds the pump's capabilities, it may be necessary to change the pump, the system requirements, or both. In many applications, the pump is either too small or too large. In these instances, it is necessary to replace the pump with one that is properly sized.

For the application where the TSH is too low and the pump is operating in run-out condition (i.e., maximum flow and minimum discharge pressure),

Table 21.2 *Common failure modes of rotary-type, positive-displacement pumps*

THE CAUSES	No liquid delivery	Insufficient discharge pressure	Insufficient capacity	Starts, but loses prime	Excessive wear	Excessive heat	Excessive vibration and noise	Excessive power demand	Motor trips	Elevated motor temperature	Elevated liquid temperature
Air leakage into suction piping or shaft seal		•	•				•			•	
Excessive discharge pressure			•		•		•	•	•		•
Excessive suction liquid temperatures			•	•							
Insufficient liquid supply		•	•	•	•		•		•		
Internal component wear	•	•	•				•				
Liquid more viscous than design								•	•	•	•
Liquid vaporizing in suction line		•	•	•			•				•
Misaligned coupling, belt drive, chain drive					•	•	•	•		•	
Motor or driver failure	•										
Pipe strain on pump casing					•	•	•	•		•	
Pump running dry	•	•			•	•	•				
Relief valve stuck open or set wrong		•	•								
Rotating element binding					•	•	•	•	•	•	
Solids or dirt in liquid					•						
Speed too low		•	•						•		
Suction filter or strainer clogged	•	•	•				•			•	
Suction piping not immersed in liquid	•	•		•							
Wrong direction of rotation	•	•								•	

Source: Integrated Systems Inc.

the system demand can be corrected by restricting the discharge flow of the pump. This approach, called false head, changes the system's head by partially closing a discharge valve to increase the back-pressure on the pump. Because the pump must follow its hydraulic curve, this forces the pump's performance back toward its BEP.

When the TSH is too great, there are two options: replace the pump or lower the system's back-pressure by eliminating line resistance due to elbows, extra valves, etc.

Table 21.3 *Common failure modes of reciprocating positive-displacement pumps*

	THE PROBLEM								
THE CAUSES	No liquid delivery	Insufficient capacity	Short packing life	Excessive wear liquid end	Excessive wear power end	Excessive heat power end	Excessive vibration and noise	Persistent knocking	Motor trips
Abrasives or corrosives in liquid			•	•					
Broken valve springs		•		•			•		
Cylinders not filling		•	•	•			•		
Drive-train problems							•		•
Excessive suction lift	•	•							
Gear drive problem							•	•	•
Improper packing selection			•						
Inadequate lubrication						•	•		•
Liquid entry into power end of pump						•			
Loose cross-head pin or crank pin								•	
Loose piston or rod								•	
Low volumetric efficiency		•	•						
Misalignment of rod or packing			•						•
Noncondensables (air) in liquid	•	•	•				•		•
Not enough suction pressure	•	•							
Obstructions in lines	•						•		•
One or more cylinders not operating		•							
Other mechanical problems: wear, rusted, etc.					•	•	•	•	
Overloading					•				•
Pump speed incorrect		•				•			
Pump valve(s) stuck open		•							
Relief or bypass valve(s) leaking		•							
Scored rod or plunger		•							•
Supply tank empty	•								
Worn cross-head or guides			•			•			
Worn valves, seats, liners, rods, or plungers	•	•		•					

Source: Integrated Systems Inc.

Positive Displacement

Positive-displacement pumps are more tolerant to variations in system demands and pressures than centrifugal pumps. However, they are still subject to a variety of common failure modes caused directly or indirectly by the process.

Rotary Type

Rotary-type, positive-displacement pumps share many common failure modes with centrifugal pumps. Both types of pumps are subject to process-induced failures caused by demands that exceed the pump's capabilities. Process-induced failures are also caused by operating methods that either result in radical changes in their operating envelope or instability in the process system.

Table 21.2 lists common failure modes for rotary-type, positive-displacement pumps. The most common failure modes of these pumps are generally attributed to problems with the suction supply. They must have a constant volume of clean liquid in order to function properly.

Reciprocating

Table 21.3 lists the common failure modes for reciprocating-type, positive-displacement pumps. Reciprocating pumps can generally withstand more abuse and variations in system demand than any other type. However, they must have a consistent supply of relatively clean liquid in order to function properly.

The weak links in the reciprocating pump's design are the inlet and discharge valves used to control pumping action. These valves are the most frequent source of failure. In most cases, valve failure is due to fatigue. The only positive way to prevent or minimize these failures is to ensure that proper maintenance is performed regularly on these components. It is important to follow the manufacturer's recommendations for valve maintenance and replacement.

Because of the close tolerances between the pistons and the cylinder walls, reciprocating pumps cannot tolerate contaminated liquid in their suction-supply system. Many of the failure modes associated with this type of pump are caused by contamination (e.g., dirt, grit, and other solids) that enters the suction-side of the pump. This problem can be prevented by the use of well-maintained inlet strainers or filters.

22 Steam Traps

Steam-supply systems are commonly used in industrial facilities as a general heat source as well as a heat source in pipe and vessel tracing lines used to prevent freeze-up in nonflow situations. Inherent with the use of steam are the problems of condensation and the accumulation of noncondensable gases in the system.

Steam traps must be used in these systems to automatically purge condensate and noncondensable gases, such as air, from the steam system. However, a steam trap should never discharge live steam. Such discharges are dangerous as well as costly.

Configuration

There are five major types of steam traps commonly used in industrial applications: inverted bucket, float and thermostatic, thermodynamic, bimetallic, and thermostatic. Each of the five major types of steam trap uses a different method to determine when and how to purge the system. As a result, each has a different configuration.

Inverted Bucket

The inverted-bucket trap, which is shown in Figure 22.1, is a mechanically actuated steam trap that uses an upside-down, or inverted, bucket as a float. The bucket is connected to the outlet valve through a mechanical linkage. The bucket sinks when condensate fills the steam trap, which opens the outlet valve and drains the bucket. It floats when steam enters the trap and closes the valve.

As a group, inverted-bucket traps can handle a wide range of steam pressures and condensate capacities. They are an economical solution for low- to medium-pressure and medium-capacity applications, such as plant heating and light processes. When used for higher-pressure and higher-capacity applications, these traps become large, expensive, and difficult to handle.

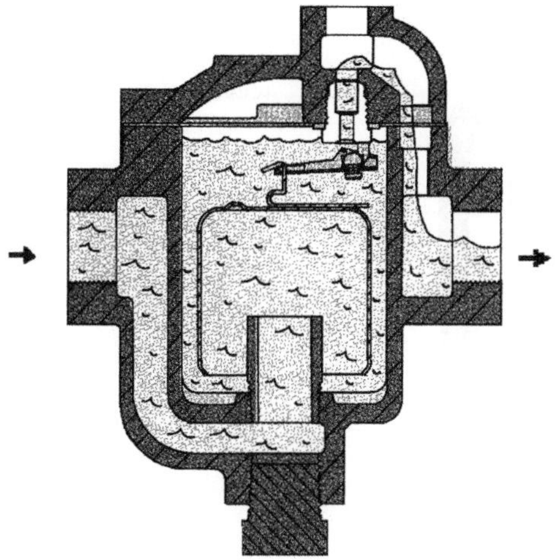

Figure 22.1 *Inverted-bucket trap*

Each specific steam trap has a finite, relatively narrow range that it can handle effectively. For example, an inverted-bucket trap designed for up to 15-psi service will fail to operate at pressures above that value. An inverted-bucket trap designed for 125-psi service will operate at lower pressures, but its capacity is so diminished that it may back up the system with unvented condensate. Therefore, it is critical to select a steam trap designed to handle the application's pressure, capacity, and size requirements.

Float and Thermostatic

The float-and-thermostatic trap shown in Figure 22.2 is a hybrid. A float similar to that found in a toilet tank operates the valve. As condensate collects in the trap, it lifts the float and opens the discharge or purge valve. This design opens the discharge only as much as necessary. Once the built-in thermostatic element purges noncondensable gases, it closes tightly when steam enters the trap. The advantage of this type of trap is that it drains condensate continuously.

Like the inverted-bucket trap, float-and-thermostatic traps as a group handle a wide range of steam pressures and condensate loads. However, each individual trap has a very narrow range of pressures and capacities. This makes

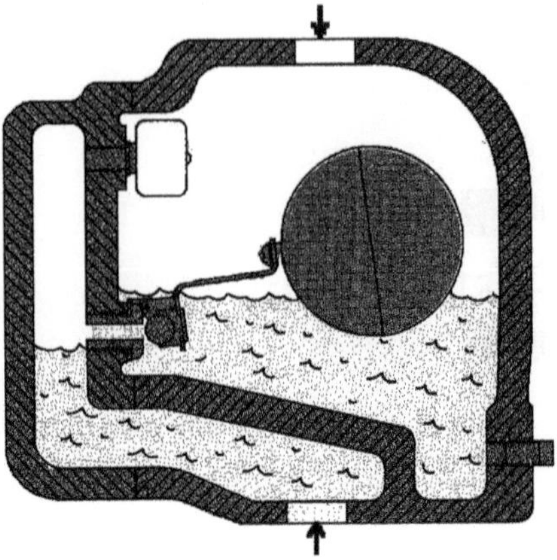

Figure 22.2 *Float-and-thermostatic trap*

it critical to select a trap that can handle the specific pressure, capacity, and size requirements of the system.

The key advantage of float-and-thermostatic traps is their ability for quick steam-system startup because they continuously purge the system of air and other noncondensable gases. One disadvantage is the sensitivity of the float ball to damage by hydraulic hammer.

Float-and-thermostatic traps are an economical solution for lighter condensate loads and lower pressures. However, when the pressure and capacity requirements increase, the physical size of the unit increases and its cost rises. It also becomes more difficult to handle.

Thermodynamic or Disk-Type

Thermodynamic, or disk-type, steam traps use a flat disk that moves between a cap and seat (see Figure 22.3). Upon startup, condensate flow raises the disk and opens the discharge port. Steam, or very hot condensate entering the trap, seats the disk. It remains seated, closing the discharge port, as long as pressure is maintained above it. Heat radiates out through the cap, thus diminishing the pressure over the disk, opening the trap to discharge condensate.

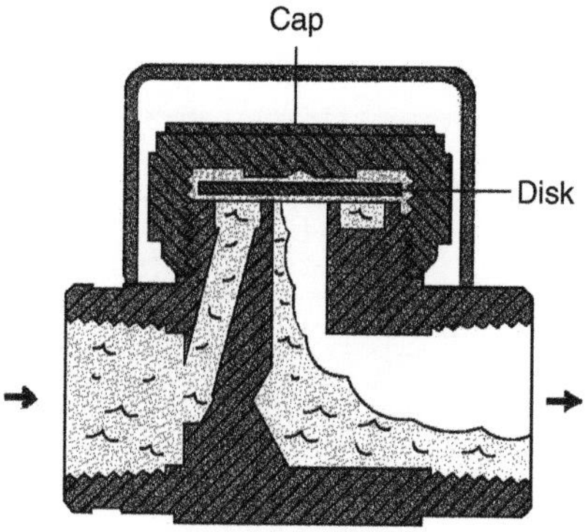

Figure 22.3 *Thermodynamic steam trap*

Wear and dirt are particular problems with a disk-type trap. Because of the large, flat seating surfaces, any particulate contamination, such as dirt or sand, will lodge between the disk and the valve seat. This prevents the valve from sealing and permits live steam to flow through the discharge port. If pressure is not maintained above the disk, the trap will cycle frequently. This wastes steam and can cause the device to fail prematurely.

The key advantage of these traps is that one trap can handle a complete range of pressures. In addition, they are relatively compact for the amount of condensate they discharge. The chief disadvantage is difficulty in handling air and other noncondensable gases.

Bimetallic

A bimetallic steam trap, which is shown in Figure 22.4, operates on the same principle as a residential-heating thermostat. A bimetallic strip, or wafer, connected to a valve disk bends or distorts when subjected to a change in temperature. When properly calibrated, the disk closes tightly against a seat when steam is present and opens when condensate, air, and other gases are present.

Two key advantages of bimetallic traps are: (1) compact size relative to their condensate load-handling capabilities, and (2) immunity to hydraulic-hammer damage.

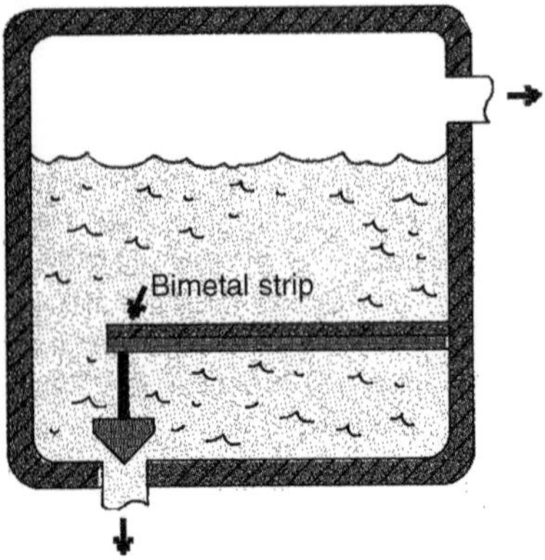

Figure 22.4 *Bimetal trap*

Their biggest disadvantage is the need for constant adjustment or calibration, which is usually done at the factory for the intended steam operating pressure. If the trap is used at a lower pressure, it may discharge live steam. If used at a higher pressure, condensate may back up into the steam system.

Thermostatic or Thermal Element

Thermostatic, or thermal-element, traps are thermally actuated using an assembly constructed of high-strength, corrosion-resistant stainless steel plates that are seam-welded together. Figure 22.5 shows this type of trap.

Upon startup, the thermal element is positioned to open the valve and purge condensate, air, and other gases. As the system warms up, heat generates pressure in the thermal element, causing it to expand and throttle the flow of hot condensate through the discharge valve. The steam that follows the hot condensate into the trap expands the thermal element with great force, which causes the trap to close. Condensate that enters the trap during system operation cools the element. As the thermal element cools, it lifts the valve off the seat and allows condensate to discharge quickly.

Thermal elements can be designed to operate at any steam temperature. In steam-tracing applications, it may be desirable to allow controlled amounts of condensate to back up in the lines in order to extract more heat from

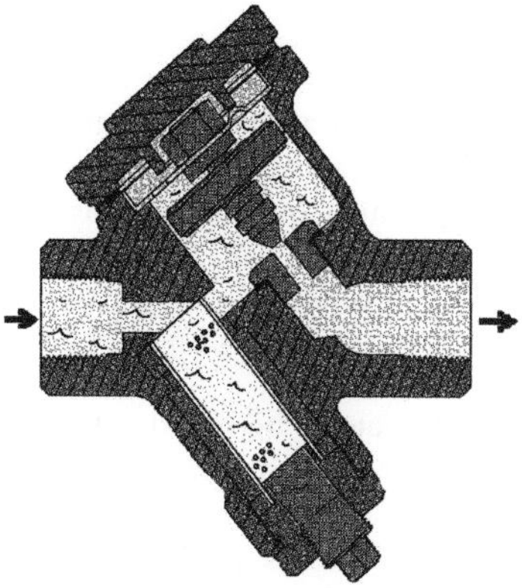

Figure 22.5 *Thermostatic trap*

the condensate. In other applications, any hint of condensate in the system is undesirable. The thermostatic trap can handle either of these conditions, but the thermal element must be properly selected to accommodate the specific temperature range of the application.

Thermostatic traps are compact, and a given trap operates over a wide range of pressures and capacities. However, they are not recommended for condensate loads over 15,000 pounds per hour.

Performance

When properly selected, installed, and maintained, steam traps are relatively trouble-free and highly efficient. The critical factors that affect efficiency include capacity and pressure ratings, steam quality, mechanical damage, and calibration.

Capacity Rating

Each type and size of steam trap has a specified capacity for the amount of condensate and noncompressible gas that it can handle. Care must be taken to ensure that the proper steam trap is selected to meet the application's capacity needs.

Pressure Rating

As discussed previously, each type of steam trap has a range of steam pressures that it can effectively handle. Therefore, each application must be carefully evaluated to determine the normal and maximum pressures that will be generated by the steam system. Traps must be selected for the worst-case scenario.

Steam Quality

Steam quality determines the amount of condensate to be handled by the steam trap. In addition to an increased volume of condensate, poor steam quality may increase the amount of particulate matter present in the condensate. High concentrations of solids directly affect the performance of steam traps. If particulate matter is trapped between the purge valve and its seat, the steam trap may not properly shut off the discharge port. This will result in live steam being continuously exhausted through the trap.

Mechanical Damage

Inverted-bucket and float-type steam traps are highly susceptible to mechanical damage. If the level arms or mechanical linkages are damaged or distorted, the trap cannot operate properly. Regular inspection and maintenance of these types of traps are essential.

Calibration

Steam traps, such as the bimetallic type, must be periodically recalibrated to ensure proper operation. All steam traps should be adjusted on a regular schedule.

Installation

Installation of steam traps is relatively straightforward. As long as they are properly sized, the only installation imperative is that they are plumb. If the trap is tilted or cocked, the bucket, float, or thermal valve will not operate properly. In addition, a nonplumb installation may prevent the condensate chamber from fully discharging accumulated liquids.

Table 22.1 *Common failure modes of steam traps*

THE CAUSES	THE PROBLEM: Trap will not discharge	Will not shut-off	Continuously blows steam	Capacity suddenly falls off	Condensate will not drain	Not enough steam heat	Traps freeze in winter	Back flow in return line
Back-pressure too high				●				
Boiler foaming or priming		●				●		
Boiler gauge reads low	●							
Bypass open or leaking	●		●					
Condensate load greater than design		●						
Condensate short-circuits					●			
Defective thermostatic elements						●		
Dirt or scale in trap			●		●			
Discharge line has long horizontal runs							●	
Flashing in return main				●				●
High-pressure traps discharge into low-pressure return								●
Incorrect fittings or connectors				●				●
Internal parts of trap broken or damaged	●	●	●		●			
Internal parts of trap plugged	●				●			
Kettles or other units increasing condensate load		●						
Leaky steam coils		●						
No cooling leg ahead of thermostatic trap						●		●
Open by-pass or vent in return line				●				
Pressure regulator out of order	●							
Process load greater than design		●						
Plugged return lines				●				
Plugged strainer, valve, or fitting ahead of trap	●							
Scored or out-of-round valve seat in trap						●		
Steam pressure too high	●							
System is air-bound					●			
Trap and piping not insulated							●	
Trap below return main				●				●
Trap blowing steam into return				●				
Trap inlet pressure too low				●	●			
Trap too small for load		●						

Source: Integrated Systems Inc.

Operating Methods

Steam traps are designed for a relatively constant volume, pressure, and condensate load. Operating practices should attempt to maintain these parameters as much as possible. Actual operating practices are determined by the process system, rather than the trap selected for a specific system.

The operator should periodically inspect them to ensure proper operation. Special attention should be given to the drain line to ensure that the trap is properly seated when not in the bleed or vent position.

Troubleshooting

A common failure mode of steam traps is failure of the sealing device (i.e., plunger, disk, or valve) to return to a leak-tight seat when in its normal operating mode. Leakage during normal operation may lead to abnormal operating costs or degradation of the process system. A single $\frac{3}{4}''$ steam trap that fails to seat properly can increase operating costs by $40,000 to $50,000 per year. Traps that fail to seat properly or are constantly in an unload position should be repaired or replaced as quickly as possible. Regular inspection and adjustment programs should be included in the standard operating procedures (SOPs).

Most of the failure modes that affect steam traps can be attributed to variations in operating parameters or improper maintenance. Table 22.1 lists the more common causes of steam trap failures.

Operation outside the trap's design envelope results in loss of efficiency and may result in premature failure. In many cases, changes in the condensate load, steam pressure or temperature, and other related parameters are the root causes of poor performance or reliability problems. Careful attention should be given to the actual versus design system parameters. Such deviations are often the root causes of problems under investigation.

Poor maintenance practices or the lack of a regular inspection program may be the primary source of steam trap problems. It is important for steam traps to be routinely inspected and repaired to assure proper operation.

23 V-Belt Drives

"Only Permanent Repairs Made Here"

V-belt drives are widely used in industry and commercial applications. V-belts are utilized to transfer energy from a driver to the driven and usually transfer one speed ratio to another through the use of different sheave sizes.

V-belts are constructed for three basic components, which vary from maker to maker:

1 Load carrying section to transfer power.

2 Rubber compression section to expand sideways in the groove.

3 Cover of cotton or synthetic fiber to resist abrasion.

Understanding the construction of V-belts assists in the understanding of belt maintenance. The standard V-belt must ride in the sheave properly. If the belt is worn or the sheave is worn, then you will have slippage of the belt and transfer of power, and speed will change resulting in a speed change to a piece of equipment. If a V-belt drive is located near oil, grease, or chemicals the V-belts could lose their capability through the deterioration of the belt material, again resulting in the reduction of energy transfer and quickly resulting in belt breakage or massive belt slippage.

Introduction

Belt drives are an important part of a conveyor system. They are used to transmit needed power from the drive unit to a portion of the conveyor system. This chapter will cover:

1 Various types of belts that are used to transmit power;

2 The advantages and disadvantage of using belt drives;

3 The correct installation procedure for belt drives;

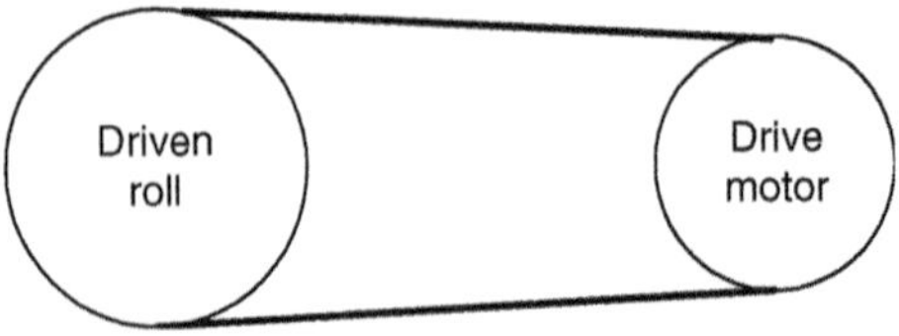

Figure 23.1 *Belt drive*

4 How to maintain belt drives;

5 How to calculate speeds and ratios that will enable you to make corrections or adjustments to belt drive speeds;

6 How to determine belt length and sheave sizes when making speed adjustments.

Belt Drives

Belt drives are used to transmit power between a drive unit and a driven unit. For example, if we have an electric motor and a contact roll on a conveyor, we need a way to transmit the power from the electric motor to the roll. This can be done easily and efficiently with a belt drive unit. See Figure 23.1.

Belt drives can consist of one or multiple belts, depending on the load that the unit must transmit.

The belts need to be the matched with the sheave type, and they must be tight enough to prevent slippage. Examples of the different belt and sheave sizes are as follows:

1 Fractional horsepower V-belts: 2L, 3L, 4L, and 5L;

2 Conventional V-belts: A, B, A-B, C, D, and E;

Conventional cogged V-belts: AX, BX, and CX;

3 Narrow V-belts: 3V, 5V, and 8V;

Narrow cogged V-belts: 3VX and 5VX;

4 Power band belts: these use the same top width designations as the above belts, but the number of bands is designated by the number preceding

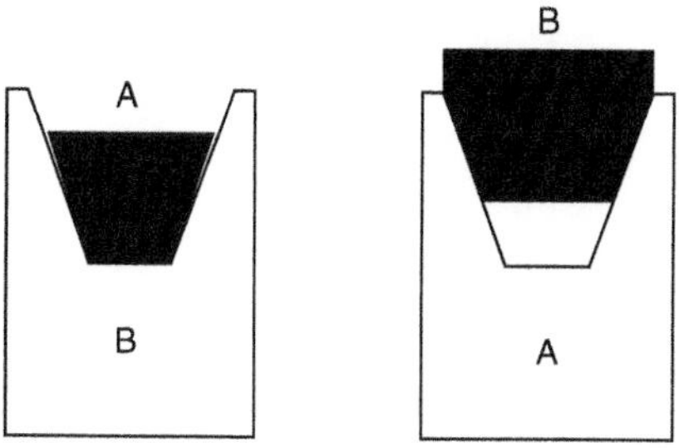

Figure 23.2 *Examples of V-belts*

the top width designation. For example, a 3-ribbed 5V belt would be labeled 3/5V;

5 Positive-drive belts: XL, L, H, XH, and XXH.

The size of the belt must match the sheave size. If they do not match, then the belt will not make proper contact with the sheave and will decrease the amount of load it can transmit. They may look something like the illustration in Figure 23.2.

Usually a set of numbers will follow the belt designation. These numbers represent the actual length of the belt in inches. On conventional belts, the length is given for the inside length of the belt, and on narrow belts it is given for the outside length. An example of this would be a 5V750 belt; the size of the belt gives it the 5V and the outside length of 75.0" gives it the 750.

More information about the specific belt dimensions can be found in the *Goodyear Power Transmission Belt Drives* manual.

Belt Selection

V-Belts

V-belts are best suited for transmitting light loads between short range sheaves. They are excellent at absorbing shock. When an overload occurs, they will act as an overload device and slip, thereby protecting

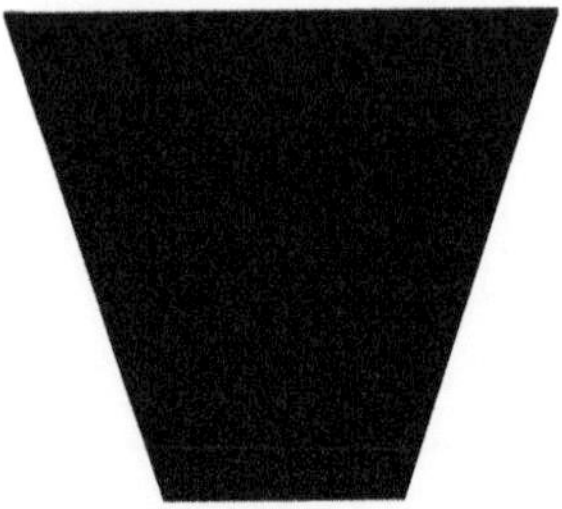

Figure 23.3 *Standard V-belt*

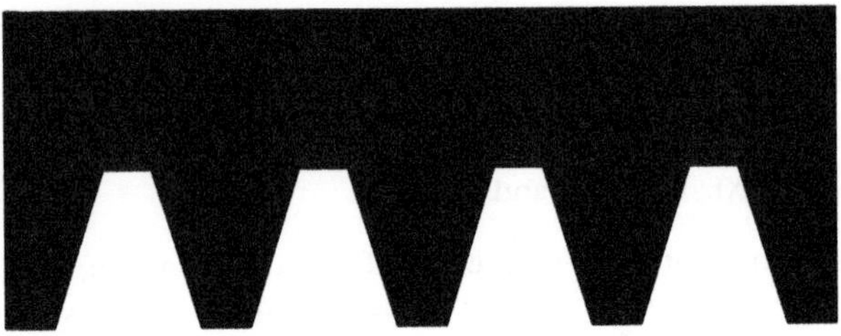

Figure 23.4 *Cogged belt*

valuable equipment. They are also much quieter than other power transmission devices such as chains.

Because of their design, they are easier to install and maintain than other belt types. Other than an occasional retensioning, V-belts are virtually maintenance free. When properly installed and maintained, V-belts will provide years of trouble-free operation. For an example, see Figure 23.3.

Cogged Belts

Cogged belts provide even longer life than conventional V-belts. Because of their design, they run cooler than conventional belts, thereby increasing the overall life of the belt. For an example, see Figure 23.4.

Joined Belts

Joined or power band belts provide a good alternative in pulsating drives where standard V-belts have a tendency to turn over. They function like a

Figure 23.5 *Joined belt (VX type)*

Figure 23.6 *Positive drive belt*

standard V-belt, with the exception that they are joined by the top fabric of the belt. These belts can be used with the standard V-belt sheaves, making selection and installation easy. For an example, see Figure 23.5.

Positive-Drive Belts

Positive-drive belts are sometimes called timing belts because they are often used in operations when timing a piece of equipment is critical. However, they are also used in applications where heavy loads cause standard V-belts to slip. They are flexible and provide the same benefit as standard V-belts, but their alignment is more critical. For an example, see Figure 23.6.

Sheaves

Sheaves are wheels with a grooved rim on which the belt rides. Sheaves are manufactured in various widths and diameters. Some have spokes, and some do not. For an example, see Figure 23.7.

Sheaves are made of cast steel for heavy-duty applications. For lighter applications, they are forged out of steel plate. Cast-iron sheaves are always used in applications where fluctuating loads are present. They provide a flywheel effect that minimizes the effects of fluctuating loads.

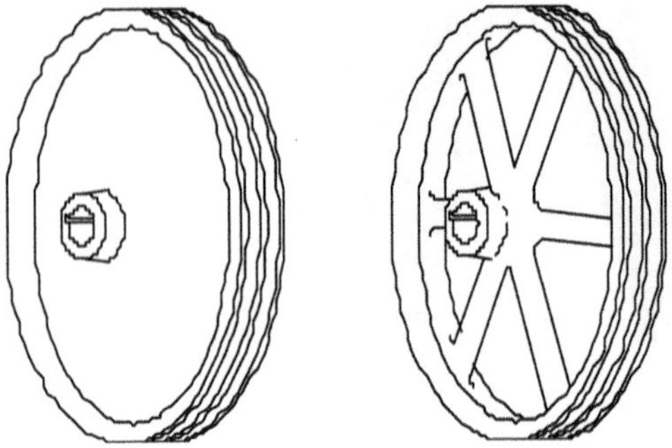

Figure 23.7 *Positive drive belt*

When they are mounted to a shaft, sheaves should be straight and have little or no wobble. For drives where the belt enters the sheave at an angle, deep-groove sheaves are available. These are especially useful when the belts must turn or twist.

Deep-groove sheaves can be used anywhere belt stability is a problem. In some cases, one drive shaft drives more than one driven shaft. When this occurs, more than one sheave can be mounted on one shaft. This is necessary only when sheaves of more than one size are needed. If the drive sheaves are the same size, one multibelt sheave can be used.

Most sheaves are balanced and capable of belt speeds of 6,000 feet per minute or less. If you note excessive vibration during operation or excessive bearing wear, you may need to balance or replace the sheaves.

Power Train Formulas

Shaft Speed

The size of the sheaves in a belt drive system determines the speed relationship between the drive and driven sheaves. For example, if the drive sheave has the same size sheave as the driven, then the speed will be equal. See Figure 23.8.

If we change the size of the driven sheave, then the speed of the shaft will also change. We know what the speed is of the electric motor and the size

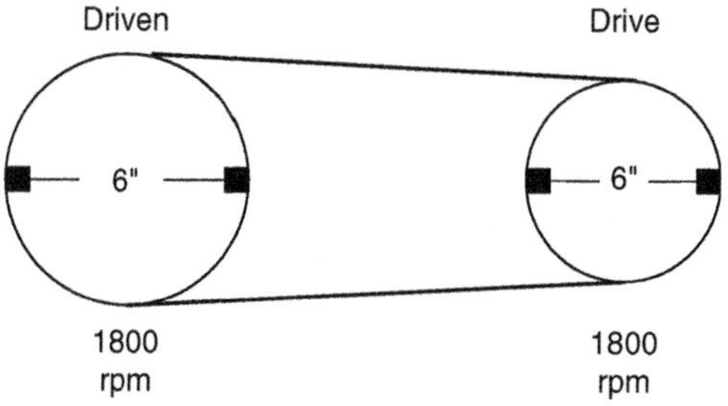

Figure 23.8 *Shaft speed*

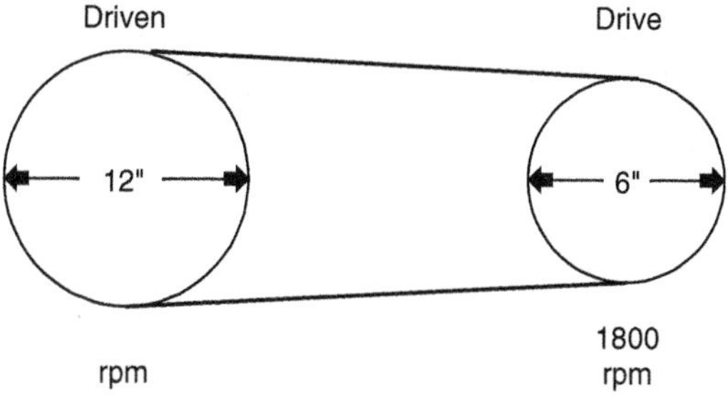

Figure 23.9 *Belt drive speed ratio*

of the sheaves, and now we can calculate the speed of the driven shaft by using the following formula (see Figure 23.9):

$$\text{Driven shaft rpm} = \frac{\text{Drive sheave diameter in inches} \times \text{drive shaft rpm}}{\text{Driven sheave diameter}}$$

$$\text{Driven shaft rpm} = \frac{6 \times 1800}{12}$$

$$900 = \frac{6 \times 1800}{12}$$

Now we understand how changing the size of a sheave will also change the shaft speed. Knowing this, we could also assume that to change the shaft rpm we must change the sheave size. The problem is, how do we know the exact size sheave that we need in order to reach the desired speed? Use the

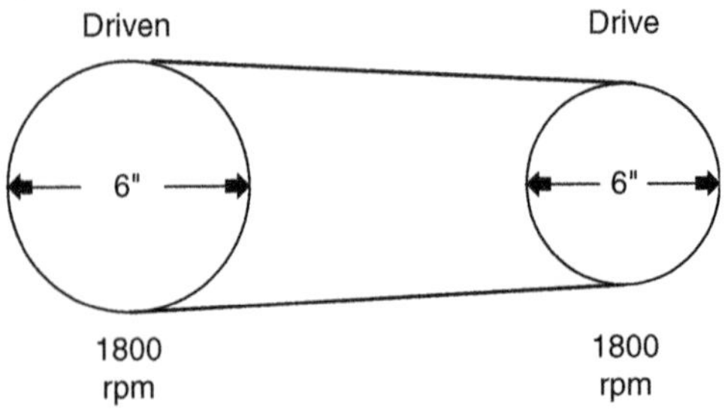

Figure 23.10 *Speed ratio*

same formula that was used to calculate shaft speed, only switch the location of the driven shaft speed and the driven sheave diameter.

$$\text{Driven shaft rpm} = \frac{\text{Drive sheave diameter in inches} \times \text{drive shaft rpm}}{\text{Driven sheave diameter}}$$

Let's change the problem to look like this:

$$\text{Driven sheave diameter} = \frac{\text{Drive sheave diameter in inches} \times \text{drive shaft rpm}}{\text{Driven shaft rpm}}$$

Let's say that we have a problem similar to the ones that we just did, but we want to change the shaft speed of the driven unit. If we know the speed we are looking for, we can use the formula above to calculate the sheave size required. See Figure 23.10

Let's change the speed of the driven shaft to 900 rpm (see Figure 23.11):

$$\text{Driven shaft rpm} = \frac{6 \times 1800}{900}$$

$$12 = \frac{6 \times 1800}{900}$$

Belt Length

Many times when a mechanic has to change out belts, the numbers on the belts cannot be read. So what should be done? Take a tape measure and wrap it around the sheaves to get the belt length? This is not a very accurate

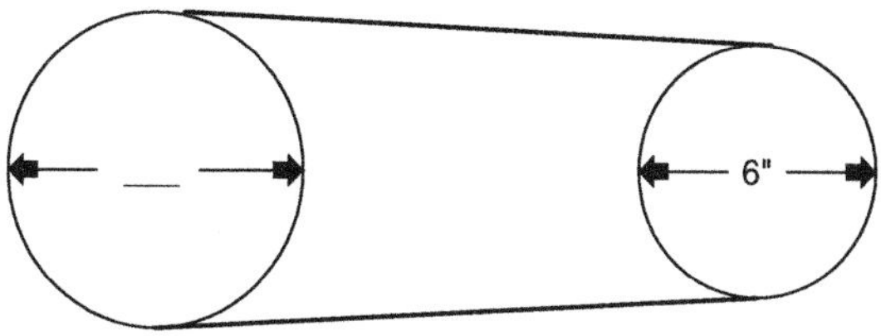

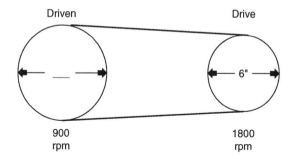

Figure 23.11 *Speed ratio calculation*

way to determine the length. So, usually the mechanic ends up taking a number of different size belts hoping to have a size that will fit.

Instead, take a couple of measurements, then use a simple formula to calculate the actual length that is needed. First, move the sheaves together until they are as close as the adjustments will allow. Then move the motor or drive out $\frac{1}{4}$ of its travel. Now you are ready to take the measurements. The following information is needed for an equation to find belt length (see Figure 23.12):

1 Diameter of the drive sheave.

2 Diameter of the driven sheave.

3 Center-to-center distance between the shafts.

Now use the following formula to solve the equation:

$$\text{Belt length} = \frac{\text{drive diameter} \times 3.14}{2} + \frac{\text{driven diameter} \times 3.14}{2} + \text{center to center} \times 2$$

Use the formula above to find the belt length.

$$\text{Belt length} = \frac{6'' \times 3.14}{2} + \frac{12'' \times 3.14}{2} + 35'' \times 2$$

$$98.26'' \text{ or } 98'' = \frac{6'' \times 3.14}{2} + \frac{12'' \times 3.14}{2} + 35'' \times 2$$

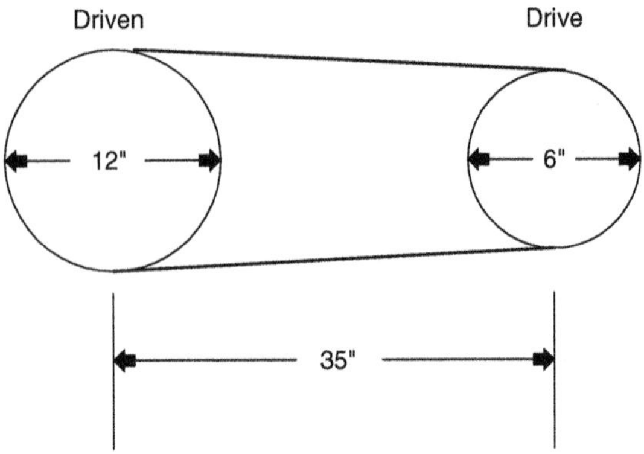

Figure 23.12 *Belt length example*

Multiple Sheaves

When calculating multiple sheave systems, think of each set of sheaves as a two-sheave system. Try to solve the following problem by only calculating two sheaves at a time.

Belt Speed

In order to calculate the speed of a belt in feet per minute (FPM), the following information is needed:

1 The diameter of the sheave that the belt is riding on.

2 The shaft rpm of the sheave.

With this information, we can use the following formula:

$$\text{FPM} = \frac{\text{diameter} \times 3.14 \times \text{rpm}}{12}$$

Use this formula to find the speed of the following belt (see Figure 23.13):

$$\text{FPM} = \frac{\text{diameter} \times 3.14 \times \text{rpm}}{12}$$

$$2826 = \frac{6'' \times 3.14 \times 1800}{1''}$$

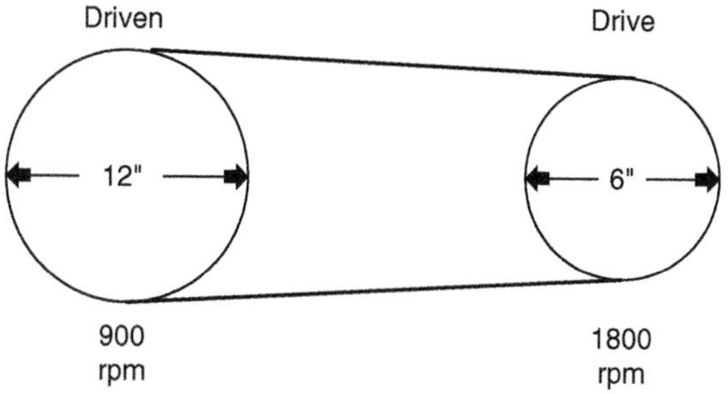

Figure 23.13 *Belt speed calculation*

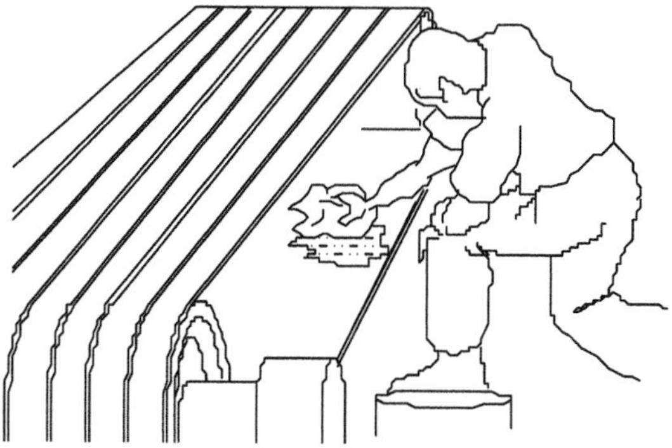

Figure 23.14 *Belt maintenance*

Belt Maintenance

Routine maintenance is essential if a belt drive is to operate properly. Belt maintenance should include regular checks of belt alignment and tension. You should also perform frequent inspections of the sheaves and shafts.

Routine maintenance will extend the life of the sheaves and belts. Belt-drive maintenance requires little time, but it must be done regularly. Keeping the belts clean and free of oil and grease will help ensure long belt life. See Figure 23.14.

When you replace a belt, always check the tension immediately after installation. Check it again after 24 hours of operation.

Never force a V-belt onto a sheave. There have been a number of injuries to fingers and hands as a result of this.

The belt should never ride in the bottom of the sheave. The sheave is deeper than the belt. The belt is made to ride near the top of the sheave. The belt may wear to the point that it is riding on the bottom of the sheave. If so, it will slip no matter how much tension is applied to the belt.

Keep used belt sets together for use on multibelt drives.

Routine preventive maintenance is essential if a belt drive is to operate properly. Belt maintenance should include regular checks of belt condition, belt alignment, and tension. You should also perform frequent inspections of the sheaves and shafts.

You may need to replace belts that are worn or damaged from overheating or contact with oil or grease. *Never replace one belt of a multibelt drive.* Belts stretch with use. If you replace one belt of a multibelt drive, it will be tighter than the others. See Figure 23.15.

A belt that is tighter than the others in a set will pull all the load. Store the old belts as a set. You may be able to use part of the set on a drive requiring fewer belts.

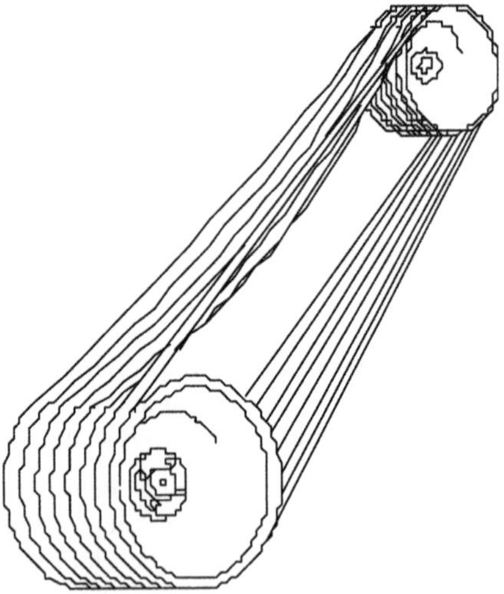

Figure 23.15 *Belt tensioning*

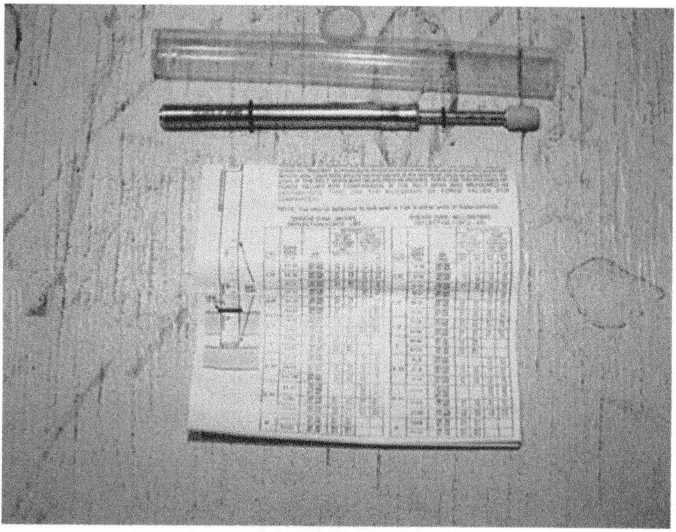

Figure 23.16 *Belt tension gauges*

Sheave and Belt Installation

Proper tools must be selected. (These must be identified on your PM inspection checklist or job plan.)

In addition to a set of basic hand tools, you will also need a reliable tension gauge with a set of belt tension tables, a set of sheave gauges, and a straightedge or string with a flashlight. See Figures 23.16 and 23.17.

When the proper procedures are followed for installing V-belts, they will yield years of trouble-free service.

Shaft and Sheave Alignment

1 The shafts must be parallel or the life of the belt will be shortened. The first step is to level the shafts; this is done by placing a level on each of the shafts. Then shim the low side until the shaft is level. See Figure 23.18.

Figure 23.17 *Sheave inspection gauges*

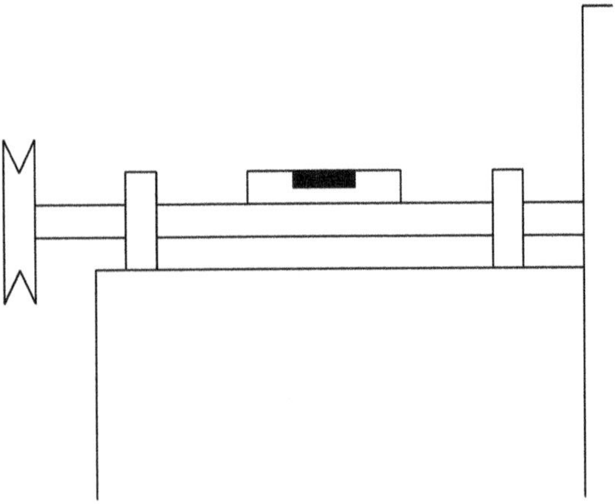

Figure 23.18 *Shaft alignment*

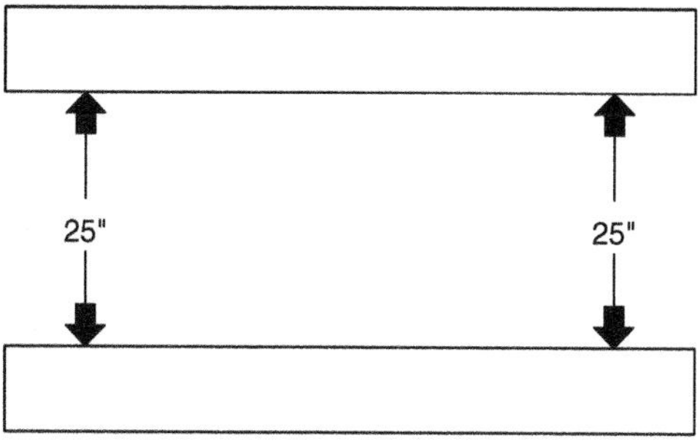

Figure 23.19 *Sheave alignment*

2 Next, make sure the shafts are parallel. This is done by measuring at different points on the shaft and adjusting the shafts until they are an equal distance apart. Make sure that the shafts are pulled in as close as possible before performing this procedure. Then you can use the jacking bolts to move the shafts apart evenly after the belt is installed. See Figure 23.19.

WARNING: Before installing a set of used sheaves, verify the size and condition of the sheaves with a sheave gauge. Select the proper gauge for the size of sheave. For example, if you have a 5V sheave that measures 14.4", use the 40-deg. gauge. Insert the gauge into the sheave groove; if you can see light on either side, the sheave is worn. Sheave gauges are also useful when the size of the sheave cannot be found stamped on it. See Figure 23.20.

3 Install the sheaves on the shafts following the manufacturer's recommendations. Locate and install the first sheave, then use a straightedge or a string to line the other one up with the one previously installed. See Figure 23.21.

Belt Installation

Install the belt on the sheaves. *Never force a belt on with a screwdriver.* This can damage the belt and could cause you to lose a finger. Next, begin increasing the distance between the sheaves by turning the jacking bolts; do this until the belt is snug but not tight. Using a belt tension gauge,

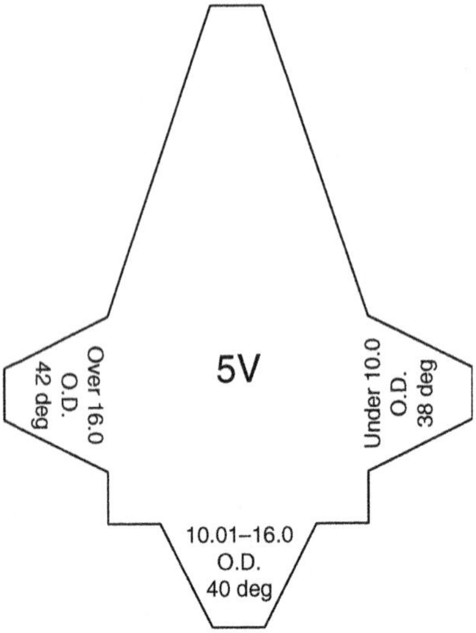

Figure 23.20 *Sheave gauge*

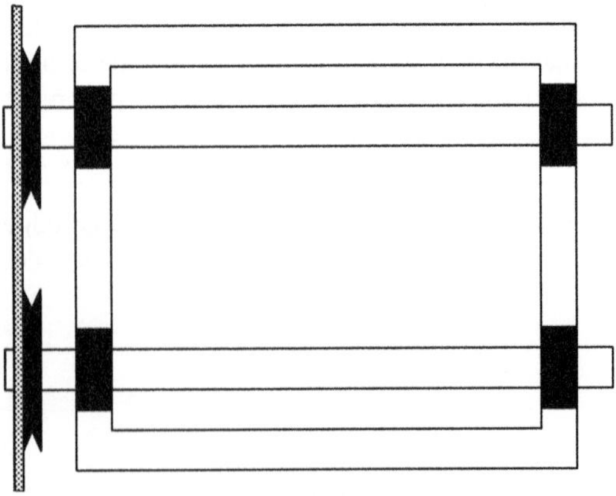

Figure 23.21 *Final alignment*

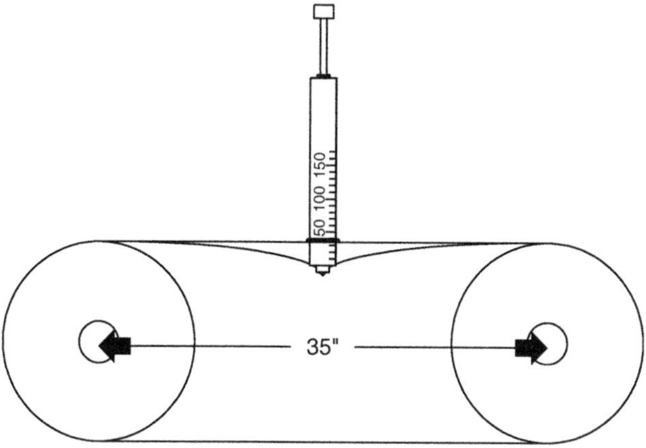

Figure 23.22 *Belt tensioning*

tighten the belts to the manufacturer's recommendation. Be sure to measure deflection and tension. This information can be found in a belt tension gauge's information sheet. See Figure 23.22.

Check for parallel and angular alignment of sheaves. The tolerance for alignment of V-belts is to within $\frac{1}{10}$" per foot of span, and for positive-drive belts to within $\frac{1}{16}$ of an inch per foot of span.

When you replace a belt, always check the tension immediately after installation. Check the tension again after 24 hours of operation.

The belt should never ride in the bottom of the sheave because the sheave is deeper than the belt. The belt is made to ride near the top of the sheave. The belt may wear to the point that it is riding on the bottom of the sheave. If so, it will slip no matter how much tension is applied to the belt.

Belt Storage

Sometimes belts are stored on shelves in their original packaging. Other times they are stacked without packaging. If possible, store them on two or more pegs to prevent distortion. Keep belts away from damp floors and high heat areas.

You may need to replace belts that are worn or damaged from overheating or contact with oil or grease. Never replace one belt of a multibelt drive. Belts stretch with use. If you replace one belt of a multibelt drive, it will be tighter than the others and thus not meet the horsepower requirements the drive was designed for.

Preventive Maintenance Procedures

Inspection (failure risks for not following the procedures below are noted along with a rating): LOW: minimal risk/low chance of failure; MEDIUM: failure is possible, and equipment not operating to specifications is highly probable; HIGH: failure will happen prematurely.

- Check belt tension using a belt tension gauge. Measure the deflection and tension for the size of the belt. (Be sure to write tension and deflection specifications for the mechanic on the PM checklist.) Set tension on belt if deficiency noted.

Risk if the procedure is not followed: MEDIUM. Belt slippage will occur, thus resulting in equipment not operating to operation specifications. Another result from slippage is for belts to break, and the consequences could be a fire or at least machine stoppage.

- Identify any type of oil, grease, or chemical within 36 inches of belts (oil leakage from gearbox, motor, bearing, or chemicals from other sources). Write a corrective maintenance work order to repair leak or eliminate source of oil, grease, or chemical from the area.

Risk if the procedure is not followed: HIGH. Belt slippage will occur, thus resulting in equipment not operating to operation specifications. Another result for slippage is for belts to break, and the consequences could be a fire or at least machine stoppage.

- Check sheave alignment. If sheaves are not in alignment, align to manufacturer's specification. (Be sure to write the specification on this procedure; mechanics should not guess on this specification.)

Risk if the procedure is not followed: MEDIUM. Rapid belt wear will occur, thus resulting in equipment not operating to specifications. The belts could break if cords in the belt, begin to break due to this misalignment.

- Check sheaves for wear. Use a sheave gauge to ensure the sheave is not worn. If worn, write a corrective maintenance work order to change the sheave at a later date.

Risk if the procedure is not followed: HIGH. Belts will slip (even though you may not hear the slippage), thus resulting in equipment not operating to specifications.

24 Maintenance Welding

Introduction

An important use of arc welding is the repair of plant machinery and equipment. In this respect, welding is an indispensable tool without which production operations would soon shut down. Fortunately, welding machines and electrodes have been developed to the point where reliable welding can be accomplished under the most adverse circumstances. Frequently, welding must be done under something less than ideal conditions, and therefore, equipment and operators for maintenance welding should be the best.

Besides making quick, on-the-spot repairs of broken machinery parts, welding offers the maintenance department a means of making many items needed to meet a particular demand promptly. Broken castings, when new ones are no longer available, can be replaced with steel weldments fashioned out of standard shapes and plates (see Figure 24.1). Special machine tools required by production for specific operations often can be designed and made for a fraction of the cost of purchasing a standard machine and adapting it to the job. Material-handling devices can be made to fit the plant's physical dimensions. Individual jib cranes can be installed. Conveyors, either rolldown or pallet-type, can be tailor-made for specific applications. Tubs and containers can be made to fit products. Grabs, hooks, and other handling equipment can be made for shipping and receiving. Jigs and fixtures, as well as other simple tooling, can be fabricated in the maintenance department as either permanent tooling or as temporary tooling for a trial lot.

The almost infinite variety of this type of welding makes it impossible to do more than suggest what can be done. Figures 24.2 through 24.5 provide just a few examples of the imaginative applications of welding technology achieved by some maintenance technicians. The welding involved should present no particular problems if the operators have the necessary training and background to provide them with a knowledge of the many welding techniques that can be used.

A maintenance crew proficient in welding can fabricate and erect many of the structures required by a plant, even to the extent of making structural

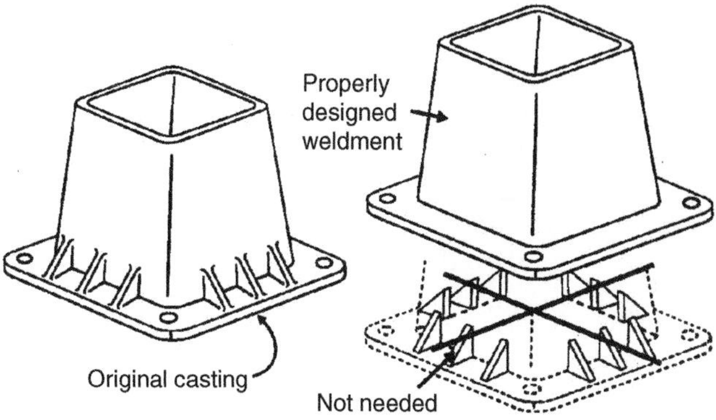

Figure 24.1

Figure 24.2

steel for a major plant expansion. Welding can be done either in the plant maintenance department or on the erection site. Structures must, of course, be adequately designed to withstand the loads to which they will be subjected. Such loads will vary from those of wind and snow in simple sheds

Figure 24.3

Figure 24.4

to dynamic loads of several tons where a crane is involved. Materials and joint designs must be selected with a knowledge of what each can do. Then the design must be executed by properly trained and qualified welders. Structural welding involves out-of-position work, so a welder must be able to make good welds under all conditions. Typical joints that are used in welded structures are shown in Figures 24.6 through 24.9.

Figure 24.5

Standard structural shapes, including pipe, which makes an excellent structural shape, can be used. Electrodes such as the E6010-11 types are often the welder's first choice for this kind of fabrication welding because of their all-position characteristics. These electrodes, which are not low-hydrogen types, may be used providing the weldability of the steel is such that neither weld cracks nor severe porosity is likely to occur.

Scrap materials often can be put to good use. When using scrap, however, it is best to weld with a low-hydrogen E7016-18 type of electrode, since the analysis of the steel is unlikely to be known, and some high-carbon steels may be encountered. Low-hydrogen electrodes minimize cracking tendencies. Structural scrap frequently comes from dismantled structures such as elevated railroads, which used rivet-quality steel that takes little or no account of the carbon content.

Shielded Metal Arc Welding (SMAW) "Stick Welding"

Shielded metal arc welding is the most widely used method of arc welding. With SMAW, often called "stick welding," an electric arc is formed between a consumable metal electrode and the work. The intense heat of the arc, which has been measured at temperatures as high as 13,000°F, melts the

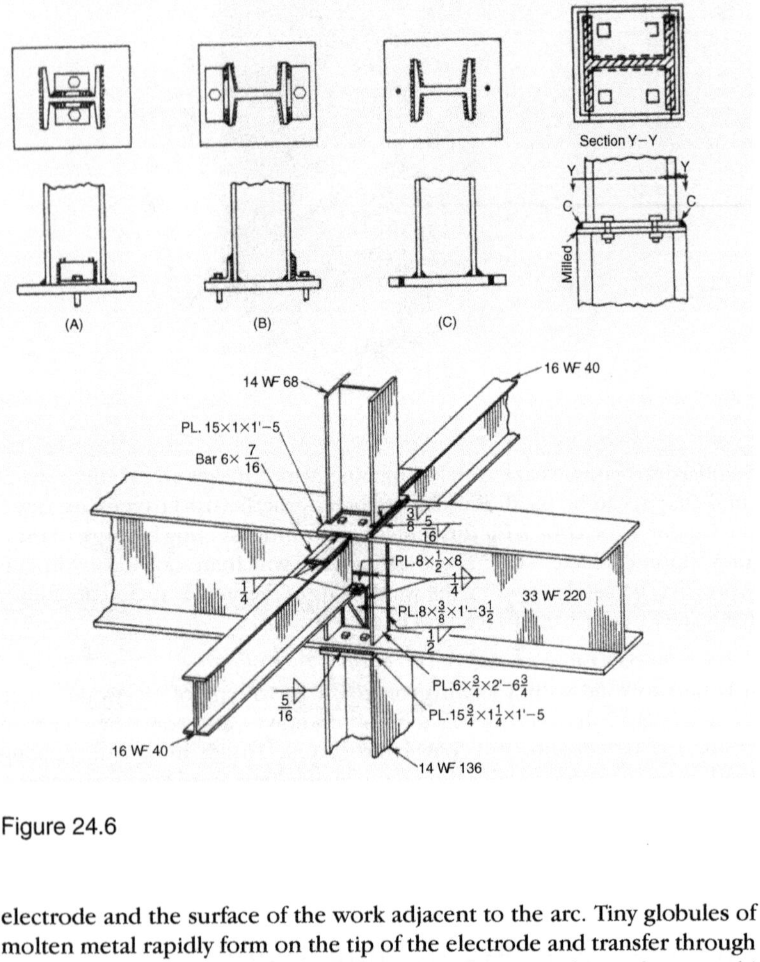

Figure 24.6

electrode and the surface of the work adjacent to the arc. Tiny globules of molten metal rapidly form on the tip of the electrode and transfer through the arc, in the "arc stream," and into the molten "weld pool" or "weld puddle" on the work's surface (see Figure 24.10).

Within the shielded metal arc welding process, electrodes are readily available in tensile strength ranges of 60,000 to 120,000 psi (see Table 24.1). In addition, if specific alloys are required to match the base metal, these, too, are readily available (see Table 24.2).

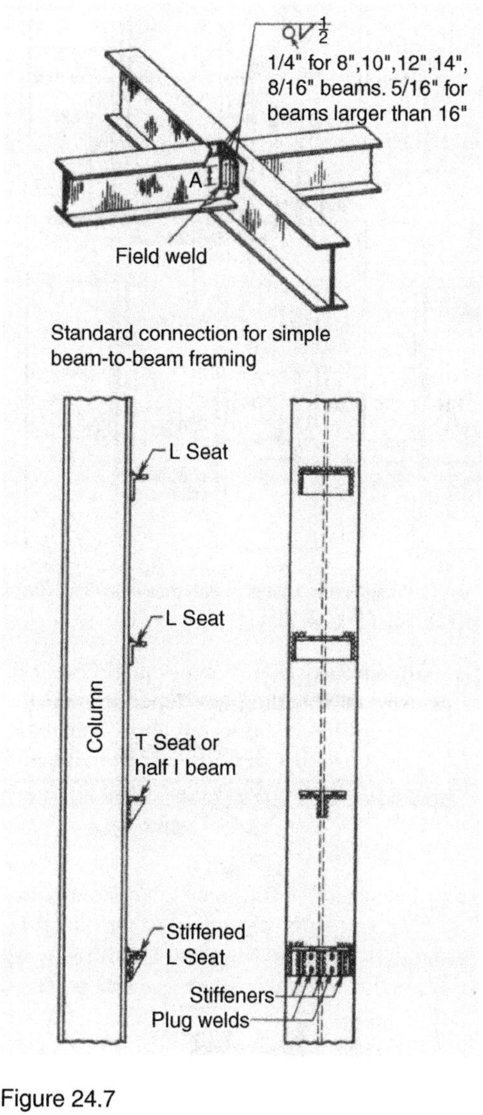
1/4" for 8",10",12",14",
8/16" beams. 5/16" for
beams larger than 16"
A
Field weld
Standard connection for simple
beam-to-beam framing
L Seat
L Seat
Column
T Seat or
half I beam
Stiffened
L Seat
Stiffeners
Plug welds

Figure 24.7

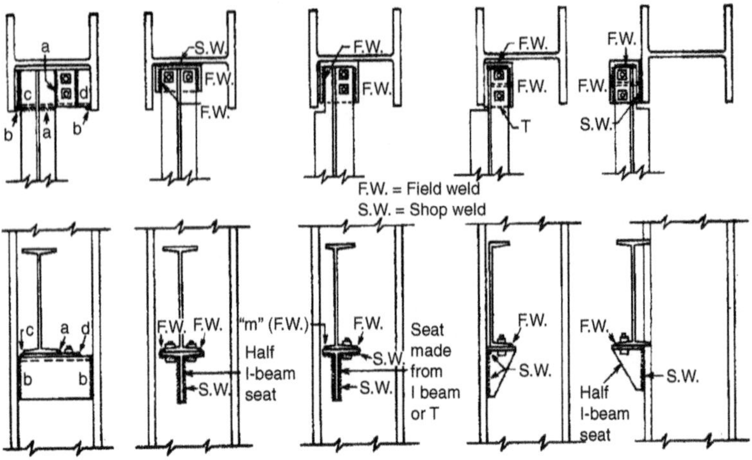

Figure 24.8

Flux-Cored Arc Welding (FCAW)

Flux-cored arc welding is generally applied as a semiautomatic process. It may be used with or without external shielding gas depending on the electrode selected. Either method utilizes a fabricated flux-cored electrode containing elements within the core that perform a scavenging and deoxidizing action on the weld metal to improve the properties of the weld.

FCAW with Gas

If gas is required with a flux-cored electrode, it is usually CO_2 or a mixture of CO_2 and another gas. These electrodes are best suited to welding relatively thick plate (not sheet metal) and for fabricating and repairing heavy weldments.

FCAW Self-Shielded

Self-shielded flux-cored electrodes, better known as Innershield, are also available. In effect, these are stick electrodes turned inside out and made into a continuous coil of tubular wire. All shielding, slagging, and

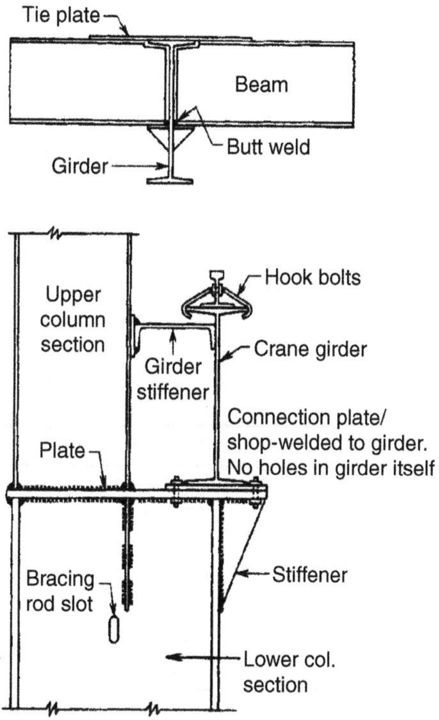

Figure 24.9

deoxidizing materials are in the core of the tubular wire. No external gas or flux is required.

Innershield electrodes offer much of the simplicity, adaptability, and uniform weld quality that account for the continuing popularity of manual welding with stick electrodes, but as a semiautomatic process, they get the job done faster. This is an open-arc process that allows the operator to place the weld metal accurately and to visually control the weld puddle. These electrodes operate in all positions: flat, vertical, horizontal, and overhead.

The electrode used for semiautomatic and fully automatic flux-cored arc welding is mechanically fed through a welding gun or welding jaws into the arc from a continuously wound coil that weighs approximately 50 pounds.

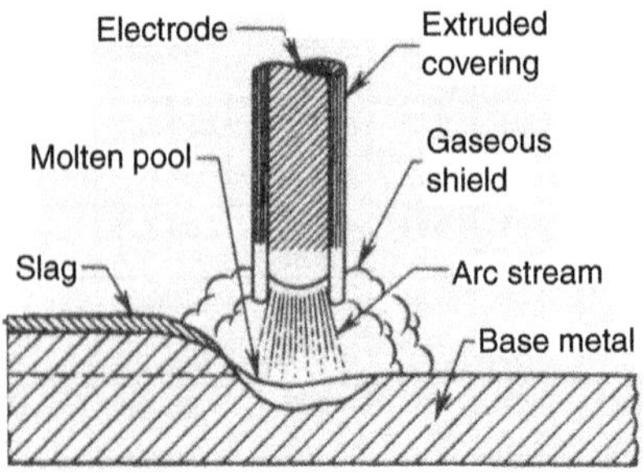

Figure 24.10

Table 24.1

AWS classification	Tensile strength, min. psi	Yield strength, min. psi
E6010-11	62,000	50,000
E7010-11	70,000	57,000
E7016-18	70,000	57,000
E8016-18	80,000	67,000
E9016-18	90,000	77,000
E10016-18	100,000	87,000
E11016-18	110,000	97,000
E12016-18	120,000	107,000

Note: E6010-11 and E7010-11 are cellulosic electrodes. All others are low-hydrogen electrodes and are better suited to welding higher-strength steels.

Only the fabricated flux-cored electrodes are suited to this method of welding, since coiling extruded flux-coated electrodes would damage the coating. In addition, metal-to-metal contact at the electrode's surface is necessary to transfer the welding current from the welding gun into the electrode. This is impossible if the electrode is covered.

A typical application of semiautomatic and fully automatic equipment for FCAW is shown in Figure 24.11. For a given cross section of electrode wire, much higher welding amperage can be applied with semiautomatic and fully automatic processes. This is because the current travels only a very short distance along the bare metal electrode, since contact between the current-carrying gun and the bare metal electrode occurs close to the arc. In manual welding, the welding current must travel the entire length of the electrode,

Table 24.2

AWS Classification[a]	Chemical composition, percent[b]								
	C	Mn	P	S	Si	Ni	Cr	Mo	V
Chemical composition, percent[b]									
E7010-Al		0.60		0.40					
E7011-Al		0.60		0.40					
E7015-Al		0.90		0.60					
E7016-Al	0.12	0.90	0.03	0.04	0.60	–	–	0.40–0.65	–
E7018-Al		0.90		0.80					
E7020-A1		0.60		0.40					
E7027-Al		1.00		0.40					
Chromium-molybdenum steel electrodes									
E8016-B1 E7017-B1	0.05 to 0.12	0.90	0.03	0.04	0.60 0.80	–	0.40–0.65	0.40–0.65	–
E8015-B2L	0.05	0.90	0.03	0.04	1.00		1.00–1.50	0.40–0.65	–
E8016-B2 E8018-B2	0.05 to 0.12	0.90	0.03	0.04	0.60	–	1.00–1.50	0.40–0.65	–
E8018-B2L	0.05	0.90	0.03	0.04	0.80	–	1.00–1.50	0.40–0.65	–
E9015-B3L	0.05	0.90	0.03	0.04	1.00	–	2.00–2.50	0.90–1.20	–
E9015-B3 E9016-B3 E9018-B3	0.05 to 0.12	0.90	0.03	0.04	0.60 0.60 0.80	–	2.00–2.50	0.90–1.20	–
E9018-B3L	0.05	0.09	0.03	0.04	0.80	–	2.00–2.50	0.90–1.20	–
E8015-B4L	0.05	0.90	0.03	0.04	1.00	–	1.75–2.25	0.40–0.65	–
B8016-B5	0.07 to 0.15	0.40 to 0.70	0.03	0.04	0.30–0.60	–	0.40–0.60	1.00–1.25	0.05
Nickel steel electrodes									
E8016-C1 E8018-C1	0.12	1.25	0.03	0.04	0.06 0.08	2.00–2.75	– –	– –	– –
E7015-C1L E7016-C1L E7018-C1L	0.05	1.25	0.03	0.04	0.50	2.00–2.75	–	–	–
E8016-C2 E8018-C2	0.12	1.25	0.03	0.04	0.60 0.80	3.00–3.75	–	–	–
E7015-C2L E7016-C2L E7018-C2L	0.05	1.25	0.03	0.04	0.50	3.00–3.75	–	–	–
E8016-C3[c] E8018-C3[c]	0.12	0.40–1.25	0.03	0.03	0.80	0.80–1.10	0.15	0.35	0.05

Continued

Table 24.2 *continued*

AWS Classification[a]	Chemical composition, percent[b]								
	C	Mn	P	S	Si	Ni	Cr	Mo	V
Nickel-molybdenum steel electrodes									
E8018-NM[d]	0.10	0.80–1.25	0.02	0.03	0.60	0.80–1.10	0.05	0.40–0.65	0.02
Manganese-molybdenum steel electrodes									
E9015-D1	0.12	1.25–1.75	0.03	0.04	0.60	–	–	0.25–0.45	–
E9018-DI					0.80				
E8016-D3	0.12	1.00–1.75	0.03	0.04	0.60	–	–	0.40–0.65	
E8018-D3					0.80				
E10015-D2	0.15	1.65–2.00	0.03	0.04	0.60		–	0.25–0.45	–
E10016-D2					0.60	–			
E10018-D2					0.80				
All other low-alloy steel electrodes[e]									
EXXIO-G[e]	–	1.00 min[f]	–	–	0.80 min[f]	0.50 min[f]	0.30 min[f]	0.20 min[f]	0.10 min[f]
EXX11-G									
EXX13-G									
EXX15-G									
EXX16-G									
EXX18-G									
E7020-G									
E9018-M[c]	0.10	0.60–1.25	0.030	0.030	0.80	1.40–1.80	0.15	0.35	0.05
E10018-M[c]	0.10	0.75–1.70	0.030	0.030	0.60	1.40–2.10	0.35	0.25–0.50	0.05
E11018-M[c]	0.10	1.30–1.80	0.030	0.030	0.60	1.25–2.50	0.40	0.25–0.50	0.05
E12018-M[c]	0.10	1.30–2.25	0.030	0.030	0.60	1.75–2.50	0.30–1.50	0.30–0.55	0.05
E12018-M1[c]	0.10	0.80–1.60	0.015	0.012	0.65	3.00–3.80	0.65	0.20–0.30	0.05
E7018-W[g]	0.12	0.40–0.70	0.025	0.025	0.40–0.70	0.20–0.40	0.15–0.30	–	0.08
E8018-W1[g]	0.12	0.50–1.30	0.03	0.04	0.35–0.80	0.40–0.80	0.45–0.70	–	–

Note: Single values shown are *maximum* percentages, except where otherwise specified.

[a] The suffixes A1, B3, C2, etc. designate the chemical composition of the electrode classification.

[b] For determining the chemical composition, DECN (electrode negative) may be used where DC, both polarities, is specified.

[c] These classifications are intended to conform to classifications covered by the military specifications for similar compositions.

[d] Copper shall be 0.10% max and aluminum shall be 0.05% max for E8018-NM electrodes.

[e] The letters "XX" used in the classification designations in this table stand for the various strength levels (70, 80, 90, 100, 110, and 120) of electrodes.

[f] In order to meet the alloy requirements of the G group, the weld deposit need have the minimum, as specified in the table, of only one of the elements listed. Additional chemical requirements may be agreed between supplier and purchaser.

[g] Copper shall be 0.30 to 0.60% for E7018-W electrodes.

Figure 24.11

and the amount of current is limited to the current-carrying capacity of the wire. The higher currents used with automatic welding result in a high weld metal deposition. This increases travel speed and reduces welding time, thereby lowering costs.

Gas-Shielded Metal Arc Welding (GMAW)

The GMAW process, sometimes called metal inert gas (MIG) welding, incorporates the automatic feeding of a continuous consumable electrode that is shielded by an externally supplied gas. Since the equipment provides for automatic control of the arc, only the travel speed, gun positioning, and guidance are controlled manually. Process control and function are achieved through the basic elements of equipment shown in Figure 24.12. The gun guides the consumable electrode and conducts the electric current and shielding gas to the workpiece. The electrode feed unit and power source are used in a system that provides automatic regulation of the arc length. The basic combination used to produce this regulation consists of a constant-voltage power source (characteristically providing an essentially flat volt-ampere curve) in conjunction with a constant-speed electrode feed unit.

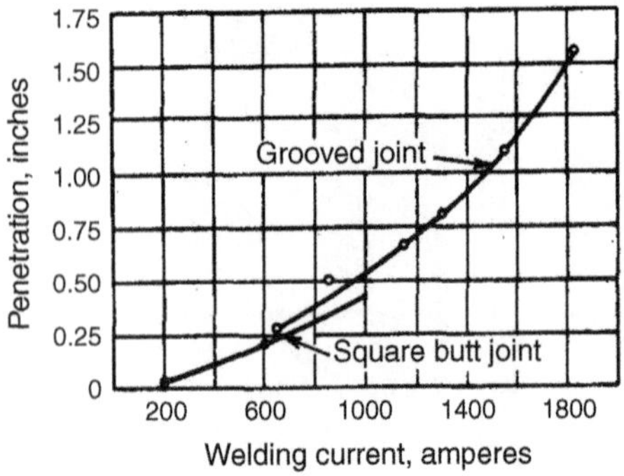

Figure 24.12

GMAW for Maintenance Welding

In terms of maintenance welding applications, GMAW has the following advantages over SMAW:

1 Can be used in all positions with the low-energy modes.

2 Produces virtually no slag to remove or be trapped in weld.

3 Requires less operator training time than SMAW.

4 Adaptable to semiautomatic or machine welding.

5 Low-hydrogen process.

6 Faster welding speeds than SMAW.

7 Suitable for welding carbon steels, alloy steels, stainless steels, aluminum, and other nonferrous metals. Table 24.3 lists recommended filler metals for GMAW.

Gas Selection for GMAW

There are many different gases and combinations of gases that can be used with the GMAW process. These choices vary with the base metal, whether a spray arc or short-circuiting arc is desired, or sometimes just according to operator preference. Recommended gas choices are given in Tables 24.4 and 24.5.

Table 24.3

Base metal type	Recommended electrode		AWS filler metal specification (use latest edition)	Electrode diameter		Current range Amperes
	Material type	Electrode classification		in.	mm	
Aluminum	1100	ER1100 or ER4043		0.030	0.8	50–175
and	3003, 3004	ER 1100 or ER5356		3/64	1.2	90–250
aluminum	5052, 5454	ER5554, ER5356	A5.10	1/16	1.6	160–350
alloys		or ER5183		3/32	2.4	225–400
	5083, 5086, 5456	ER5556 or ER5356		1/8	3.2	350–475
	6061, 6063	ER4043 or ER5356				
Magnesium	AZ10A	ERAZ61A, ERAZ92A				
alloys	AZ3IB,AZ61A,			0.040	1.0	150–300[2]
	AZ80A	ERAZ61A, ERAZ92A		3/64	1.2	160–320[2]
	ZE10A	ERAZ61A, ERAZ92A		1/16	1.6	210–400[2]
	ZK21A	ERAZ61A, ERAZ92A		3/32	2.4	320–510[2]
	AZ63A, AZ81A,		A5.19	1/8	3.2	400–600[2]
	AZ91C	ERAZ92A				
	AZ92A, AM100A	ERAZ92A				
	HK31A, HM21A,					
	HM31A	EREZ33A				
	LA141A	EREZ33A				
Copper and	Silicon Bronze	ERCuSi-A				
copper	Deoxidized copper	ERCu		0.035	0.9	150–300
alloys	Cu-Ni alloys	ERCuNi	A5.7	0.045	1.2	200–400
	Aluminum bronze	ERCuA1-A1, A2 or A3		1/16	1.6	250–450
	Phosphor bronze	ERCuSn-A		3/32	2.4	350–550
Nickel and				0.020	0.5	–
nickel	Monel[3] Alloy 400	ERNiCu-7		0.030	0.8	–
alloys	Inconel[3] Alloy 600	ERNiCrFe-5	A5.14	0.035	0.9	100–160
				0.045	1.2	150–260
				1/16	1.6	100–400
Titanium	Commercially	Use a filler metal one		0.030	0.8	–
and	pure	or two grades lower		0.035	0.9	–
titanium	Ti-0.15 Pd	ERTi-0.2 Pd	A5.16	0.045	1.2	–
alloys	Ti-5A1–2.5Sn	ERTi-5A1–2.5Sn or				
		commercially pure				
Austenitic	Type 201	ER308		0.020	0.5	–
stainless	Types 301, 302,			0.025	0.6	–
steels	304 & 308	ER308		0.030	0.8	75–150
	Type 304L	ER308L		0.035	0.9	100–160
	Type 310	ER310	A5.9	0.045	1.2	140–310
	Type 316	ER316		1/16	1.6	280–450
	Type 321	ER321		5/64	2.0	–
	Type 347	ER347		3/32	2.4	–
				7/64	2.8	–
				8	3.2	–

Table 24.3 *continued*

Base metal type	Recommended electrode		AWS filler metal specification (use latest edition)	Electrode diameter		Current range Amperes
	Material type	Electrode classification		in.	mm	
Steel	Hot rolled or cold-drawn plain carbon steels	ER70S-3 or ER70S-1 ER70S-2, ER70S-4 ER70S-5. ER70S-6	A5.18	0.020 0.025 0.030 0.035 0.045 0.052 1/16 5/64 3/32 1/8	0.5 0.6 0.8 0.9 1.2 1.3 1.6 2.0 2.4 3.2	- - 40-220 60-280 125-380 260-460 275-450 - - -
Steel	Higher strength carbon steels and some low alloy steels	ER80S-D2 ER80S-Nil ER100S-G	A5.28	0.035 0.045 1/16 5/64 3/32 1/8 5/32	0.9 1.2 1.6 2.0 2.4 3.2 4.0	60-280 125-380 275-450 - - - -

[2]Spray transfer mode.
[3]Trademark-International Nickel Co.

Table 24.4

Metal	Shielding gas	Advantages
Aluminum	Argon	0 to 1 in. (0 to 25 mm) thick: best metal transfer and arc stability; least spatter.
	35% argon + 65% helium	1 to 3 in. (25 to 76 mm) thick: higher heat input than straight argon; improved fusion characteristics with 5XXX series Al-Mg alloys.
	25% argon + 75% helium	Over 3 in. (76 mm) thick: highest heat input; minimizes porosity.
Magnesium	Argon	Excellent cleaning action.
Carbon steel	Argon + 1–5% oxygen	Improves arc stability; produces a more fluid and controllable weld puddle; good coalescence and bead contour, minimizes undercutting; permits higher speeds than pure argon.
	Argon + 3–10% CO_2	Good bead shape; minimizes spatter; reduces chance of cold lapping; cut not weld out of position.
Low-alloy steel	Argon + 2% oxygen	Minimizes undercutting; provides good toughness.

Continued

Table 24.4 *continued*

Metal	Shielding gas	Advantages
Stainless steel	Argon + 1% oxygen	Improves arc stability; produces a more fluid and controllable weld puddle, good coalescence and bead contour, minimizes undercutting on heavier stainless steels.
	Argon + 2% oxygen	Provides better arc stability, coalescence, and welding speed than 1 percent oxygen mixture for thinner stainless steel materials.
Copper, nickel, and their alloys	Argon	Provides good wetting; decreases fluidity of weld metal for thickness up to 1/8 in. (3.2 mm).
	Argon + helium	Higher heat inputs of 50 & 75 percent helium mixtures offset high heat dissipation of heavier gages.
Titanium	Argon	Good arc stability; minimum weld contamination; inert gas backing is required to prevent air contamination on back of weld area.

Table 24.5

Metal	Shielding gas	Advantages
Carbon steel	75% argon + 25% CO_2	Less than 1/8 in. (3.2 mm) thick: high welding speeds without burn-thru; minimum distortion and spatter.
	75% argon + 25% CO_2	More than 1/8 in. (3.2 mm) thick: minimum spatter; clean weld appearance; good puddle control in vertical and overhead positions.
	CO_2	Deeper penetration; faster welding speeds.
Stainless steel	90% helium + 7.5% argon + 2.5% CO_2	No effect on corrosion resistance; small heat-affected zone; no undercutting; minimum distortion.
Low alloy steel	60-70% helium + 25–35% argon + 4–5% CO_2	Minimum reactivity; excellent toughness; excellent arc stability, wetting characteristics, and bead contour, little spatter.
	75% argon + 25% CO_2	Fair toughness; excellent arc stability, wetting characteristics, and bead contour; little spatter.
Aluminum, copper, magnesium, nickel, and their alloys	Argon & argon + helium	Argon satisfactory on sheet metal; argon-helium preferred on thicker sheet material [over 1/8 in. (3.2 mm)].

Gas Tungsten Arc Welding (GTAW)

The gas tungsten arc welding (GTAW) process, also referred to as the tungsten inert gas (TIG) process, derives the heat for welding from an electric arc established between a tungsten electrode and the part to be welded (Figure 24.13). The arc zone must be filled with an inert gas to protect the tungsten electrode and molten metal from oxidation and to provide a conducting path for the arc current. The process was developed in 1941 primarily to provide a suitable means for welding magnesium and aluminum, where it was necessary to have a process superior to the shielded metal arc (stick electrode) process. Since that time, GTAW has been refined and has been used to weld almost all metals and alloys.

The GTAW process requires a gas- or water-cooled torch to hold the tungsten electrode; the torch is connected to the weld power supply by a power cable. In the lower-current gas-cooled torches (Figure 24.14), the power cable is inside the gas hose, which also provides insulation for the conductor. Water-cooled torches (Figure 24.15) require three hoses: one for the water supply, one for the water return, and one for the gas supply. The power cable is

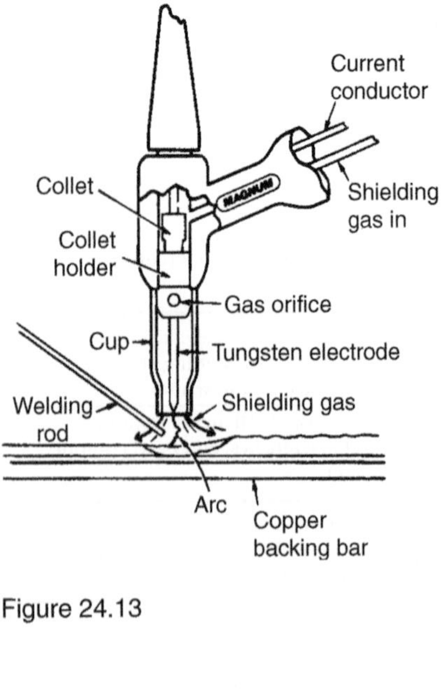

Figure 24.13

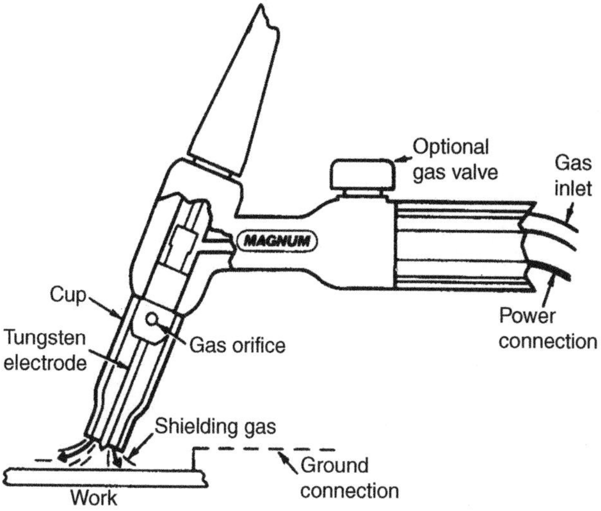

Figure 24.14

usually located in the water-return hose. Water cooling of the power cable allows use of a smaller conductor than that used in a gas-cooled torch of the same current rating.

Applicability of GTAW

The GTAW process is capable of producing very high-quality welds in almost all metals and alloys. However, it produces the lowest metal deposition rate of all the arc welding processes. Therefore, it normally would not be used on steel, where a high deposition rate is required and very high quality usually is not necessary. The GTAW process can be used for making root passes on carbon and low-alloy steel piping with consumable insert rings or with added filler metal. The remainder of the groove would be filled using the coated-electrode process or one of the semiautomatic processes such as GMAW (with solid wire) or FCAW (with flux-cored wire).

A constant-current or drooping-characteristic power supply is required for GTAW, either DC or AC and with or without pulsing capabilities. For water-cooled torches, a water cooler circulator is preferred over the use of tap water.

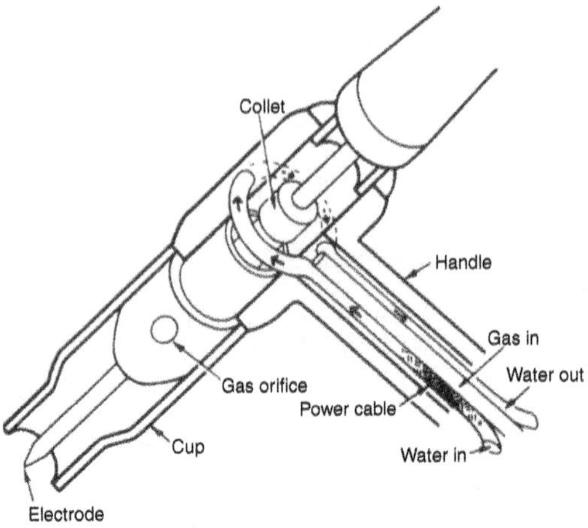

Figure 24.15

For automatic or machine welding, additional equipment is required to provide a means of moving the part in relation to the torch and feeding the wire into the weld pool. A fully automatic system may require a programmer consisting of a microprocessor to control weld current, travel speed, and filler wire feed rate. An inert-gas supply (argon, helium, or a mixture of these), including pressure regulators, flowmeters, and hoses, is required for this process. The gases may be supplied from cylinders or liquid containers. A schematic diagram of a complete gas tungsten arc welding arrangement is shown in Figure 24.16.

GTAW would be used for those alloys for which high-quality welds and freedom from atmospheric contamination are critical. Examples of these are the reactive and refractory metals such as titanium, zirconium, and columbium, where very small amounts of oxygen, nitrogen, and hydrogen can cause loss of ductility and corrosion resistance. It can be used on stainless steels and nickel-base superalloys, where welds exhibiting high quality with respect to porosity and fissuring are required. The GTAW process is well suited for welding thin sheet and foil of all weldable metals because it can be controlled at the very low amperages (2 to 5 amperes) required for these thicknesses. GTAW would not be used for welding the very low-melting metals, such as tin-lead solders and zinc-base alloys, because the high temperature of the arc would be difficult to control.

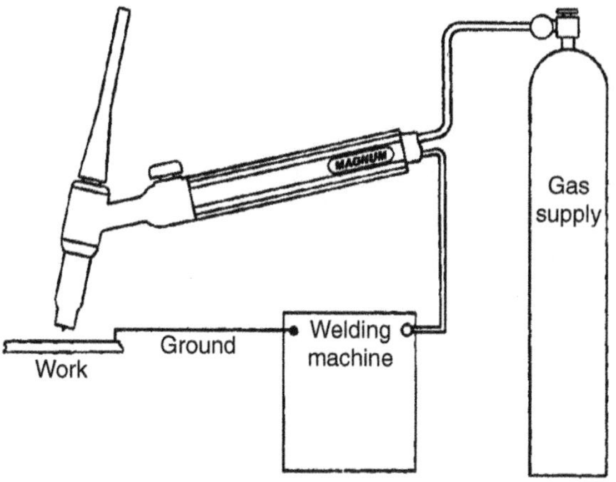

Figure 24.16 *Complete gas tungsten arc welding arrangement*

Advantages and Disadvantages of GTAW

The main advantage of GTAW is that high-quality welds can be made in all weldable metals and alloys except the very low-melting alloys. This is because the inert gas surrounding the arc and weld zone protects the hot metal from contamination. Another major advantage is that filler metal can be added to the weld pool independently of the arc current. With other arc welding processes, the rate of filler metal addition controls the arc current. Additional advantages are very low spatter, portability in the manual mode, and adaptability to a variety of automatic and semiautomatic applications.

The main disadvantage of GTAW is the low filler metal deposition rate. Further disadvantages are that it requires greater operator skill and is generally more costly than other arc welding processes.

Principles of Operating GTAW

In the GTAW process, an electric arc is established in an inert-gas atmosphere between a tungsten electrode and the metal to be welded. The arc is surrounded by the inert gas, which may be argon, helium, or a mixture of these two. The heat developed in the arc is the product of the arc current times the arc voltage, where approximately 70% of the heat is generated at

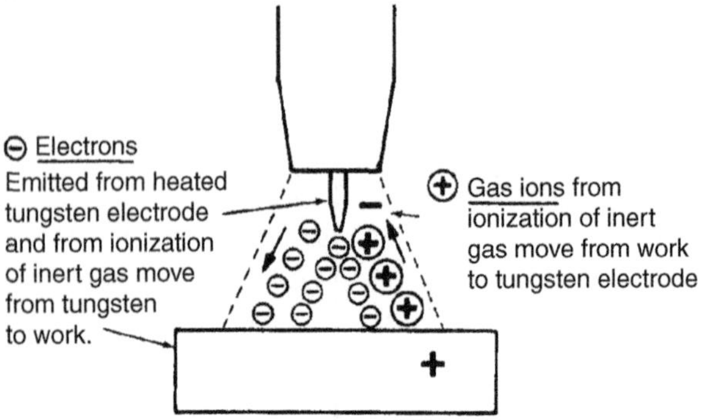
⊖ Electrons
Emitted from heated tungsten electrode and from ionization of inert gas move from tungsten to work.
⊕ Gas ions from ionization of inert gas move from work to tungsten electrode
−
+

Figure 24.17

the positive terminal of the arc. Arc current is carried primarily by electrons (Figure 24.17), which are emitted by the heated negative terminal (cathode) and obtained by ionization of the gas atoms. These electrons are attracted to the positive terminal (anode), where they generate approximately 70% of the arc heat. A smaller portion of the arc current is carried by positive gas ions which are attracted to the negative terminal (cathode), where they generate approximately 30% of the arc heat. The cathode loses heat by the emission of electrons, and this energy is transferred as heat when the electrons deposit or condense on the anode. This is one reason why a significantly greater amount of heat is developed at the anode than at the cathode.

The voltage across an arc is made up of three components: the cathode voltage, the arc column voltage, and the anode voltage. In general, the total voltage of the gas tungsten arc will increase with arc length (Figure 24.18), although current and shielding gas have effects on voltage, which will be discussed later. The total arc voltage can be measured readily, but attempts to measure the cathode and anode voltages accurately have been unsuccessful. However, if the total arc voltage is plotted against arc length and extrapolated to zero arc length, a voltage that approximates the sum of cathode voltage plus anode voltage is obtained. The total cathode plus anode voltage determined in this manner is between 7 and 10 volts for a tungsten cathode in argon. Since the greater amount of heat is generated at the anode, the GTAW process is normally operated with the tungsten electrode or cathode negative (negative polarity) and the work or anode positive. This puts the heat where it is needed, at the work.

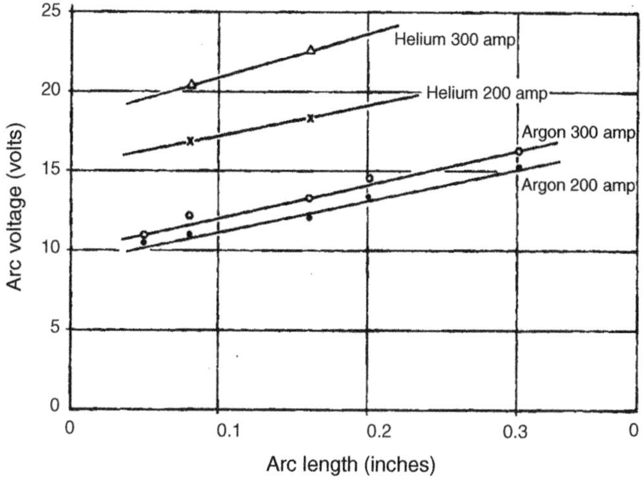

Figure 24.18

Polarity and GTAW

The GTAW process can be operated in three different modes: electrode-negative (straight) polarity, electrode-positive (reverse) polarity, or AC (Figure 24.19). In the electrode-negative mode, the greatest amount of heat is developed at the work. For this reason, electrode-negative (straight) polarity is used with GTAW for welding most metals. Electrode-negative (straight) polarity has one disadvantage—it does not provide cleaning action on the work surface. This is of little consequence for most metals, because their oxides decompose or melt under the heat of the arc so that molten metal deposits will wet the joint surfaces. However, the oxides of aluminum and magnesium are very stable and have melting points well above that of the metal. They would not be removed by the arc heat and would remain on the metal surface and restrict wetting.

In the electrode-positive (reverse) polarity mode, cleaning action takes place on the work surface by the impact of gas ions. This removes a thin oxide layer while the surface is under the cover of an inert gas, allowing molten metal to wet the surface before more oxide can form.

When AC gas tungsten arc welding aluminum, rectification occurs, and more current will flow when the electrode is negative (Fig. 24.20). This condition

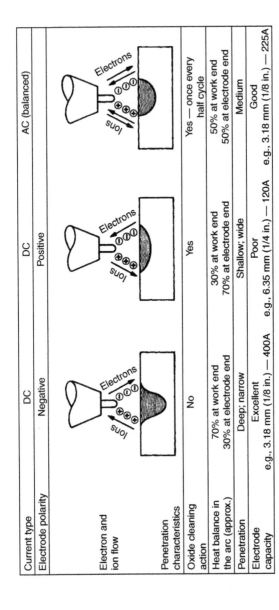

Current type	DC	DC	AC (balanced)
Electrode polarity	Negative	Positive	
Electron and ion flow Penetration characteristics	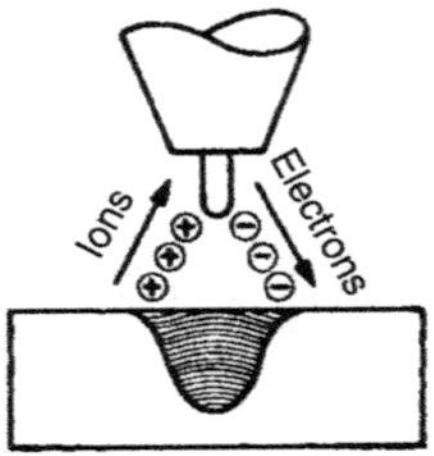	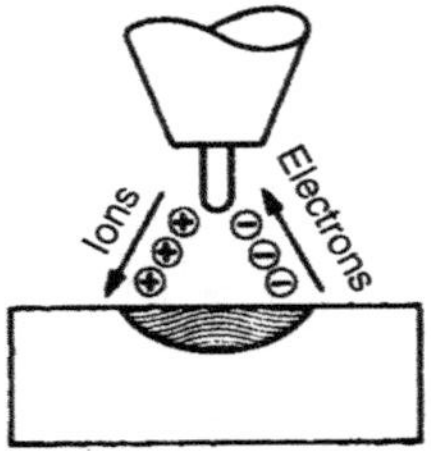	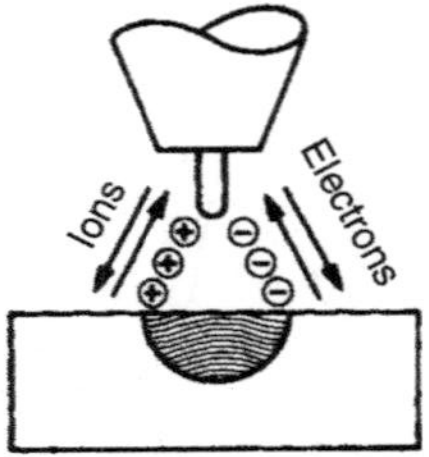
Oxide cleaning action	No	Yes	Yes — once every half cycle
Heat balance in the arc (approx.)	70% at work end 30% at electrode end	30% at work end 70% at electrode end	50% at work end 50% at electrode end
Penetration	Deep; narrow	Shallow; wide	Medium
Electrode capacity	Excellent e.g., 3.18 mm (1/8 in.) — 400A	Poor e.g., 6.35 mm (1/4 in.) — 120A	Good e.g., 3.18 mm (1/8 in.) — 225A

Figure 24.19

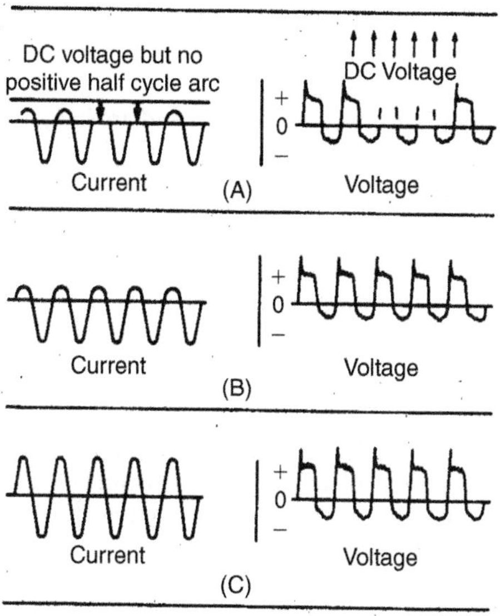

Figure 24.20

exists because the clean aluminum surface does not emit electrons as readily as the hot tungsten electrode. It will occur with standard AC welding power supplies. More advanced GTA welders incorporate circuits that can balance the negative- and positive-polarity half-cycles. Generally, this balanced condition is desirable for welding aluminum. The newest GTA power supplies include solid-state control boards, which allow adjustment of the AC current so as to favor either the positive- or negative-polarity half-cycle. These power supplies also chop the tip of the positive and negative half-cycles to produce a squarewave AC rather than a sinusoidal AC. When maximum cleaning is desired, the electrode-positive mode is favored; when maximum heat is desired, the electrode-negative mode is favored.

GTAW Shielding Gases and Flow Rates

Any of the inert gases could be used for GTAW. However, only helium (atomic weight 4) and argon (atomic weight 40) are used commercially, because they are much more plentiful and much less costly than the other inert gases. Typical flow rates are 15 to 40 cubic feet per hour (cfh).

Argon is used more extensively than helium for GTAW because:

1 It produces a smoother, quieter arc action;

2 It operates at a lower arc voltage for any given current and arc length;

3 There is greater cleaning action in the welding of materials such as aluminum and magnesium in the AC mode;

4 Argon is more available and lower in cost than helium;

5 Good shielding can be obtained with lower flow rates;

6 Argon is more resistant to arc zone contamination by cross drafts;

7 The arc is easier to start in argon.

The density of argon is approximately 1.3 times that of air and 10 times that of helium. For this reason, argon will blanket a weld area and be more resistant than helium to cross drafts. Helium, being much lighter than air, tends to rise rapidly and cause turbulence, which will bring air into the arc atmosphere. Since helium costs about three times as much as argon, and its required flow rate is two to three times that for argon, the cost of helium used as a shielding gas can be as much as nine times that of argon.

Although either helium or argon can be used successfully for most GTAW applications, argon is selected most frequently because of the smoother arc operation and lower overall cost. Argon is preferred for welding thin sheet to prevent melt-through. Helium is preferred for welding thick materials and materials of high thermal conductivity such as copper and aluminum.

Electrode Material for GTAW

In selecting electrodes for GTAW, five factors must be considered: material, size, tip shape, electrode holder, and nozzle. Electrodes for GTAW are classified as pure tungsten, tungsten containing 1 or 2% thoria, tungsten containing 0.15 to 0.4% zirconia, and tungsten that contains an internal lateral segment of thoriated tungsten. The internal segment runs the full length of the electrode and contains 1 or 2% thoria. Overall, these electrodes contain 0.35 to 0.55% thoria. All tungsten electrodes are normally available in diameters from 0.010 to 0.250" and lengths from 3" to 24". Chemical composition requirements for these electrodes are given in AWS A5.12, "Specification for Tungsten Arc Welding Electrodes."

Pure tungsten electrodes, which are 99.5% pure, are the least expensive but also have the lowest current-carrying capacity on AC power and a low resistance to contamination. Tungsten electrodes containing 1 or 2% thoria have greater electron emissivity than pure tungsten and, therefore, greater current-carrying capacity and longer life. Arc starting is easier, and the arc is more stable, which helps make the electrodes more resistant to contamination from the base metal. These electrodes maintain a well sharpened point for welding steel.

Tungsten electrodes containing zirconia have properties in between those of pure tungsten and thoriated tungsten electrodes with regard to arc starting and current-carrying capacity. These electrodes are recommended for AC welding of aluminum over pure tungsten or thoriated tungsten electrodes because they retain a balled end during welding and have a high resistance to contamination. Another advantage of the tungsten-zirconia electrodes is their freedom from the radioactive element thorium, which, although not harmful in the levels used in electrodes, is of concern to some welders.

GTAW Electrode Size and Tip Shape

The electrode material, size, and tip shape (Figure 24.21) will depend on the welding application, material, thickness, type of joint, and quantity. Electrodes used for AC or electrode-positive polarity will be of larger diameter than those used for electrode-negative polarity.

The total length of an electrode will be limited by the length that can be accommodated by the GTAW torch. Longer lengths allow for more redressing of the tip than short lengths and are therefore more economical. The extension of the electrode from the collet or holder determines the heating and voltage drop in the electrode. Since this heat is of no value to the weld, the electrode extension should be kept as short as necessary to provide access to the joint.

It is recommended that electrodes to be used for DC negative-polarity welding be of the 2% thoria type and be ground to a truncated conical tip. Excessive current will cause the electrode to overheat and melt. Too low a current will permit cathode bombardment and erosion caused by the low operating temperature and resulting arc instability. Although a sharp point on the tip promotes easy arc starting, it is not recommended because it will melt and form a small ball on the end.

For AC and DC electrode-positive welding, the desirable electrode tip shape is a hemisphere of the same diameter as the electrode. This tip shape on the larger electrodes required for AC and DC electrode-positive welding provides a stable surface within the operating current range. Zirconia-type

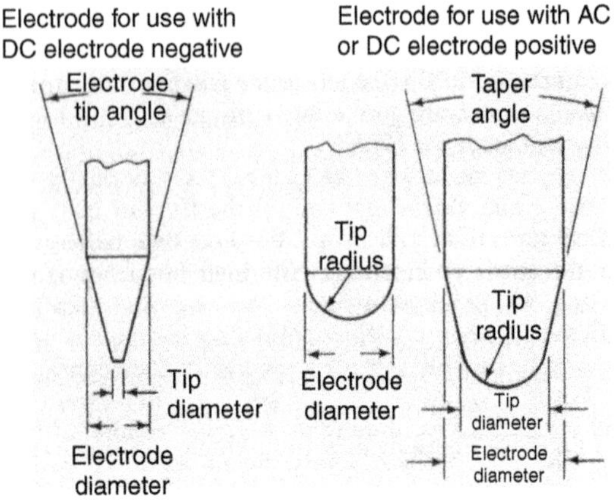

Figure 24.21

electrodes are preferred for AC and DC electrode-positive operation because they have a higher current-carrying capacity than the pure tungsten electrodes, yet they will readily form a molten ball under standard operating conditions. Thoriated electrodes do not ball readily and, therefore, are not recommended for AC or DC electrode-positive welding.

The degree of taper on the electrode tip affects weld penetration, where the smaller taper angles tend to reduce the width of the weld bead and thus increase penetration. When preparing the tip angle on an electrode, grinding should be done parallel to the length of the electrode. Special machines are available for grinding electrodes. These can be set to accurately grind any angle required.

GTAW Electrode Holders and Gas Nozzles

Electrode holders usually consist of a two-piece collet made to fit each standard-sized tungsten electrode. These holders and the part of the GTAW torch into which they fit must be capable of handling the required welding current without overheating. These holders are made of a hardenable copper alloy.

The function of the gas nozzle is to direct the flow of inert gas around the holder and electrode and then to the weld area. The nozzles are made of a hard, heat-resistant material such as ceramic and are available in various sizes and shapes. Large sizes give a more complete inert gas coverage of the weld area but may be too big to fit into restricted areas. Small nozzles can provide adequate gas coverage in restricted areas where features of the component help keep the inert gas at the joint. Most nozzles have internal threads that screw over threads on the electrode holder. Some nozzles are fitted with a washer-like device that consists of several layers of fine-wire screen or porous powder metal. These units provide a nonturbulent or lamellar gas flow from the torch, which results in improved inert gas coverage at a greater distance from the nozzle. In machine or automatic welding, more complete gas coverage may be provided by backup gas shielding from the fixture and a trailer shield attached to the torch.

Characteristics of GTAW Power Supplies

Power supplies for use with GTAW should be of the constant-current, drooping-voltage type (Figure 24.22). They may have other optional features such as up slope, down slope, pulsing, and current programming capabilities. Constant-voltage power supplies should not be used for GTAW.

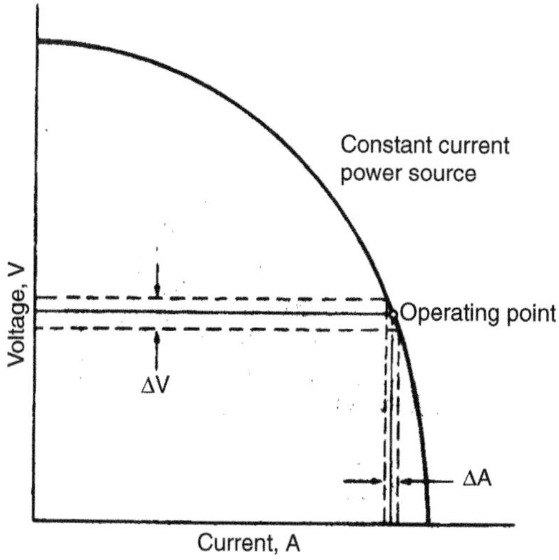

Figure 24.22

The power supply may be a single-phase transformer-rectifier, which also can supply AC for welding aluminum. Engine generator-type power supplies are usually driven by a gasoline or diesel engine and will produce DC with either a constant-current, drooping-voltage or constant-voltage characteristics. Engine alternator power supplies will produce AC for GTAW. A power supply capable of operating on either constant current or constant voltage should be set for the constant-current mode for GTAW.

Power supplies made specifically for GTAW normally will include a high-frequency source for arc starting and valves that control the flow of inert gas and cooling water for the torch. Timers allow the valves to be opened a short time before the arc is initiated and closed a short time after the arc is extinguished. The high frequency is necessary for arc starting instead of torch starting, where tungsten contamination of the weld is likely. It should be possible to set the high frequency for arc starting only, or for continuous operation in the AC mode.

Power supplies should include a secondary contactor and a means of controlling arc current remotely. For manual welding, a foot pedal would perform these functions of operating the contactor and controlling weld current. A power supply with a single current range is desirable because it allows the welder to vary the arc current between minimum and maximum without changing a range switch.

The more advanced power supplies incorporate features that permit pulsing the current in the DC mode with essentially square pulses. Both background and pulse peak current can be adjusted, as well as pulse duration and pulsing frequency (Figure 24.23). In the AC mode, the basic 60-Hz sine wave can be modified to produce a rectangular wave. Other controls permit the AC wave to be balanced or varied to favor the positive or negative half-cycles. This feature is particularly useful when welding aluminum and magnesium, where the control can be set to favor the positive half-cycle for maximum cleaning. In the DC mode, the pulsing capability allows welds to be made in thin material, root passes, and overhead with less chance of melt-through or droop.

GTAW Torches

A torch for GTAW must perform the following functions:

1 Hold the tungsten electrode so that it can be manipulated along the weld path.

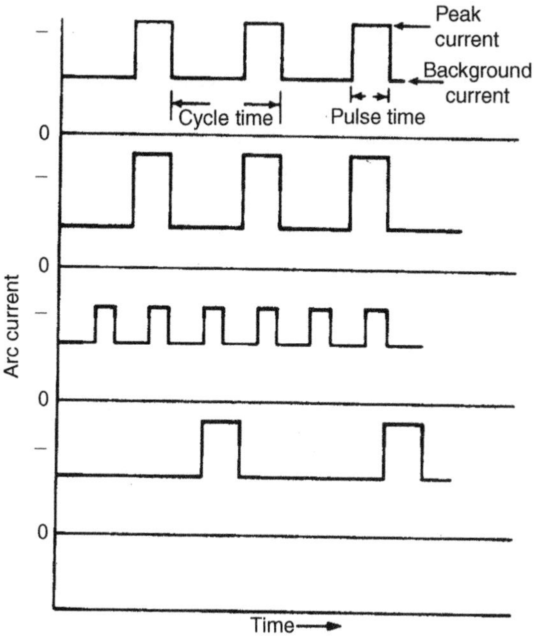

Figure 24.23

2 Provide an electrical connection to the electrode.

3 Provide inert-gas coverage of the electrode tip, arc, and hot weld zone.

4 Insulate the electrode and electrical connections from the operator or mounting bracket.

Typical GTAW torches consist of a metallic body, a collet holder, a collet, and a tightening cap to hold the tungsten electrode. The electrical cable is connected to the torch body, which is enclosed in a plastic insulating outer sheath. For manual torches, a handle is connected to the sheath. Power, gas, and water connections pass through the handle or, in the case of automatic operation, through the top of the torch. In the smaller, low-current torches, the electrode, collet, and internal components are cooled by the inert-gas flow. Larger, high-current torches are water-cooled and require connections to tap water and a drain or to a water cooler circulator. A cooler circulator with distilled or deionized water is preferred to prevent buildup of mineral deposits from tap water inside the torch.

Inert gas flows through the torch body and through holes in the collet holder to the arc end of the torch. A cup or nozzle is fitted over the arc end of the torch to direct inert gas over the electrode and the weld pool. The nozzles normally screw onto the torch and are made of a hard, heat-resistant ceramic. Some are made of high-temperature glass such as Vicor and are pressed on over a compressible plastic taper. Some nozzles can be fitted with an insert washer made up of several layers of fine-wire screen sometimes called a gas lens. This produces a lamellar rather than turbulent flow of inert gas to increase the efficiency of shielding.

On most manual GTAW torches, the handle is fixed at an angle of approximately 70 degrees to the torch body. Some makes of torches have a flexible neck between the handle and torch body which allows the angle between the handle and the torch body to be adjusted over a range from about 50 degrees to 90 degrees.

Manual GTAW Techniques

To become proficient in manual gas tungsten arc welding, the welder must develop skills in manipulating the torch with one hand, while controlling weld current with a foot pedal or thumb control and feeding filler metal with the other hand (see Figure 24.24). Before welding is started on any job, a rough idea of the welding conditions, such as filler material, current, shielding gas, etc., is needed.

Establishing Welding Parameters for GTAW

The material, thickness, joint design, and service requirements will determine the weld current, inert gas, voltage, and travel speed. This information may be available in a "welding procedure specification" (WPS) or from handbook data on the material and thickness. If welding parameters are not provided in a WPS, the information given in Tables 24.6, 24.7, and 24.8 can be used as starting-point parameters for carbon and low-alloy steels, stainless steels, and aluminum. These should be considered starting values; final values should be established by running a number of test parts.

Gas Tungsten Arc Starting Methods

The gas tungsten arc may be started by touching the work with the electrode, by a superimposed high frequency pulse, or by a high-voltage pulse.

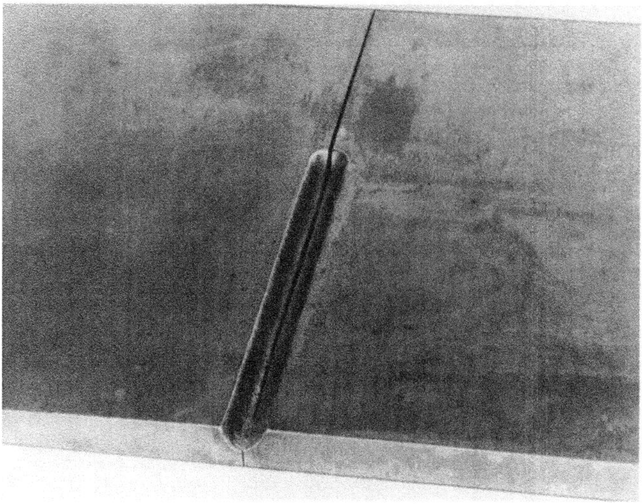

Figure 24.24

The touch method is not recommended for critical work because there is a strong possibility of tungsten contamination with this technique. Most weld power supplies intended for GTAW contain a high-frequency generator (usually a spark-gap oscillator), which superimposes the high-frequency pulse on the main weld power circuit. When welding with DC electrode-negative or -positive, the high-frequency switch should be set in the HT start position. When the welder presses the foot pedal to start welding, a timer is activated, which starts the high-frequency pulse and stops it when the arc initiates. Once started, the arc will continue after the high-frequency pulse stops as long as the power and proper arc gap are maintained. When welding with AC, the switch should be set in the HF continuous position to ensure that the arc restarts after voltage reversal on each half-cycle. High-frequency generators on welders produce frequencies in the radio communications range. Therefore, manufacturers of power supplies must certify that the radio frequency radiation from the power supply does not exceed limitations established by the Federal Communications Commission (FCC). The allowable radiation may be harmful to some computer and microprocessor systems and to communications systems. These possibilities for interference should be investigated before high-frequency starting is used. Installation instructions provided with the power supply should be studied and followed carefully.

Table 24.6

Welder size	60-Hz input voltage	Ampere rating	3 wires in conduit or 3-conductor cable, Type R	Grounding conductor
200	230 460 575	44 22 18	8 12 12	8 10 14
300	230 460 575	62 31 25	6 10 10	8 10 12
400	230 460 575	78 39 31	6 8 10	6 8 10
600	230 460 575	124 62 50	2 6 8	6 8 8
900	230 460 575	158 79 63	1 6 4	3 6 8

Table 24.7

		Amp input		Wire size (3 in conduit)			Wire size (3 in free air)		
Welder	Volts input	With condsr.	Without condsr.	With condsr.	Without condsr.	Ground conduct.	With condsr.	Without condsr.	Ground conduct.
300	200 440 550	84 42 38	104 52 42	2 6 8	1 6 6	1 6 6	4 8 10	4 8 8	4 8 8
400	220 440 550	115 57.5 46	143 71.5 57.2	0 4 8	00 3 4	00 3 4	3 6 8	1 6 6	1 6 6
500	220 440 550	148 74 61	180 90 72	000 3 4	0000 2 3	0000 2 3	1 6 6	0 4 6	0 4 6

Table 24.8

Machine size, amp	Cable sizes for lengths (electrode plus ground)		
	Up to 50 ft	50–100 ft	100–250 ft
200	2	2	1/0
300	1/0	1/0	3/0
400	2/0	2/0	4/0*
600	3/0	3/0	4/0*
900	Automatic application only		

*Recommended longest length of 4/0 cable for 400-amp welder, 150 ft; for 600-amp welder, 100 ft. For greater distances, cable size should be increased; this may be a question of cost-consider ease of handling versus moving of welder closer to work.

Oxyacetylene Cutting

Steel can be cut with great accuracy using an oxyacetylene torch (see Figure 24.25). However, not all metals cut as readily as steel. Cast iron, stainless steel, manganese steels, and nonferrous materials cannot be cut and shaped satisfactorily with the oxyacetylene process because of their reluctance to oxidize. In these cases, plasma arc cutting is recommended.

The cutting of steel is a chemical action. The oxygen combines readily with the iron to form iron oxide. In cast iron, this action is hindered by the presence of carbon in graphite form, so cast iron cannot be cut as readily as steel. Higher temperatures are necessary, and cutting is slower. In steel, the action starts at bright-red heat, whereas in cast iron, the temperature must be nearer the melting point in order to obtain a sufficient reaction.

Because of the very high temperature, the speed of cutting is usually fairly high. However, since the process is essentially one of melting without any great action, tending to force the molten metal out of the cut, some provision must be made for permitting the metal to flow readily away from the cut. This is usually done by starting at a point from which the molten metal can

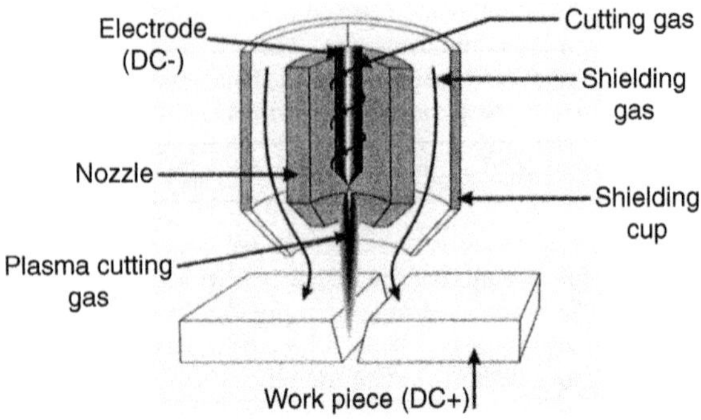

Figure 24.25

flow readily. This method is followed until the desired amount of metal has been melted away.

Air-Carbon Arc Cutting and Gouging

Air-carbon arc cutting (CAC-A) is a physical means of removing base metal or weld metal using a carbon electrode, an electric arc, and compressed air (see Figure 24.26). In the air-carbon arc process, the intense heat of the arc between the carbon electrode and the workpiece melts a portion of the base metal or weld. Simultaneously, a jet of air is passed through the arc of sufficient volume and velocity to blow away the molten material. This sequence can be repeated until the required groove or cut has been obtained. Since CAC-A does not depend on oxidation to maintain the cut, it is capable of cutting metals that oxyacetylene cutting will not cut. It is used to cut carbon steel, stainless steel, many copper alloys, and cast iron.

Arc gouging can be used to remove material approximately five times as fast as chipping. Depth of cut can be closely controlled, and welding slag does not deflect or hamper the cutting action, as it would with cutting tools. Gouging equipment generally costs less to operate than chipping hammers or gas cutting torches. An arc-gouged surface is clean and smooth and usually can be welded without further preparation. Drawbacks of the process include the fact that it requires large volumes of compressed air, and it is not as good as other processes for through-cutting.

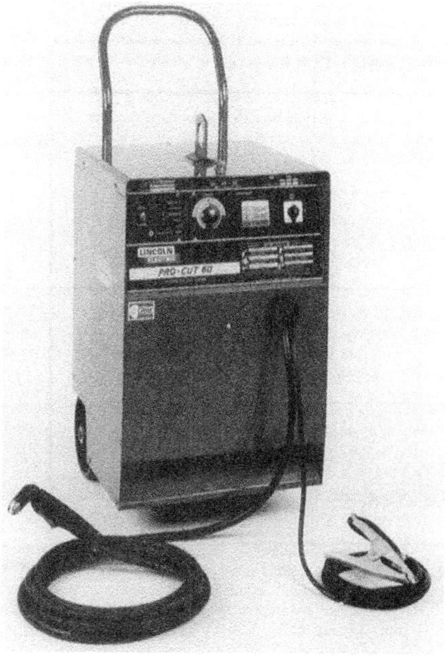

Figure 24.26

Electrode type and air supply specifications for arc gouging are outlined in Tables 24.9 and 24.10.

Applications

The CAC-A process may be used to provide a suitable bevel or groove when preparing plates for welding (see Fig. 24.27). It also may be used to back-gouge a seam prior to welding the side. CAC-A provides an excellent means of removing defective welds or misplaced welds and has many applications in metal fabrication, casting finishing, construction, mining, and general repair and maintenance. When using CAC-A, normal safety precautions must be taken and, in addition, ear plugs must be worn.

Power Sources

While it is possible to arc air gouge with AC, this is not the preferred method. A DC power source of sufficient capacity and a minimum of 60 open circuit

Table 24.9

Material	Electrode	Power
Steel	DC	DCEP
	AC	AC
Stainless steel	DC	DCEP
	AC	AC
Iron (cast iron, ductile iron, malleable iron)	AC	AC or DCEN
	DC	DCEP (high-amperage)
Copper alloys	AC	AC or DCEN
	DC	DCEP
Nickel alloys	AC	AC or DCEN

Source: *AWS Handbook*, 6th ed., Section 3A.
Note: AC is not the preferred method.

Table 24.10

Maximum Electrode Size (in.)	Application	Pressure (psi)	Consumption (cfm)
1/4	Intermittent-duty, manual torch	40	3
1/4	Intermittent-duty, manual torch	80	9
3/8	General-purpose	80	16
3/4	Heavy-duty	80	29
5/8	Semiautomatic mechanized torch	80	25

volts, either rectifier or motor generator, will give the best results. With DC, it is operated with DCEP (electrode-positive). Arc voltages normally range from about 35 to 56 volts. Table 24.11 lists recommended power sources, and Table 24.12 lists suggested current ranges. It is recommended that the power source have overload protection in the output circuit. High current surges of short duration occur with arc gouging, and these surges can overload the power source.

(a)

Conditions	SMAW (Stick)	GMAW (mig)	FCAW		GTAW (TIG)	SAW (Sub arc)
			Gas	Self shielded		
Carbon steel a. Sheet metal	★★	★★★★		★★	★★★★	
b. Plate	★★★	★★★	★★★★	★★★★		★★★★
Stainless a. Sheet metal	★	★★★★			★★★★	
b. Plate	★★	★★★			★★	★★★
Aluminum a. Sheet metal					★★★★	
b. Plate		★★★			★★	
Nonferrous Copper, Bronze, etc					★★★★	
High dep. rate	★	★★	★★★	★★★		★★★★
Portability of equipment	★★★	★★★	★★★	★★★	★★	
User friendly	★	★★	★★	★	★★★★	★
Min. weld cleaning	★	★★★	★★	★	★★★★	★
All position welding	★★★★	★★★★	★★	★★	★★★★	

Rating ★★★★ Excellent
★★★ Very good
★★ Good
★ Fair
(none) Not recommended

Figure 24.27

(b)

Weldment:____________________ Quality Category:

Drawing No.____________________ Critical ☐

Functional ☐

Process	Electrode Size/Type	Polarity	Volts	WFS/AMP (either)	ESO	Shield. Gas/Flux
SMAW						
GMAW						
SAW						
TIG						
FCAW						

Sketch of the Joint

Special instructions:____

Welder:____________________ Proced. Apprvd by:____________

Welder:____________________ Date:____________

Figure 24.27 *continued*

Table 24.11

Type of current	Type of power source	Remarks
DC	Variable-voltage motor-generator, rectifier or resistor-grid equipment	Recommended for all electrode sizes
DC	Constant-voltage motor-generator, or rectifier	Recommended only for electrodes above 1/4-in. diameter
AC	Transformer	Should be used only with AC electrodes
AC-DC		

Table 24.12

Type of electrode and power	Maximum and minimum current (amp)					
	Electrode size (in.)					
	5/32	3/16	1/4	5/16	3/8	1/2
DC electrodes, DECP power	90–150	150–200	200–400	250–450	350–600	600–1000
AC electrodes, AC power	—	150–200	200–300	—	300–500	400–600
AC electrodes, DCEN power	—	150–180	200–250	—	300–400	400–500

Plasma Arc Cutting

Plasma arc cutting has become an essential requirement for any properly equipped maintenance department. It provides the best, fastest, and cheapest method of cutting carbon or alloy steel, stainless steel, aluminum, nonferrous metals, and cast iron. In fact, it will cut any conductive material. The cuts are clean and precise, with very little dross or slag to remove. The heat is so concentrated within the immediate area of the cut that very little distortion takes place (see Figure 24.28). On gauge thickness material, the speed is limited only by the skill of the operator.

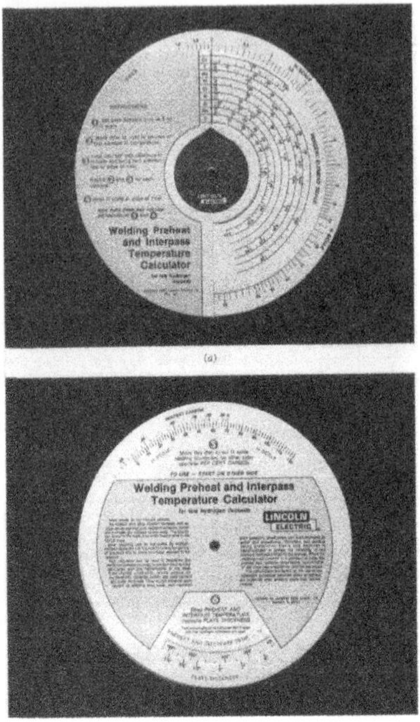

Figure 24.28

Plasma arc cutting operates on the principle of passing an electric arc through a quantity of gas through a restricted outlet. The electric arc heats the gas as it travels through the arc. This turns the gas into a plasma that is extremely hot. It is the heat in the plasma that heats the metal. A typical power source for plasma arc cutting is about the size of a small transformer welder and comes equipped with a special torch, as shown in Figure 24.29.

Plasma arc torch consumables include the electrode and the orifice. The electrode is copper, with a small hafnium insert at the center tip. The arc emanates from the hafnium, which gradually erodes with use, and this requires electrode replacement periodically. The torch tip contains the orifice which constricts the plasma arc. Various sized orifices are available, and the smaller diameters operate at lower amperages and produce a more constricted narrow arc than the larger diameters. The orifices also

Figure 24.29

gradually wear with use and must be replaced when the arc becomes too wide.

The plasma arc cutting torch can be used for arc gouging. The main changes are that the tip orifice is larger than for cutting, and the torch is held at an angle of about 30 degrees from horizontal rather than at 90 degrees, as in cutting. Plasma gouging can be used on all metals and is particularly suitable for aluminum and stainless steels, where oxyacetylene cutting is ineffective and carbon arc gouging tends to cause carbon contamination.

Welding Procedures

Much of the welding done by maintenance welders does not normally require detailed written welding procedures. The judgment and skill of the welders generally are sufficient to get the job done properly. However, there are some maintenance applications that demand the attention of a welding engineer or a supervisor, someone who knows more than the typical welder about the service conditions of the weldment or perhaps the weldability of it. Or there may be metallurgical factors or high-stress service requirements that must be given special consideration. A distinction should be made between "casual" and "critical" welding. When an application is considered critical, a welding engineer or a properly qualified supervisor should provide the welding operator with a detailed written procedure specification (WPS) providing all the information needed to make the weld properly.

Qualification of Welders

If the nature of the welding is critical and requires a written procedure, then equal consideration should be given to making sure the operator is qualified to do the kind of welding called for in the procedure. This can be done by having the operator make test welds that simulate the real thing. These welds can then be examined either destructively or nondestructively to see if the welder has demonstrated the required skills.

Note: Some welding may fall under local, state, or federal code requirements that do not permit welding to be done by in-house maintenance welders unless they have been certified.

Plasma Arc Welding

Plasma arc welding exists in several forms. The basic principle is that of an arc or jet created by the electrical heating of a plasma-forming gas (such as argon with additions of helium or hydrogen) to such a high temperature that its molecules become ionized atoms possessing extremely high energy. When properly controlled, this process results in very high melting temperatures. Plasma arc holds the potential solution to the easier joining of many hard-to-weld materials. Another application is the depositing of materials having high melting temperatures to produce surfaces of high resistance to extreme wear, corrosion, or temperature. As discussed earlier, when plasma arc technology is applied to metal cutting, it achieves unusually high speeds and has become an essential tool for a variety of maintenance applications.

Base Metals

The Carbon Steels

Carbon steels are widely used in all types of manufacturing. The weldability of the different types (low, medium, and high) varies considerably. The preferred analysis range of the common elements found in carbon steels is shown in Table 24.13. Welding metals whose elements vary above or below the range usually calls for special welding procedures.

Low-Carbon Steels (0.10 to 0.30% Carbon)

Steels of low-carbon content represent the bulk of the carbon steel tonnage used by industry. These steels are usually more ductile and easier to form than higher-carbon steels. For this reason, low-carbon steels are used in most applications requiring considerable cold forming, such as stampings and rolled or bent shapes in bar stock, structural shapes, or sheet. Steels with less than 0.13% carbon and 0.30% manganese have a slightly greater tendency to internal porosity than steels of higher carbon and manganese content.

Medium-Carbon Steels (0.31 to 0.45%)

The increased carbon content in medium-carbon steel usually raises the tensile strength, hardness, and wear resistance of the material. These steels are

Table 24.13

	Low, %	Preferred, %	High, %
Carbon	0.06	0.10 to 0.25	0.35
Manganese	0.30	0.35 to 0.80	1.40
Silicon		0.10 or under	0.30 max
Sulfur		0.035 or under	0.05 max
Phosphorus		0.03 or under	0.04 max

selectively used by manufacturers of railroad equipment, farm machinery, construction machinery, material-handling equipment, and other similar products. The medium-carbon steels can be welded successfully with the E60XX electrode if certain simple precautions are taken and the cooling rate is controlled to prevent excessive hardness.

High-Carbon Steels (0.46% and Higher)

The high-carbon steels are generally used in a hardened condition. This group includes most of the steels used in tools for forming, shaping, and cutting. Tools used in metalworking, woodworking, mining, and farming, such as lathe tools, drills, dies, knives, scraper blades, and plowshares, are typical examples. The high-carbon steels are often described as being "difficult to weld" and are not suited to mild steel welding procedures. Usually, low-hydrogen electrodes or processes are required, and controlled welding procedures, including preheating and postheating, are needed to produce crack-free welds.

The higher the carbon content of the steel, the harder the material becomes when it is quenched from above the critical temperature. Welding raises steel above the critical temperature, and the cold mass of metal surrounding the weld area creates a quench effect. Hardness and the absence of ductility result in cracking as the weld cools and contracts. Preheating from 300°F to 600°F and slow cooling will usually prevent cracking. Figure 24.30 shows a calculator for determining preheat and interpass temperatures.

For steels in the higher carbon ranges (over 0.30%), special electrodes are recommended. The lime ferritic low-hydrogen electrodes (E7016 or E7018) can be used to good advantage in overcoming the cracking tendencies of high-carbon steels. A 308 stainless steel electrode also can be used to give good physical properties to a weld in high-carbon steel.

Cast Iron

Cast iron is a complex alloy with a very high carbon content. Quickly cooled cast iron is harder and more brittle than slowly cooled cast iron. The metal also naturally exhibits low ductility, which results in considerable strain on parts of a casting when one local area is heated. The brittleness and the uneven contraction and expansion of cast iron are the principal concerns when welding it.

Each job must be analyzed to predetermine the effect of welding heat so that corresponding procedures can be adopted. Welds can be deposited in

Figure 24.30

short lengths, allowing each to cool. Peening of the weld metal while it is red hot may be used to stretch the weld deposit. Steel, cast iron, carbon, or nonferrous electrodes may be used. All oil, dirt, and foreign matter must be removed from the joint before welding. With steel electrodes, intermittent welds no longer than 3" should be used with light peening. To reduce contraction, the work should never be allowed to get too hot in one spot. Preheating will help to reduce hardening of the deposit to make it more machinable.

For the most machinable welds, a nonferrous alloy rod should be used. A two-layer deposit will have a softer fusion zone than a single-layer deposit. When it is practical, heating of the entire casting to a dull red heat is recommended in order to further soften the fusion zone and burn out dirt and foreign matter. When the weld is in a deep groove, it is general practice to use a steel electrode for welding cast iron to fill up the joint to within approximately $\frac{1}{8}$" of the surface and then finish the weld with the more machinable nonferrous deposit, usually a 95 to 98% nickel electrode.

The Alloy Steels

High-Tensile, Low-Alloy Steels

These steels are finding increasing use in metal fabricating because their higher strength levels permit the use of thinner sections, thereby saving metal and reducing weight. They are made with a number of different alloys and can be readily welded with specially designed electrodes that produce excellent welds of the same mechanical properties as the base metal. However, it is not necessary to have a core wire of exactly the same composition as the steel.

Stainless Steels

Electrodes are made to match various types of stainless steels so that corrosion-resistance properties are not destroyed in welding. The most commonly used types of stainless steels for welded structures are the 304, 308, 309, and 310 groups. Group 304 stainless, with a maximum carbon content of 0.8%, is commonly specified for weldments.

Welding procedures are much the same as for welding mild steel, except that one must take into account the higher electrical resistance of stainless steels and reduce the current accordingly. It is important to work carefully, cleaning all edges of foreign material. Light-gauge work must be clamped firmly to prevent distortion and buckling. Small-diameter and short electrodes should be used to prevent loss of chromium and undue overheating of the electrode. The weld deposit should be approximately the same analysis as the plate (see Tables 24.14 and 24.15).

Stainless Clad Steel

The significant precautions in welding this material are in joint design, including edge preparation, procedure, and choice of electrode. The electrode should be of the correct analysis for the cladding being welded. The joint must be prepared and welded to prevent dilution of the clad surface by the steel backing material. The backing material is welded with a mild steel electrode but in multiple passes to prevent excessive penetration into the cladding. The clad side is also welded in small passes to prevent penetration

Table 24.14

AISI type	Composition* (%)			
	Carbon	Chromium	Nickel	Other†
201	0.15	16.0–18.0	3.5–5.5	0.25 N, 5.5–7.5 Mn, 0.060 P
202	0.15	17.0–19.0	4.0–6.0	0.25 N, 7.5–10.0 Mn, 0.060 P
301	0.15	16.0–18.0	6.0–8.0	–
302	0.15	17.0–19.0	8.0–10.0	–
302B	0.15	17.0–19.0	8.0–10.0	2.0–3.0 Si
303	0.15	17.0–19.0	8.0–10.0	0.20 P, 0.15 S (min), 0.60 Mo (opt)
303Se	0.15	17.0–19.0	8.0–10.0	0.20 P, 0.06 S, 0.15 Se (min)
304	0.08	18.0–20.0	8.0–12.0	–
304L	0.03	18.0–20.0	8.0–12.0	–
305	0.12	17.0–19.0	10.0–13.0	–
308	0.08	19.0–21.0	10.0–12.0	–
309	0.20	22.0–24.0	12.0–15.0	–
309S	0.08	22.0–24.0	12.0–15.0	–
310	0.25	24.0–26.0	19.0–22.0	1.5 Si
310S	0.08	24.0–26.0	19.0–22.0	1.5 Si
314	0.25	23.0–26.0	19.0–22.0	1.5–3.0 Si
316	0.06	16.0–18.0	10.0–14.0	2.0–3.0 Mo
316L	0.04	16.0–18.0	10.0–14.0	2.0–3.0 Mo
317	0.06	18.0–20.0	11.0–15.0	3.0–4.0 Mo
321	0.06	17.0–19.0	9.0–12.0	Ti (5 × %C min)
347	0.08	17.0–19.0	9.0–13.0	Cb + Ta (10 × %C min)
348	0.08	17.0–19.0	9.0–13.0	Cb + Ta (10 × %C min but 0.10 Ta max), 0.20 Co

*Single values denote maximum percentage unless otherwise noted.
†Unless otherwise noted, other elements of all alloys listed include maximum contents of 2.0% Mn, 1.0% Si, 0.045% P, and 0.030% S. Balance is Fe.

into the backing material and resulting dilution of the stainless joint. When one is welding thin-gauge material and it is necessary to make the weld in one pass, a 309 stainless electrode should be used for the steel side as well as for the stainless side. The design and preparation of the joint can do much to prevent iron pickup, as well as reduce the labor costs of making the joint.

Table 24.15

AISI type	Composition*(%)			
	Carbon	Chromium	Manganese	Other†
405	0.08	11.5–14.5	1.0	0.1-0.3 Al
430	0.12	14.0–18.0	1.0	–
430F	0.12	14.0–18.0	1.25	0.060 P, 0.15 S(min), 0.60 Mo (opt)
430FSe	0.12	14.0–18.0	1.25	0.060 P, 0.060 S, 0.15 Se (min)
442	0.20	18.0–23.0	1.0	–
446	0.20	23.0–27.0	1.5	0.25 N

*Single values denote maximum percentage unless otherwise noted.
†Unless otherwise noted, other elements of all alloys listed include maximum contents of 1.0% Si, 0.040% P, and 0.030% S. Balance is Fe.

Straight Chromium Steels

The intense air-hardening property of these steels, which is proportional to the carbon and chromium content, is the chief consideration in establishing welding procedures. Considerable care must be taken to keep the work warm during welding, and it must be annealed afterward; otherwise, the welds and the areas adjacent to the welds will be brittle. It is a good idea to consult steel suppliers for specific details of proper heat treatment.

High-Manganese Steels

High-manganese steels (11 to 14% Mn) are very tough and are work-hardening, which makes them ideally suited for surfaces that must resist abrasion or wear as well as shock. When building up parts made of high-manganese steel, an electrode of similar analysis should be used.

The Nonferrous Metals

Aluminum

Most fusion welding of aluminum alloys is done with either the gas metal arc (GMAW) process or the gas tungsten arc (GTAW) process. In either case, inert-gas shielding is used.

With GMAW, the electrode is aluminum filler fed continuously from a reel into the weld pool. This action propels the filler metal across the arc to the workpiece in line with the axis of the electrode, regardless of the orientation of the electrode. Because of this, and because of aluminum's qualities of density, surface tension, and cooling rate, horizontal, vertical, and overhead welds can be made with relative ease. High deposition rates are practical, producing less distortion, greater weld strength, and lower welding costs than can be attained with other fusion welding processes.

GTAW uses a nonconsumable tungsten electrode, with aluminum alloy filler material added separately, either from a handheld rod or from a reel. Alternating current (AC) is preferred by many users for both manual and automatic gas tungsten arc welding of aluminum because AC GTAW achieves an efficient balance between penetration and cleaning.

Copper and Copper Alloys

Copper and its alloys can be welded with shielded metal arc, gas-shielded carbon arc, or gas tungsten arc welding. Of all these, gas-shielded arc welding with an inert gas is preferred. Decrease in tensile strength as temperature rises and a high coefficient of contraction may make welding of copper complicated. Preheating usually is necessary on thicker sections because of the high heat conductivity of the metal. Keeping the work hot and pointing the electrode at an angle so that the flame is directed back over the work will aid in permitting gases to escape. It is also advisable to put as much metal down per bead as is practical.

Control of Distortion

The heat of welding can distort the base metal; this sometimes becomes a problem in welding sheet metal or unrestrained large sections. The following suggestions will help in overcoming problems of distortion:

1 Reduce the effective shrinkage force.

- a Avoid overwelding. Use as little weld metal as possible by taking advantage of the penetrating effect of the arc force.
- b Use correct edge preparation and fit-up to obtain required fusion at the root of the weld.
- c Use fewer passes.
- d Place welds near a neutral axis.
- e Use intermittent welds.
- f Use back-step welding method.

2 Make shrinkage forces work to minimize distortion.

  a Preset the parts so that when the weld shrinks, they will be in the correct position after cooling.
  b Space parts to allow for shrinkage.
  c Prebend parts so that contraction will pull the parts into alignment.

3 Balance shrinkage forces with other forces (where natural rigidity of parts is insufficient to resist contraction).

  a Balance one force with another by correct welding sequence so that contraction caused by the weld counteracts the forces of welds previously made.
  b Peen beads to stretch weld metal. Care must be taken so that weld metal is not damaged.
  c Use jigs and fixtures to hold the work in a rigid position with sufficient strength to prevent parts from distorting. Fixtures can actually cause weld metal to stretch, preventing distortion.

Special Applications

Sheet Metal Welding

Plant maintenance frequently calls for sheet metal welding. The principles of good welding practice apply in welding sheet metal as elsewhere, but welding thin-gauge metals poses the specific challenges of potential distortion and/or burn-through. Special attention should therefore be given to all the factors involved in controlling distortion: the speed of welding, the choice of proper joints, good fit-up, position, proper current selection, use of clamping devices and fixtures, number of passes, and sequence of beads.

Good welding practice normally calls for the highest arc speeds and the highest currents within the limits of good weld appearance. In sheet metal work, however, there is always the limitation imposed by the threat of burn-through. As the gap in the work increases in size, the current must be decreased to prevent burn-through; this, of course, will reduce welding speeds. A clamping fixture will improve the fit-up of joints, making higher speeds possible. If equipped with a copper backing strip, the clamping fixture will make welding easier by decreasing the tendency to burn-through and also removing some of the heat that can cause warpage. Where possible, sheet metal joints should be welded downhill at about a 45 degrees angle with the same currents that are used in the flat position or slightly higher. Tables 24.16 and 24.17 offer guides to the selection of proper current, voltage, and electrodes for the various types of joints used with 20- to 8-gauge sheet metal.

Hard Surfacing

The building up of a layer of metal or a metal surface by electric arc welding, commonly known as hard surfacing, has important and useful applications in equipment maintenance. These may include restoring worn cutting edges and teeth on excavators, building up worn shafts with low- or medium-carbon deposits, lining a carbon-steel bin or chute with a stainless steel corrosion-resistant alloy deposit, putting a tool-steel cutting edge on a medium-carbon steel base, and applying wear-resistant surfaces to metal machine parts of all kinds. The dragline bucket shown in Fig. 24.31 is being returned to "new" condition by rebuilding and hard surfacing. Arc weld surfacing techniques include, but are not limited to, hard surfacing. There are many buildup applications that do not require hard surfacing. Excluding the effects of corrosion, wear of machinery parts results from various combinations of abrasion and impact. Abrasive wear results from one material scratching another, and impact wear results from one material hitting another.

Resisting Abrasive Wear

Abrasive wear is resisted by materials with a high scratch hardness. Sand quickly wears metals with a low scratch hardness, but under the same conditions, it will wear a metal of high scratch hardness very slowly. Scratch hardness, however, is not necessarily measured by standard hardness tests. Brinell and Rockwell hardness tests are not reliable measures for determining the abrasive wear resistance of a metal. A hard-surfacing material of the chromium carbide type may have a hardness of 50 Rockwell C. Sand will wear this material at a slower rate than it will a steel hardened to 60 Rockwell C. The sand will scratch all the way across the surface of the steel. On the surfacing alloy, the scratch will progress through the matrix material and then stop when the sand grain comes up against one of the microscopic crystals of chromium carbide, which has a higher scratch hardness than sand. If two metals of the same type have the same kind of microscopic constituents, however, the metal having the higher Rockwell hardness will be more resistant to abrasive wear.

Table 24.16

Type of welded joint	20 ga			18 ga			16 ga			14 ga			12 ga			10 ga			8 ga		
	F*	V*	O*	F	V	O	F	V	O	F	V	O	F	V	O	F	V	O	F	V	O
Plain butt	30†	30†	30†	40†	40†	40†	70†	70†	70†	85†	80	85†	115	110	110	135	120	115	190	130	120
Lap	40†	40†	40†	60†	60†	60†	100	100	100	130	130	130	135	120	120	155	130	120	165	140	120
Fillet				40†	40†	40†	70†	70†	70†	100	90	85	150	140	120	160	150	130	160	160	130
Corner	40†	40†	40†	60†	60†	60†	90†	90†	90†	90	80	75	125	110	110	140	130	125	175	130	125
Edge	40†	40†	40†	60†	60†	60†	80†	80†	80†	110	80	80	145	110	110	150	120	120	160	120	120

*F—fat position; V—vertical; O—overhead.
†Electrode negative, work positive.

Table 24.17

Type of welded joint	20 ga			18 ga			16 ga			14 ga			12 ga			10 ga			8 ga		
	F*	V*	O*	F	V	O	F	V	O	F	V	O	F	V	O	F	V	O	F	V	O
Plain butt	3/32	3/32	3/32	8/32	8/32	8/32	1/8	1/8	1/8	1/8	1/8	1/8	5/32	5/32	5/32	5/32	5/32	5/32	8/16	5/32	5/32
Lap	3/32	3/32	3/32	8/32	8/32	8/32	1/8	1/8	1/8	5/32	5/32	5/32	5/32	5/32	5/32	3/16	3/16	5/32	3/16	3/16	5/32
Fillet				8/32	8/32	8/32	1/8	1/8	1/8	1/8	1/8	1/8	5/32	5/32	5/32	3/16	5/32	5/32	3/16	5/32	5/32
Corner	3/32	3/32	3/32	8/32	8/32	8/32	1/8	1/8	1/8	1/8	1/8	1/8	3/16	5/32	5/32	3/16	5/32	5/32	3/16	5/32	5/32
Edge	3/32	3/32	3/32	8/32	8/32	8/32	1/8	1/8	1/8	1/8	1/8	1/8	3/16	5/32	5/32	3/16	5/32	5/32	3/16	5/32	5/32

*F—fat position; V—vertical; O—overhead.

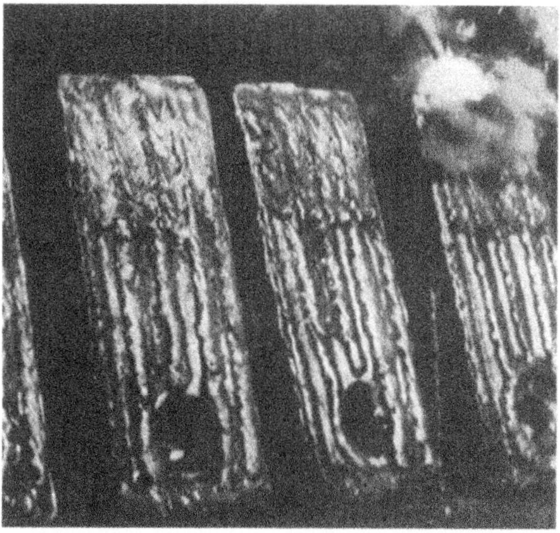

Figure 24.31

Resisting Impact Wear

Whereas abrasive wear is resisted by the surface properties of a metal, impact wear is resisted by the properties of the metal beneath the surface. To resist impact, a tough material is used, one that does not readily bend, break, chip, or crack. It yields so as to distribute or absorb the load created by impact, and the ultimate strength of the metal is not exceeded. Included in impact wear is that caused by bending or compression at low velocity without impact, resulting in loss of metal by cracking, chipping, upsetting, flowing, or crushing.

Types of Surfacing Electrodes

Many different kinds of surfacing electrodes are available. The problem is to find the best one to do a given job. Yet because service conditions vary so widely, no universal standard can be established for determining the ability of the surfacing to resist impact or abrasion. Furthermore, there is no ideal surfacing material that resists impact and abrasion equally well. In manufacturing the surfacing electrodes, it is necessary to sacrifice one quality somewhat to gain the other. High impact resistance is gained by sacrificing abrasion resistance, and vice versa.

Price is no index of electrode quality. An expensive electrode ingredient does not necessarily impart wear resistance. Therefore, the user of surfacing materials must rely on a combination of the manufacturer's recommendations and the user's own tests to select the best surfacing material for a particular purpose.

Choosing Hard-Facing Material

The chart shown in Figure 24.32 lists the relative characteristics of manual hard-facing materials. This chart is a guide to selecting the two items in the following list:

1 The hard-facing electrode best suited for a job not hard-faced before.

2 A more suitable hard-facing electrode for a job where the present material has not produced the desired results.

Example 1

Application: Dragline bucket tooth, as shown in Fig. 24.32.

Service: Sandy gravel with some good-sized rocks.

Maximum wear that can be economically obtained is the goal of most hard-facing applications. The material chosen should rate as highly as possible in the resistance-to-abrasion column, unless some other characteristics shown in the other columns make it unsuited for this particular application.

First, consider the tungsten carbide types. Notice that they are composed of very hard particles in a softer and less abrasion-resistant matrix. Although such material is the best for resisting sliding abrasion on hard material, in sand the matrix is apt to scour out slightly, and then the brittle particles are exposed. These particles are rated poor in impact resistance, and they may break and spall off when they encounter the rocks.

Next best in terms of abrasion, as listed in the chart, is the high-chromium carbide type shown in the electrode size column to be a powder. It can be applied only in a thin layer and also is not rated high in impact resistance. This makes it of dubious use in this rocky soil.

The rod-type high-chromium carbides also rate very high in abrasion resistance but do not rate high in impact resistance. However, the second does show sufficient impact rating to be considered if two or three different materials are to be tested in a field test. Given the possibility that it has enough

(a)

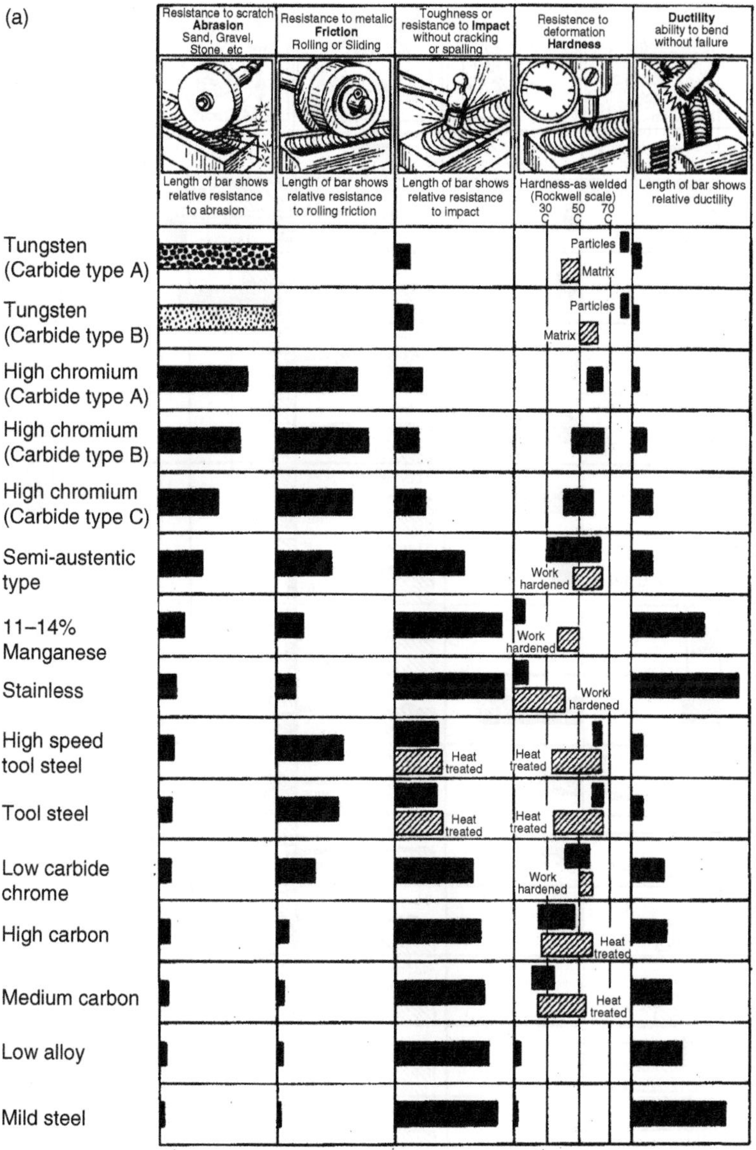

Figure 24.32

(b)

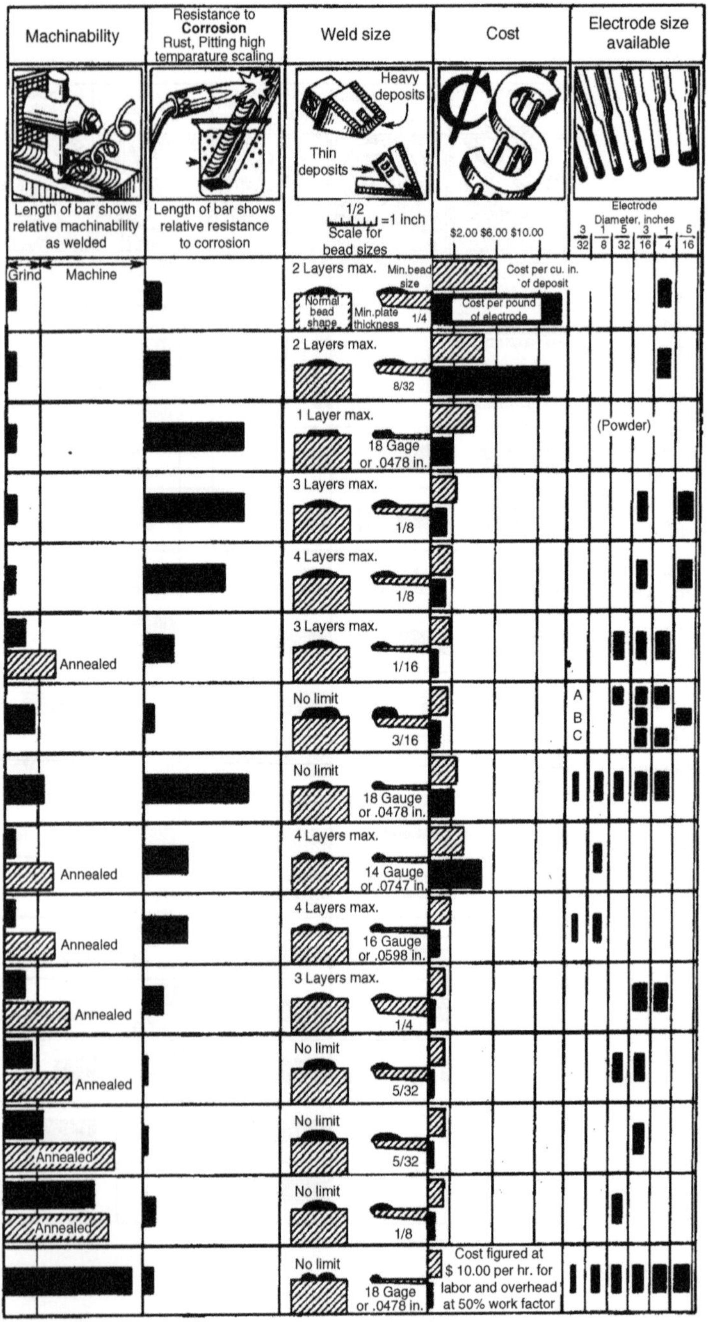

Figure 24.32 *continued*

impact resistance to do this job, there may be reluctance to pass up its very good wearing properties.

Nevertheless, the semiaustenitic type is balanced in both abrasion and impact resistance. It is much better in resistance to impact than the materials that rate higher in abrasion resistance. Therefore, the semiaustenitic is the first choice for this job, considering that the added impact resistance of the austenitic type is not necessary, since the impact in this application is not extreme.

Example 2

Application: Same dragline tooth used in Example 1.

Service: Soil changed to clay and shale.

The semiaustenitic type selected in the first example stands up well, but the teeth wear only half as long as the bucket lip. With double the wear on the teeth, only half the downtime periods would be needed for resurfacing, and both teeth and bucket could be done together. Since impact wear is now negligible with the new soil conditions, a material higher in the abrasion column should be considered. A good selection would be the first high-chromium carbide rod, which could give twice the wear by controlling the size of bead applied while still staying within a reasonable cost range.

Example 3

Application: Same dragline tooth used as in Examples 1 and 2.

Service: Soil changed to obtain large rocks.

If the earth contains many hard and large rocks, and the teeth are failing because of spalling under impact, one should move down the abrasion-resisting column to a more impact-resistant material, such as the semiaustenitic type.

These examples demonstrate that where a dragline operates in all kinds of soils, a material that is resistant to both impact and abrasion, such as a semiaustenitic type, is the best choice. When the same type of reasoning is used to check the important characteristics, an appropriate material can be chosen for any application. If, for any reason, the first choice does not prove satisfactory, it is usually easy to improve the next application by choosing a material that is rated higher in the characteristic that was lacking. Where failures occur because of cracking or spalling, it usually indicates that a material higher in impact or ductility rating should be used. Where normal wear alone seems too rapid, a material with a higher abrasion rating is indicated.

Check Welding Procedure

Often, hard-facing failures due to cracking or spalling may be caused by improper welding procedures. Before changing the hard-surfacing material, consider whether or not the material has been properly applied. For almost any hard-facing application, very good results can be obtained by following these precautions:

1 Do not apply hard-surfacing material over cracked or porous areas. Remove any defective areas down to sound base metal.

2 Preheat. Preheating to 400°F to 500°F improves the resistance to cracking and spalling. This minimum temperature should be maintained until welding is completed. The exception to the rule is 11 to 14% manganese steel, which should be kept cool.

3 Cool slowly. If possible, allow the finished weldment to cool under an insulating material such as lime or sand.

4 Do not apply more than the recommended number of layers.

When more than normal buildup is required, apply intermediate layers of either medium carbon or stainless steel. This will provide a good bond to the base metal and will eliminate excessively thick layers of hard-surfacing material that might otherwise spall off. Stainless steel is also an excellent choice for intermediate layers on manganese steels or for hard-to-weld steels where preheating is not practical.

Check Before the Part Is Completely Worn

Whenever possible, examine a surfaced part when it is only partly worn. Examination of a part after it is completely worn is unsatisfactory. Did the surface crumble off, or was it scratched off? Is a tougher surface needed, or is additional abrasion resistance required? Should a heavier layer of surfacing be used? Should surfacing be reduced? All these questions can be answered by examination of a partially worn part and with a knowledge of the surfacing costs and service requirements.

When it is impossible to analyze the service conditions thoroughly in advance, it is always on the safe side to choose a material tougher than

Figure 24.33

is thought to be required. A tough material will not spall or chip off and will offer some resistance to abrasion. A hard, abrasion-resistant material is more susceptible to chipping, and surfacing material does no good if it falls off.

After some experience with surfacing materials, various combinations of materials can be tried to improve product performance. For example, on a part which is normally surfaced with a tough, semiaustenitic electrode, it may be possible to get additional abrasion resistance without sacrificing resistance to cracking. A little of the powdered chromium carbide material can be fused to critical areas where additional protection is needed.

Many badly worn parts are first built up to almost finished size with a high-carbon electrode, then surfaced with an austenitic rod, and finally a few beads of chromium carbide deposit are placed in spots requiring maximum protection against abrasion. Regardless of the circumstances, a careful analysis of the surfacing problem will be well worthwhile. Examples of jobs are shown in Figures 24.33 to 24.35.

Hard Surfacing with SAW

The submerged arc process offers several advantages for hard surfacing. The greater uniformity of the surface makes for better wearing qualities. The speed of SAW creates major economies in hard-surfacing areas that require the deposition of large amounts of metal. These areas may be either flat or curved surfaces. Mixer bottom plates, scraper blades, fan blades, chutes,

Figure 24.34

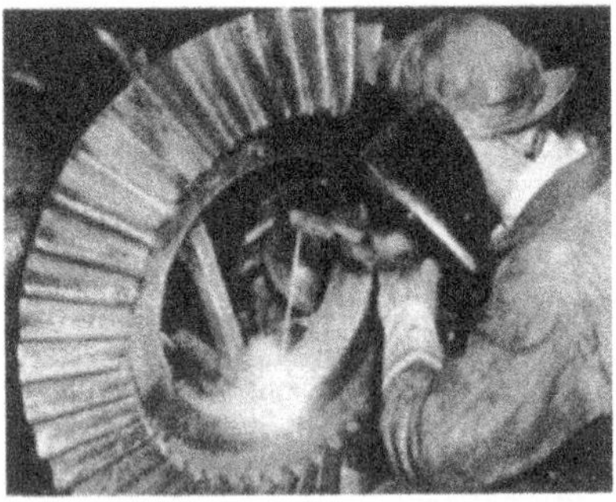

Figure 24.35

and refinery vessels are examples of the flat plate to be surfaced. Shafts, blooming mill spindles, skelp rolls, crane wheels, tractor idlers and rollers, and rams are examples of cylindrical surfaces (Figures 24.36 to 24.39).

The process can be used with either fully automatic or semiautomatic equipment depending on the economics of the application. Fully automatic equipment can be quickly fitted with auxiliary accessories, resulting in more economical metal deposition. An oscillating device can be added to an automatic head to create a bead up to 3" wide in a single pass. Another attachment permits the feeding of two electrode wires through a single head and a single contact jaw. Both of these attachments are useful in hard surfacing.

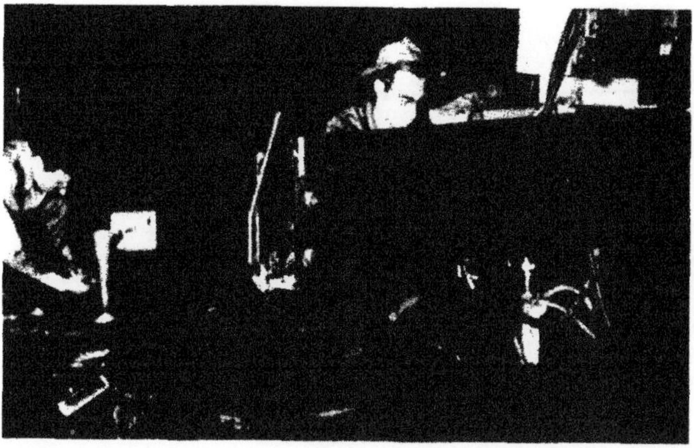

Figure 24.36

Figure 24.37

Hard surfacing with a submerged arc can be done with several different types of materials. The hard-surfacing deposit can be created by using solid alloy wires and a neutral granular flux. It also can be created by using a solid mild-steel wire and an agglomerated-alloy flux, the alloys being added

Figure 24.38

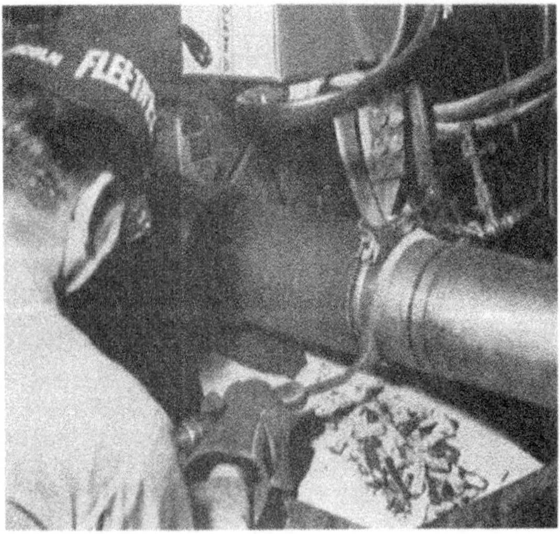

Figure 24.39

to the deposit through the flux rather than through the wire. Also available are tubular wires that contain alloying material in the hollow portion of the mild-steel tube. All the methods have specific advantages. With SAW, considerable variation in the hard-surfacing deposit can be made by changing the welding procedure to control admixture and the heat-treatment effect of the welding cycle. Methods and procedures should be established with the help of qualified engineers.

Selection and Maintenance of Equipment

Machines

Satisfactory welding can be accomplished with either alternating or direct welding current. Each type of current, however, has particular advantages that make it best suited for certain types of welding and welding conditions. The chief advantage of alternating current is its elimination of arc blow, which may be encountered when welding on heavy plate or into a corner. The magnetic fields set up in the plate deflect the path of the arc. Alternating current tends to minimize this deflection and also will increase the speed of welding with larger electrodes, over $\frac{3}{16}$" diameter, and with the iron powder type of electrodes.

The chief advantages of direct current are the stability of the arc and the fact that the current output of the motor-generator type of welder will remain constant in spite of variations in the input voltage that affect a transformer-type welder. Direct current, therefore, is a more versatile welding current. Certain electrodes, such as stainless, require a very stable arc; these electrodes operate much better with direct current. Direct current, because of its stability, is also better for sheet metal welding, where the danger of burn-through is present. The DC arc also can be more readily varied to meet different welding conditions. A wider range of control over both voltage and current permits closer adjustment of the arc for difficult welding conditions, such as might be encountered in vertical or overhead welding. Because of its versatility, direct current should be available for maintenance welding.

Direct-current welders (Figs. 24.40 and 24.41) are made either as motor-generator sets or as transformer-rectifier sets. Motor-generator sets are powered by AC or DC motors. Generators are also powered by small air-cooled gasoline engines (Figure 24.42). The advantage of this type of set

Figure 24.40

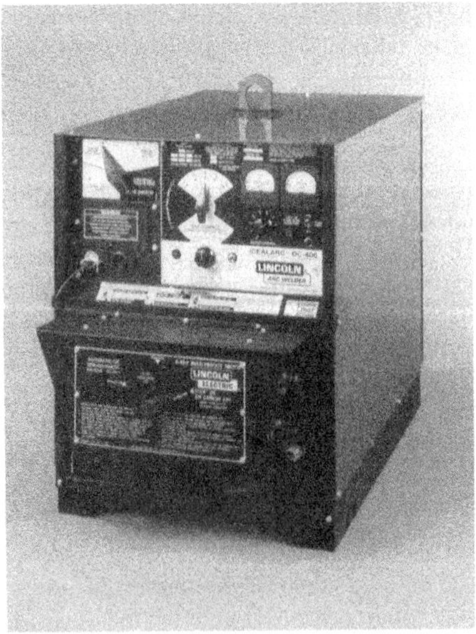

Figure 24.41

Figure 24.42

is that for on-the-spot maintenance welding, it is not necessary to string electric power lines to the job site. Engine-driven welders powered by gasoline engines are also available and come in larger sizes than the air-cooled engine sets (Figure 24.43). These are suitable where the size of the plant maintenance operation warrants a larger welder.

For most general maintenance welding, a 250-amp output capacity is ample. Several manufacturers make compact, portable machines especially for this type of welding. Higher amperages may be required in particular applications; for these, heavy-duty machines should be used.

Another type of welding machine is one that produces both alternating and direct welding current, either of which is available at the flip of a switch (Figure 24.44). This is ideal for maintenance welding, since it makes any kind of welding arc available, offering complete flexibility.

Figure 24.43

Accessory Equipment

The varied and severe service demands made on maintenance welding equipment require that the best in accessories be used. Most maintenance welders make racks or other storage conveniences, which they attach directly to the welding machine to facilitate storing and transporting electrodes and accessories. While these arrangements will vary to suit individual tastes and needs, the end result is to have everything immediately available for use.

A fire extinguisher is an essential accessory. Many electrode holders are available, but only a few combine all the desirable features. The operator holds the electrode clamped in a holder, and the current from the welding set passes through the holder to the electrode. The clamping device should be designed to hold the electrode securely in position yet permit the quick and easy exchange of electrodes. It should be light in weight, properly balanced and easy to handle, yet sturdy enough to withstand rough use. It should be designed to remain cool enough to be handled comfortably (see Figure 24.45).

Figure 24.44

Figure 24.45

Face or head shields are generally constructed of some kind of pressed fiber insulating material, usually black to reduce glare. The shield should be light in weight and comfortable to wear. The glass windows in the shield should be of a material that absorbs infrared rays, ultraviolet rays, and most visible rays emanating from the arc. The welding lens should be protected from molten metal spatter and breakage by a chemically treated clear "nonspatter" glass covering the exposed side of the lens. The operator should always wear a protective shield when welding and should never look at the arc with the naked eye. When a new lens is put into the shield, care should be taken to make sure no light leaks in around the glass. If practical, the welding room should be painted a dead black or some other dark color to prevent the reflection of light and glare. Others working around the welding area can be easily shielded from light and sparks by the use of portable screens.

Special goggles are used by welders' helpers, foremen, supervisors, inspectors, and others working close to a welding arc to protect their eyes from occasional flashes. A good set of goggles has an adjustable elastic head band and is lightweight, cool, well ventilated, and comfortable. Clear cover glasses and tinted lenses in various shades are available for this type of goggle.

During the arc welding process, some sparks and globules of molten metal are thrown out from the arc. For protection from possible burns, the operator is advised to wear an apron of leather or other protective material. Some operators also wear spats or leggings and sleevelets of leather or other fire-resistant material. Some sort of protection should be provided for the operator's ankles and feet, since a globule of molten metal can cause a painful burn before it can be extracted from the shoe. A gauntlet type of glove, preferably made of leather, is generally used by operators to protect their hands from the arc rays, spatters of molten metal, sparks, etc. Gloves also provide protection when the operator is handling the work.

Other tools of value in any shop where welding is done include wire brushes for cleaning the welds, cold chisels for chipping, clamps for holding work in position for welding, wedges, and, where work is large or heavy, a crane or chain block. A drill, air hammer, and grinder are also valuable accessories.

Installation of Equipment

Good welding begins with the proper installation of equipment. Installations should be made in locations that are as clean as possible, and there should be provisions for a continuous supply of clean air for ventilation. It is important to provide separate enclosures if the atmosphere is excessively moist or contains corrosive vapors. If welding must be done where the ambient temperature is high, place the equipment in a different location. Sets operated outdoors should be equipped with protection against inclement weather.

When installing welding equipment, consider the following:

1 Contact the local power company to ensure an adequate supply of electric power.

2 Provide an adequate and level support for the equipment.

3 Protect adequately against mechanical abuse and atmospheric conditions.

4 Provide fresh air for ventilation and cooling.

5 Electrically ground the frame of the welder.

6 Check electrical connections to make sure they are clean and mechanically tight.

7 The fuses for a motor-generator welder should be of the "high lag" type and be rated two or three times the input-current rating of the welder.

8 Provide welding leads of sufficient capacity to handle the required current.

9 Check the set before operating it to make sure that no parts are visibly loose or in poor condition.

Equipment Operation and Maintenance

The following precautions will do much to ensure maximum service and performance from arc welding equipment.

Keep the Machine Clean and Cool

Because of the large volume of air pulled through welders by the fans in order to keep the machines cool, the greatest enemies of continuous efficient performance are airborne dust and abrasive materials. Machines that are exposed to ordinary dust should be blown out at least once a week with dry, clean compressed air at a pressure not exceeding 30 psi. Higher pressures may damage windings.

In foundries or machine shops, where cast iron or steel dust is present, vacuum cleaning should be substituted for compressed air. Compressed air under high pressure tends to drive the abrasive dust into the windings.

Abrasive material in the atmosphere grooves and pits the commutator and wears out brushes. Greasy dirt or lint-laden dust quickly clogs air passages between coils and causes them to overheat. Since resistance of the coils is raised and the conductivity lowered by heat, it reduces efficiency and can result in burned-out coils if the machine is not protected against overload. Overheating makes the insulation between coils dry and brittle. Neither the air intake nor the exhaust vents should be blocked, because this will interrupt the flow of air through the machine. The welder covers should be kept on; removing them destroys the proper path of ventilation.

Do Not Abuse the Machine

Never leave the electrode grounded to the work. This can create a "dead" short circuit. The machine is forced to generate much higher current than it was designed for, which can result in a burned-out machine.

Do Not Work the Machine Over Its Rated Capacity

A 200-amp machine will not do the work of a 400-amp machine. Operating above capacity causes overheating, which can destroy the insulation or melt the solder in the commutator connections.

Use extreme care in operating a machine on a steady load other than arc welding, such as thawing water pipes, supplying current for lighting, running motors, charging batteries, or operating heating equipment. For example, a DC machine, NEMA-rated 300 amp to 40 volts or 12 kW, should not be used for any continuous load greater than 9.6 kW and not more than 240 amp. This precaution applies to machines with a duty cycle of at least 60%. Machines with lower load-factor ratings must be operated at still lower percentages of the rated load.

Do Not Handle Roughly

A welder is a precisely aligned and balanced machine. Mechanical abuse, rough handling, or severe shock may disturb the alignment and balance of the machine, resulting in serious trouble. Misalignment can cause bearing failure, bracket failure, unbalanced air gap, or unbalance in the armature.

Never pry on the ventilating fan or commutator to try to move the armature. To do so will damage the fan or commutator. If the armature is jammed, inspect the unit for the cause of the trouble. Check for dirt or foreign particles between the armature and frames. Inspect the banding wire on the armature. Look for a frozen bearing.

Do not neglect the engine if the welder is an engine-driven unit. It deteriorates rapidly if not properly cared for. Follow the engine manufacturer's recommendations. Change the oil regularly. Keep air filters and oil strainers clean. Do not allow grease and oil from the engine to leak back into the generator. Grease quickly accumulates dirt and dust, clogging the air passages between the coils.

Maintain the Machine Regularly

Bearings

The ball bearings in modern welders have sufficient grease to last the life of the machine under normal conditions. Under severe conditions—heavy use or a dirty location—the bearings should be greased about once a year. An ounce of grease a year is sufficient for each bearing. A pad of grease approximately one cubic inch in volume weighs close to 1 ounce. Dirt is responsible for more bearing failures than any other cause. This dirt may get into the grease cup when it is removed to refill, or it may get into the grease in its original container. Before the grease cup or pipe plug is removed, it is important to wipe it absolutely clean. A piece of dirt no larger than the period at the end of this sentence may cause a bearing to fail in a short time. Even small particles of grit that float around in the factory atmosphere are dangerous.

If too little grease is applied, bearings fail. If the grease is too light, it will run out. Grease containing solid materials may ruin antifriction bearings. Rancid grease will not lubricate. Dirty grease or dirty fittings or pipes can cause bearing failures.

Generally, bearings do not need inspection. They are sealed against dirt and should not be opened. If bearings must be pulled, it should be done using a special puller designed to act against the inner race.

Never clean new bearings before installing them. Handle them with care. Put them in place by driving against the inner race. Make sure that they fit squarely against the shoulders.

Brackets or End Bolts

If it becomes necessary to remove a bracket, to replace a bearing, or to disassemble the machine, do so by removing the bolts and tapping lightly and evenly with a babbitt hammer all around the outside diameter of the bracket ring. Do not drive off with a heavy steel hammer. The bearing may become worn over size, caused by the pounding of the bearing when the armature is out of balance. The bearing should slide into the housing with a light drive fit. Replace the bracket if the housing is over size.

Brushes and Brush Holders

Set brush holders approximately $\frac{1}{32}$" to $\frac{3}{32}$" above the surface of the commutator. If brush holders have been removed, be certain that they are set squarely in the rocker slot when replaced. Do not force the brush holder into the slot by driving on the insulation. Check to ensure that the brush holder insulation is squarely set. Tighten brush holders firmly. When properly set, they are parallel to the mica segments between commutator bars. Use the grade of brushes recommended by the manufacturer of the welding set. Brushes that are too hard or too soft may damage the commutator. Brushes will be damaged by excessive clearance in the brush holder or uneven brush spring pressure. High commutator bars, high mica segments, excessive brush spring pressure, and abrasive dust will also wear out brushes rapidly.

Inspect brushes and holders regularly. A brush may wear down and lose spring tension. It will then start to arc, with damage to the commutator and other brushes. Keep the brush contact surface of the holder clean and free from pit marks. Brushes must be able to move freely in the holder. Replace them when the pigtails are within $\frac{1}{8}$" of the commutator or when the limit of spring travel is reached.

New brushes must be sanded in to conform to the shape of the commutator. This may be done by stoning the commutator with a stone or by using fine sandpaper (not emery cloth or paper). Place the sandpaper under the brush, and move it back and forth while holding the brush down in the normal position under slight pressure with the fingers. See that the brush holders and springs seat squarely and firmly against the brushes and that the pigtails are fastened securely.

Commutators

Commutators normally need little care. They will build up a surface film of brown copper oxide, which is highly conductive, hard, and smooth. This surface helps to protect the commutator. Do not try to keep a commutator bright and shiny by constant stoning. The brown copper oxide film prevents the buildup of a black abrasive oxide film that has high resistance and causes excessive brush and commutator wear. Wipe clean occasionally with a rag or canvas to remove grease discoloration from fumes or other unnatural film. If brushes are chattering because of high bars, high mica, or grooves, stone by hand or remove and turn in a lathe, if necessary.

Most commutator trouble starts because the wrong grade of brushes is used. Brushes that contain too much abrasive material or have too high a copper content usually scratch the commutator and prevent the desired surface film from building up. A brush that is too soft may smudge the surface with the same result as far as surface film is concerned. In general, brushes that have a low voltage drop will give poor commutation. Conversely, a brush with high voltage drop commutates better but may cause overheating of the commutator surface.

If the commutator is burned, it may be dressed down by pressing a commutator stone against the surface with the brushes raised. If the surface is badly pitted or out of round, the armature must be removed from the machine and the commutator turned in a lathe. It is good practice for the commutator to run within a radial tolerance of 0.003". The mica separating the bars of the commutator is undercut to a depth of $\frac{1}{32}$" to $\frac{1}{16}$". Mica exposed at the commutator surface causes brush and commutator wear and poor commutation. If the mica is even with the surface, undercut it. When the commutator is operating properly, there is very little visible sparking. The brush surface is shiny and smooth, with no evidence of scratches.

Generator Frame

The generator frame and coils need no attention other than inspection to ensure tight connections and cleanliness. Blow out dust and dirt with compressed air. Grease may be cleaned off with naphtha. Keep air gaps between armature and pole pieces clean and even.

Armature

The armature must be kept clean to ensure proper balance. Unbalance in the set will pound out the bearings and wear the bearing housing oversize. Blow out the armature regularly with clean, dry compressed air. Clean out the inside of the armature thoroughly by attaching a long pipe to the compressed air line and reaching into the armature coils.

Motor Stator

Keep the stator clean and free from grease. When reconnecting it for use on another voltage, solder all connections. If the set is to be used frequently on different voltages, it may save time to place lugs on the ends of all the stator leads. This eliminates the necessity for loosening and resoldering to make connections, since the lugs may be safely joined with a screw, nut, and lock washer.

Exciter Generator

If the machine has a separate exciter generator, its armature, coils, brushes, and brush holders will need the same general care recommended for the welder set. Keep the covers over the exciter armature, since the commutator can be damaged easily.

Controls

Inspect the controls frequently to ensure that the ground and electrode cables are connected tightly to the output terminals. Loose connections cause arcing that destroys the insulation around the terminals and burns them. Do not bump or hit the control handles—it damages the controls, resulting in poor electrical contacts. If the handles are tight or jammed, inspect them for the cause. Check the contact fingers of the magnetic starting switch regularly. Keep the fingers free from deep pits or other defects that will interfere with a smooth, sliding contact. Copper fingers may be filed lightly. All fingers should make contact simultaneously. Keep the switch clean and free from dust. Blow out the entire control box with low-pressure compressed air.

Connections of the leads from the motor stator to the switch must be tight. Keep the lugs in a vertical position. The line voltage is high enough to jump between the lugs on the stator leads if they are allowed to become loose and cocked to one side or the other. Keep the cover on the control box at all times.

Condensers

Condensers may be placed in an AC welder to raise the power factor. When condensers fail, it is not readily apparent from the appearance of the condenser. Consequently, to check a condenser, one should see if the input current reading corresponds to the nameplate amperes at the rated input voltage and with the welder drawing the rated output load current. If the reading is 10 to 20% more, at least one condenser has failed. Caution: Never touch the condenser terminals without first disconnecting the welder from the input power source; then discharge the condenser by touching the two terminals with an insulated screwdriver.

Delay Relays

The delay relay contacts may be cleaned by passing a cloth soaked in naphtha between them. Do not force the contact arms or use any abrasives to clean the points. Do not file the silver contacts. The pilot relay is enclosed in a dust-proof box and should need no attention. Relays are usually adjusted at the factory and should not be tampered with unless faulty operation is obvious. Table 24.18, a troubleshooting chart, may prove to be a great timesaver.

Table 24.18

Trouble	Cause	Remedy
Welder will not start (Starter not operating)	Power circuit dead	Check voltage
	Broken power lead	Repair
	Wrong supply voltage	Check name plate against supply
	Open power switches	Close
	Blown fuses	Replace
	Overload relay tripped	Let set cool. Remove cause of overloading
	Open circuit to starter button	Repair
	Defective operating coil	Replace
	Mechanical obstruction in contactor	Remove
Welder will not start (Starter operating)	Wrong motor connections	Check connection diagram
	Wrong supply voltage	Check name plate against supply
	Rotor stuck	Try turning by hand
	Power circuit single-phased	Replace fuse; repair open line
	Starter single-phased	Check contact of starter tips
	Poor motor connection	Tighten
	Open circuit in windings	Repair
Starter operates and blows fuse	Fuse too small	Should be two to three times rated motor current
	Short circuit in motor connections	Check starter and motor leads for insulation from ground and from each other
Welder starts but will not deliver welding current	Wrong direction of rotation	Check connection diagram
	Brushes worn or missing	Check that all brushes bear on commutator with sufficient tension
	Brush connections loose	Tighten
	Open field circuit	Check connection to rheostat, resistor, and auxiliary brush studs
	Series field and armature circuit open	Check with test lamp or bell ringer

Hazard	Factors to consider	Precaution summary
Electric shock can kill 	• Wetness • Welder in or workpiece • Confined space • Electrode holder and cable insulation	• Insulate welder from workpiece and ground using dry insulation. Rubber mat or dry wood. • Wear dry, *hole-free* gloves. (Change as necessary to keep dry.) • Do not touch electrically “hot” parts or electrode with bare skin or wet clothing. • If wet area and welder cannot be insulated from workpiece with dry insulation, use a semiautomatic, constant-voltage welder or stick welder with voltage redusing device. • Keep electrode holder and cable insulation in good condition. Do not use if insulation damaged or missing.
Fumes and gases can be dangerous 	• Confined area • Positioning of welder’s head • Lack of general ventilation • Electrode types, i.e., manganese, chromium, etc., see MSDS • Base metal coatings, galvanize, paint	• Use ventillation or exhaust to keep air breathing zone clear, comfortable. • Use helmet and positioning of head to minimize fume in breathing zone. • Read warnings on electrode container and material safety data sheet (MSDS) for electrode. • Provide additional ventilation/exhaust where special ventillation requirements exist. • Use special care when welding in a confined area. • Do not weld unless ventillation is adequate.

Figure 24.46

Welding sparks can cause fire or explosion 	• Containers which have held combustibiles • Flammable materials	• Do not weld on containers which have held combustible materials (unless strict AWS F4.1 procedures are followed). Check before welding. • Remove flammable materiels from welding area or shield from sparks, heat. • Keep a fire watch in area during and after welding. • Keep a fire extinguisher in the welding area. • Wear fire retardent clothing and hat. Use earplugs when welding overhead
Arc rays can burn eyes and skin 	• Process: gas-shielded arc most severe	• Select a filter lens which is comfortable for you while welding. • Always use helmet when welding. • Provide nonflammable shielding to protect others. • Wear clothing which protects skin while welding.
Confined space 	• Metal enclosure • Wetness • Restricted entry • Heavier than air gas • Welder inside or on workpiece	• Carefully evaluvate adequacy of ventillation especially where electrode requires special ventillation or where gas may displace breathing air. • If basic electric shock precautions cannot be follwed to insulate welder from work and electrode, use semiautomatic, constant-voltage equipment with cold electrode or stick welder with voltage reducing device. • Provide welder helper and method of welder retrieval from outside enclosure.

Figure 24.46 *continued*

Safety

Arc welding can be done safely, provided that sufficient measures are taken to protect the operator from the potential hazards. If the proper measures are ignored or overlooked, welding operators can be exposed to such dangers as electrical shock and overexposure to radiation, fumes and gases, and fire and explosion, any of which could cause severe injury or even death. With the diversification of the welding that may be done by maintenance departments, it is vitally important that the appropriate safety measures be evaluated on a job-by-job basis and that they be rigidly enforced.

A quick guide to welding safety is provided in Figure 24.46. All the potential hazards, as well as the proper safety measures, may be found in ANSI Z-49.1, published by the American National Standards Institute and the American Welding Society. A similar publication, "Arc Welding Safety," is available from the Lincoln Electric Company.

Appendix A

Written Assessment Answers

	Safety	Lubrication	Bearings	Chain Drives	Belt Drives	Hydraulics	Couplings
1.	C	A	A	C	D	D	A
2.	A	C	A	A	A	B	A
3.	C	B	A	B	A	D	A
4.	A	C	D	C	A	C	A
5.	C	A	D	C	C	A	D
6.	A	C	D	A	C	C	C
7.	B	B	A	A	C	C	C
8.	C	D	B	A	D	B	A
9.	A	C	A	B	B	A	C
10.	D	A	A	B	C	B	C
11.	B	C	C	B	C	D	D
12.	C	B	A	C	C	B	C
13.	D	C	D	C	C	A	C
14.	A	B	B	D	A	A	A
15.	B	B	B	D	D	C	B
16.	B	A	A	A	A	A	B
17.	A	C	C	A	D		
18.	D	A	A	C			
19.	A	B	D	A			
20.	B	C	C	D			

Index